Praise for *Einstein's Wa*

"Stanley's is a superb book, one that scientists, hist
general public will enjoy in equal measure. It is wr
Those wary of technical jargon will be delighted by Stanley's lucid explanations. With almost all books written in a generalist vein, there is some worry about what might be lost—however much else is gained—by not dwelling on the details provided in the academic work on which they are based. *Einstein's War*, however, is that very rare work from which I came away understanding the scholarly literature better for having had its context presented to me in gripping and readable prose." —*American Scientist*

"A thrilling history of the development of the theory of relativity . . . a superb account of Einstein's and Eddington's spectacularly successful struggles to work and survive under miserable wartime conditions."

—*Kirkus Reviews*, starred review

"Dissecting the debate over whether philosophical attraction to Einstein's views caused Eddington to skew his Principe data, Stanley affirms both the empirical integrity and the political bravery of this Briton's confirmation of a German's theory. The international human drama in epoch-making science."

—*Booklist*, starred review

"Impressive work of popular science . . . Well told . . . Delivers a wider, and still relevant, message that how science is performed is inextricable from other aspects of people's lives." —*Publishers Weekly*, starred review

"Detailed and readable . . . It is especially revealing about Einstein's scientific work and private life leading up to the momentous events of 1919." —*Nature*

"Fast moving and engaging." —*New York Journal of Books*

"Fans of popular science, Einstein, physics, and World War I will find this to be entertaining and informative." —*Library Journal*

"[Stanley] skillfully interweaves the lives on Einstein and Eddington into a readable narrative." —*Science*

"[Stanley] succeeds in wrapping up the global, national, and scientific politics of an era in a compelling story of one man's wild theory, lucidly sketched, and its experimental confirmation in the unlikeliest and most exotic circumstances."

—*Spectator*

"This beautifully written, moving account captures the heady thrills and crushing setbacks of one of the great intellectual adventures of modern times." —David Kaiser, MIT, author of *How the Hippies Saved Physics*

"Profoundly absorbing . . . One of the greatest epics of scientific history—the grueling intellectual labor of Albert Einstein in Berlin, on one side of the dividing line that the First World War had drawn across Europe . . . and, on the other side of it, the efforts of Arthur Eddington in Cambridge, one of Einstein's few supporters. . . . It was Eddington's achievement that at last established Einstein's for the world at large; and all this . . . amidst the intellectual indifference and near-universal xenophobia of their fellow scientists on both sides. An amazing story."

—Michael Frayn, author of the Tony Award–winning play *Copenhagen*

"Einstein, at thirty-five, was living and working in the heart of Berlin when the guns of World War I began firing. The bloody, seemingly endless conflict provoked everything he detested: violence, nationalism, and the herd-thinking that drove individuals into nothing. Matt Stanley gives us the scientist-philosopher and activist during these bloody years, tacking back and forth so effectively between the development of general relativity, plans to test it, and the lopsided battle Einstein fought with the majority of German intellectuals who lined up happily behind the battle lines. A great read about Einstein and the War that Made More Wars."

—Peter Galison, Harvard University, author of *Einstein's Clocks, Poincaré's Maps*

"Even if you know a lot about the history of relativity—even if you know the old stories about Sir Arthur Eddington's voyage in 1919 to try to prove Albert Einstein's theories correct—you probably haven't pondered just how unlikely the Einstein/Eddington pairing really was. At a time when the mere hint of fraternization with the enemy could land you in jail as a spy, a Briton embraced the ideas of an enemy scientist and helped launch the legend of arguably the greatest physicist of modern times. A fascinating story."

—Charles Seife, author of *Zero: The Biography of a Dangerous Idea*

"How an obscure German physicist became the first superstar of science . . . Stanley lets us share that excitement a hundred years later in this entertaining and gripping book. It's a must-read if you ever wondered how Einstein became 'Einstein.'" —Manjit Kumar, author of *Quantum*

EINSTEIN'S WAR

How Relativity Triumphed Amid the Vicious Nationalism of World War I

MATTHEW STANLEY

DUTTON

DUTTON
An imprint of Penguin Random House LLC
penguinrandomhouse.com

Previously published as a Dutton hardcover edition in 2019

First trade paperback printing: May 2020

THE LIBRARY OF CONGRESS HAS CATALOGUED THE HARDCOVER EDITION AS FOLLOWS:

Names: Stanley, Matthew, 1975– author.
Title: Einstein's war : the birth of relativity amid the vicious nationalism of World War I / Matthew Stanley.
Description: New York, New York : Dutton, an imprint of Penguin Random House LLC, [2019] | Includes bibliographical references and index.
Identifiers: LCCN 2019001286 | ISBN 9781524745417 (hardcover ; alk. paper) | ISBN 1524745413 (hardcover ; alk. paper) | ISBN 9781524745431 (ebook) |
Subjects: LCSH: Relativity (Physics)—History. | Science—Social aspects. | Einstein, Albert, 1879–1955. | Eddington, Arthur Stanley, Sir, 1882–1944. | World War, 1914–1918—Science.
Classification: LCC QC173.52 .S73 2019 | DDC 530.1109/041—dc23
LC record available at https://lccn.loc.gov/2019001286

Dutton trade paperback ISBN: 9781524745424

Printed in the United States of America
10 9 8 7 6 5 4 3 2 1

BOOK DESIGN BY NEUWIRTH & ASSOCIATES

CONTENTS

EINSTEIN'S WAR

PROLOGUE

"The Devil howling, 'Ho! . . .'"

THEY HAD BEEN allies for five years but were meeting for the first time. On a Friday afternoon in June 1921 a shabby scientist, wearing shoes without socks, shuffled down a hallway in London. He described himself as having a "pale face and long hair, and a tiny start of a paunch." Someone watching would notice "an awkward gait, and a cigar in the mouth . . . and a pen in pocket or hand. But crooked legs and warts he does not have, and so is quite handsome." No one on the planet would have needed an introduction to him: his lively mane and bushy mustache marked him unmistakably as Albert Einstein, the most famous thinker in the world.

Waiting for him was a neatly dressed, square-jawed Englishman. He had a lean, athletic build and a thoughtful, penetrating gaze. Arthur Stanley Eddington had just been elected president of the Royal Astronomical Society (colloquially known as the RAS), but it was not his expertise about the hearts of stars and movements of galaxies that bonded him to Einstein. He had literally journeyed across the world, suffered condemnation and the threat of prison, and battled against centuries of scientific tradition to help make this unruly visitor from Berlin into an icon of science. He was the one who had showed Einstein's relativity to be true, and who had spread that gospel around the world.

They shook hands in Burlington House, a grand Palladian structure that sat at the inner ring of British science. Not long before, it would have been unthinkable for someone who answered to "Herr Professor" to set foot there. The building had seen ferocious arguments about whether German research even counted as real science, and whether Germans could ever be allowed back into the scientific community after the horrors of the Great War. But now the main room filled to bursting with British scientists eagerly waiting to hear Einstein's every word.

THIS IS THE story of a theory, general relativity, hailed at its birth as "one of the greatest—perhaps *the* greatest—of achievements in the history of human thought." It remains, a century later, one of the essential pillars upholding our understanding of the universe. Relativity not only explained the movements of galaxies through space, predicted black holes, and defined the grand scales of the cosmos, it forced us to question the most basic ways we experience the world around us. Einstein warned that time and space were not what they seemed, that the most fundamental tools we use to make sense of reality were *warped*. Gravity bent light, twins aged at different rates, stars were askew in the heavens, matter and energy were strange shadows of each other. We saw only a distorted, partial view of the true four-dimensional universe. The truth of things was accessible only to those who could grapple with complicated mathematics and philosophical paradoxes.

Einstein's extraordinary claims, and Einstein himself, burst into public view in 1919. His theory had been complete for four years and he had been hard at work on it for almost a decade before that. But his work was still little known outside a small group of theoretical physicists. Why? This is also a story of science's journey from peace to war and back again.

The First World War, the industrialized murder that racked Europe from 1914 to 1918, coincided with Einstein's most productive years (fifty-nine publications). His struggle with his own theory was inextricably tied to the course of *the* World War, as it was known at the time. Scientists seeking to confirm his ideas were arrested as spies. Technical journals were banned as enemy propaganda. Colleagues died in the trenches. And most frustratingly, Einstein was separated from his most crucial ally by barbed wire and U-boats. This ally was Eddington, who would go on to convince the world of the truth of relativity.

Eddington took up the cause of that German physicist—whom he had never met—to show how science could triumph over nationalism and hatred. The international institutions of science had been shattered by the war and had left Einstein isolated, but Eddington realized that relativity could be the key to restoring exactly those networks. In 1919, when Europe was still in chaos from the war, Eddington led a globe-spanning expedition to catch a fleeting solar eclipse. It was a rare opportunity to confirm Einstein's bold prediction that *light* has *weight*. It was the results of this expedition—the proof of relativity, as many saw it—that put Einstein on front pages around the world. The pandemonium of the Great War was the background against which this new sage appeared. His scientific revolution depended on battles both intellectual and political, fought from Berlin to London to the very edge of the universe.

The story of Einstein's relativity was a deeply human one. Far from being an abstract development of ideas through pure rationality or a spark of inspiration from one person's mind, this was a messy adventure that combined friendship, hatred, and politics. While Einstein never held a rifle or fired a shot, the war shaped his life and work for years. Falling ill from wartime starvation, unable to send simple letters to his most important colleagues, Einstein was constantly reminded he was living in a state under siege. The war hobbled him but also created the conditions under which his theory came to fruition. The pressures of that shattering conflict pushed together specific people and ideas at just the right moments of serendipity to create the relativity revolution. Einstein's emergence as a history-making figure could have been quite different: the practice of science, and what it even meant to *be* a scientist, became deeply enmeshed in questions of politics and empire. All of this molded not only the creation of relativity but the way the world came to first meet Einstein.

One of the most extraordinary features about Einstein's ascent was the way ordinary nonscientists engaged with the mysteries of relativity. The mathematics and concepts of the theory were hardly accessible; nonetheless apostles like Eddington convinced huge swaths of the public that they should care deeply about its implications. One newspaper of the time tried to describe how the proverbial "man in the street" reacted to this scientific revolution: It "disturbs fundamentally his basic conceptions of the universe and even of his own mind. It challenges somehow the absolute nature of his thought." What did it mean to live in a universe where things were "relative"? What did science say about our everyday experience? What could we truly know about the world? The implications were mind-bending and the explanations often opaque. The Archbishop of Canterbury complained that the more he read about relativity, the less he understood. The theory can be tough to swallow whole. But in this story we walk alongside Einstein as he develops relativity from the first—we can watch it built brick by brick, from thought experiments to radical concepts to experimental confirmation, instead of needing to grasp it all at once.

Beyond the philosophical puzzles raised by the content of the theory, there were further obstacles. Science already had a theory that explained the nature of the universe: the one presented more than two centuries before by Sir Isaac Newton. Newton's ideas had answered almost every question asked of them, and they underlay everything that was known at the beginning of the twentieth century. Following his system was what it meant to do science. Alexander Pope's classic couplet captured this reverence:

> Nature and Nature's laws lay hid in Night:
> God said, "Let Newton be!" and all was light.

Einstein and Arthur Eddington around the time they met.
EMILIO SEGRÈ VISUAL ARCHIVES

And now this disheveled German sought to replace Newton? The British poet J. C. Squire offered an addendum to Pope:

> It did not last: the Devil howling "Ho!
> Let Einstein be!" restored the status quo.

So Einstein and Eddington not only had to convince the world that relativity was true, they had to make the case that it was *more true* than Newton—so extraordinary that it deserved to knock Newton from his pedestal.

This was Einstein's personal war. Not only the creation of his universal theory but the struggle to make it known, to persuade friends and enemies of its importance. And his passage from scientific outsider to the new author of our reality not only took place against the backdrop of Armageddon but was intertwined with it. Without the Great War, relativity would not be as we know it, and Einstein's name would not be synonymous with genius. Two wars, entangled, that changed the world.

CHAPTER 1

The World of Science Before the War

"Thank you. I've completely solved the problem."

THE EINSTEINS WERE early adopters. The brothers Jakob and Hermann ran a small company at the cutting edge of technological innovation: electrification. They brought electrical lights to the public streets of southern Germany, becoming part of that country's extraordinary growth at the end of the nineteenth century. Only unified in 1871 in the wake of its overwhelming victory in the Franco-Prussian War, Germany was still a young nation. The formerly feuding twenty-five principalities and kingdoms had become a world empire with an enormous army and a streamlined economy whose intellectual and cultural institutions dominated the Continent. There seemed to be no better model for modernity. When Mark Twain visited in 1878, he wrote, "What a paradise this land is! What clean clothes, what good faces, what tranquil contentment, what prosperity, what genuine freedom, what superb government! And I am so happy, for I am responsible for none of it. I am only here to enjoy."

A year after Twain's visit, Hermann Einstein's wife, Pauline, delivered their first child, Albert. They were living in Ulm at the time, a modest city on the banks of the Danube. Ulm (whose urban motto declares "the people of Ulm are mathematicians") did not hold the

Einsteins' attention for long, and they quickly moved to Munich, the great metropolis of southern Germany. The young Albert was late to talk. He had the habit of saying sentences to himself over and over to get them right before actually speaking. He was also known for his terrible tantrums in which "his face would turn completely yellow, the tip of his nose snow-white." During one of these episodes, his sister reported, he struck her in the head with a garden hoe.

The Einsteins were Jewish but almost completely secular. In Imperial Germany there were few legal restrictions on Jews but plenty of anti-Semitic boundaries. Like many assimilationist families they dove deeply into German secular culture. Hermann read Schiller and Heine aloud to the children in the evenings. Pauline, a skilled pianist, hoped Albert would be a music partner and started him on violin lessons at age six. In a sign of things to come, the boy deeply disliked the mechanical, repetitive drills he was taught and studied music only reluctantly. It was not until years later that he discovered a passion for Mozart sonatas and threw himself into studying the violin. As he recalled years later, "I believe altogether that love is a better teacher than a sense of duty—at least for me."

Unfortunately for him, the German education of the day was much more focused on duty than love. The schools—he attended the nearest one to his home, which happened to be a Catholic institution—were highly disciplined and militaristic. He was always unhappy being told what to do and became an antagonist of nearly all his teachers. He threw a chair at one tutor and had the annoying habit of addressing his instructors with the casual "*Du*" instead of the formal "*Sie*." Family legend reports that one teacher dressed him down by saying, "Your mere presence here undermines the class's respect for me." Albert was just fine with that situation.

He had few friends and was self-reliant from an early age (he walked the busiest streets of Munich alone at age four). A favorite entertainment was building houses of cards. Contrary to common legend, Albert's grades were fine. The instruction at school focused largely on classical languages, which was not to his taste, so much of his lasting education happened at home. As an older man, Albert recalled the definitive moment that set him toward a love of science. At age four or five his gadget-loving father brought him the gift of a compass. The simple consistency with which the needle always pointed north struck him deeply. He was entranced by these invisible forces, absolutely consistent and reliable. What were they? Were there more? Could they be understood? If so, how?

A brief phase of religious observance at age eleven (neither stimulated nor supported by his family) was swiftly eclipsed the next year by the discovery of what he came to call his "sacred little geometry book." The Einsteins regularly invited a poor medical student, Max Talmud, over for dinner. Talmud and Uncle Jakob brought Albert popular science and mathematics books, which Einstein credited with spurring him toward freethinking. The pivotal geometry text was Euclid's classic *Elements*, the foundation for two thousand years of European mathematical education. It is structured around a series of indisputable premises (for example, two points define a line) that are then developed steadily and rigorously into sophisticated deductions (for example, the Pythagorean theorem). Albert was stunned by the way the complex could be built from the simple, and how conclusions "could be proved with such certainty that any doubt was ruled out. This clarity and certainty made an indescribable impression on me." This became his model for how to think about the natural world: start with one clear, powerful idea and then deduce the consequences, hopefully producing something useful along the way.

Particularly important for Einstein was that this knowledge seemed to be beyond the individual. It was true in a profound, transcendental way. Mathematics and science, then, became a way for him to escape "the fetters of the merely personal." As a teenager he declared he wanted to become a theoretical physicist precisely because of the independence it would give him—from society, from convention, from tradition, from authority. He also admitted that his lack of "practical sense" meant theory would be better than application.

Einstein was perfectly willing to muddle through the expectations of a rigid school system, except for one frightening consequence awaiting him after graduation. Like all German men, he would have to serve in the army. School was bad enough, and he did not think he could survive actual military life. So in an amazing bit of lateral thinking, in 1894 he convinced a family friend to diagnose him with "neurasthenic exhaustion." This malady was the classic nineteenth-century ailment of an overworked brain and an exhausted nervous system. Einstein levered his diagnosis into dismissal from his school before graduation. He went even further and formally gave up his German citizenship. The imperial state would hold no power over him anymore.

Predictably, his family was none too happy about their suddenly uncredentialed and unemployed son. Luckily there was an excellent college, the Federal Swiss Polytechnic (abbreviated ETH in German), that did not require a high school diploma for entry. Albert convinced his family he could study on his own for the entrance exam. After an initial failure and another year of study, he succeeded. He found Switzerland, and Zurich in particular, to be a much more liberal place than Munich, and he thrived in the new environment.

Einstein's general school habits had not noticeably improved, though. If he did not find a course compelling, he often skipped lectures. These were frequently his mathematics classes, despite the ETH having some of the finest mathematicians in central Europe. One of those instructors was Hermann Minkowski, who disparaged Einstein as a "lazy dog." Fortunately Albert became close friends with Marcel Grossmann, a mathematics student who took diligent notes. Einstein would read the notes by himself and show up for exams, earning both decent grades and an official reprimand for neglect. He looked back on this arrangement with gratitude for Grossmann's well-organized notebooks. "I would rather not speculate how I might have fared without them."

Even his physics classes barely held his attention. They largely focused on well-established science and avoided the exciting new work being done in electrodynamics and the study of heat. It was not so different from science education today. The classes were not really about preparing students to do new science; they were about ensuring mastery of the old. Memorization of concepts, endless sample problems to solve, repeating classic experiments. He and his friends had to read up on modern developments on their own initiative.

Physics was loosely divided into experimental and theoretical branches, though most physicists did some of each, and students like Einstein had to demonstrate proficiency in both. Experimentation was typically done in laboratories—specially designed spaces filled (hopefully) with special instruments used to measure electrical currents, or perhaps the thermal properties of a metal, or the viscosity of a gas. You might be searching for a new phenomenon, finding varieties of old ones, or establishing a better number for, say, the speed of sound in glass. This required patience, steady hands, and a good relationship with machines.

Doing theory required little equipment beyond chalkboards, ink, and paper. The work there was largely conceptual—finding the patterns, usually mathematical in form, that shaped the natural world. The equations that emerged would then (hopefully) explain the variety of the physical world through a few elegant concepts like gravity or inertia. A theorist sought something they could never touch—the laws of nature. Here, one required a good relationship with the intangibles such as ideas, numbers, and mathematical beauty. Einstein had a clear preference for theory over experiment. He wanted to know the principles that made the universe function, that explained *why* things happened. That said, he also deeply enjoyed his work in the laboratory—he liked seeing concepts play out in a tangible way.

Although Einstein could both run an experiment and derive an equation, it seemed that he was hardly on the road to scientific success. Most important, his general disrespect for authority began to hobble him. When he approached his physics professor H. F. Weber about designing a novel experiment, Weber shut him down immediately: "You are a smart boy . . . but you have one great fault: you do not let yourself be told anything." Albert was a largely unremarkable student, more known for his "roaring, booming, friendly, all-enveloping laughter" than his scientific skills. There were few signs of this changing.

THE IMPERIAL AMBITION that chased Einstein out of Germany was aimed at a specific rival: Great Britain. The British Empire stretched around the globe, centered in the vast metropolis of London. Its network of colonial possessions brought resources, markets, and, most important of all, vast prestige. Germany, or more specifically, Kaiser Wilhelm II, was deeply envious. Ironically, Wilhelm was actually the grandson of Queen Victoria and spoke fluent English. He would happily recount his childhood memories of playing at her seaside estate. Despite (or perhaps because of) this, Wilhelm marked Britain as his chief rival. His nation shared long land borders with two large, well-armed, and unsympathetic neighbors—France and Russia—but much of his focus remained on the handful of islands in the North Sea.

Technically known as the United Kingdom of Great Britain and Ireland, the country was a hodgepodge of ethnicities (Scots, Welsh, Irish) under the domination of the English majority. Indeed, foreigners and patriots often referred to "England" when they meant the entire British nation, a practice that survives to this day. Britain thought of itself as a liberal state, primarily in the classical sense of a limited central government with a largely unregulated free-market economy. But it was also testing the waters of liberalism in the modern sense of toleration of different beliefs and practices. The nineteenth century saw the weakening of the centuries-old established Church, under which only professed Anglicans could hold political office, attend universities, and be full members of British society.

Roman Catholics and Protestant "Dissenters" such as Baptists, Unitarians, and Quakers had been literal second-class citizens. The Quakers in particular had largely been content to respond to this by having limited interaction with wider British society and keeping to their own tight-knit communities. Many Quaker families could trace themselves back to the turbulent seventeenth century, when their founder, George Fox, and his followers were routinely imprisoned and assaulted for their belief in direct mystical contact with God and a radical egalitarianism. The name "Quaker" was originally an insult, referring to the shaking brought on by religious devotion; they were formally known as the Religious Society of Friends, casually referring to one another simply as "Friends."

One of these Quaker families, the Eddingtons, welcomed a new baby in 1882, naming him Arthur Stanley. Only two years later his schoolteacher father, Arthur Henry Eddington, died in a typhoid epidemic. The young Eddington grew up very close to his mother, Sarah Ann, and his sister, Winifred, in the lovely southwest of England. The green hills and valleys there were the setting of the Arthurian legends and provided many adventures for the young boy.

Stanley, as he was known to his family, showed talent for mathematics very early. He learned the 24 x 24 multiplication table before he could read. He tried to count the number of words in the Bible (he made it through Genesis) and the stars in the sky. These eye-straining activities may have led to his need for glasses, which he received at age twelve. The sheer joy of being able to see things clearly meant that he spent much of that year simply staring closely at trees and stone walls. He was fascinated by the natural world, writing articles about Jupiter for a school newspaper or giving lectures about the moon to an audience consisting of the housemaid in the attic. One childhood essay talked about total eclipses, and how "some of the greatest astronomers in the world" would carry out expeditions to observe them.

Sarah Ann belonged to a generation of Quakers known for their conservative views about, well, everything. Alcohol, the theatre, and tobacco were all forbidden. The Quakers were already famously austere—no priests, no rituals, their meeting halls unadorned by so much as a crucifix, their worship conducted in silence punctuated only by moments of divine inspiration. But most had gone even further, focusing their religious practice inward and disdaining participation in politics or the modern world.

Arthur Stanley was part of the first generation to reject this. The so-called Quaker Renaissance presented a new interpretation of their sect's theology. Quakers had always contended that every human had a direct connection to God known as the Inner Light (sometimes Inward Light). Everyone was able to commune personally with the divine without the need of institutions or priests. The presence of this Inner Light was also the justification for their pacifism—violence against another person was violence against God. To the older Friends this simply meant declining to fight; to the younger Quakers this meant an obligation to *wage peace.*

So when Eddington went to study science in Manchester and, later, Cambridge, it was an assertion that his religious beliefs had an important role to play in the modern world. It was a conscious choice to represent the Friends. He was taught by mentors who asserted that science and religion were both critical to modernity in the world and had no inherent conflict. There was a long British tradition, particularly at Cambridge, of science and religion mutually supporting each other. The Quakers were distinctive in calling this out as essential for the new socially diverse, technologically sophisticated world that the twentieth century promised.

Despite this commitment to participation in modern British life, Eddington still stood out. He retained many of his mother's puritan habits and wore a humble cap to class in an era of ubiquitous bowler hats. He was known to be modest, polite, and reserved—stereotypical for a Quaker. His family was fairly poor, which meant that his higher education was contingent on winning a steady stream of competitive awards, grants, and competitions. He first went to study physics at Manchester University, where he worked with Arthur Schuster, a German immigrant doing cutting-edge laboratory experiments. He thrived there, where a lively and modern Quaker community helped him find his place in the world.

Everything changed in December 1901, when Eddington learned that he had been granted £75 to study the next year at Trinity College Cambridge, the home of Newton, Maxwell, and Tennyson. This was a change of enormous significance. From industrial, lower-class Manchester, the pacifist Dissenter would move to the very heart of refined, Anglican, imperial England. He was one of the first Quakers to make this transition, an event inconceivable a generation before.

At Cambridge, Eddington displayed immense powers of concentration and dedication studying physics and mathematics. Nonetheless, he ever so gradually began to loosen up. At first he took up solitary pursuits, such as cycling. He gradually discovered a passion for fine and not-so-fine literature; both *The Rubáiyát* and *Alice's Adventures in Wonderland* were equally inspirational (perhaps some combination of the two led to his passion for whimsical rhyming). Later he joined the chess club, read with a Shakespeare society, and took up tennis and golf with more enthusiasm than skill. One hockey teammate recalled a game in which Eddington "indiscriminately bruised the shins of friend and foe alike as he relentlessly but somewhat myopically chased the ball up and down the field." He was remarkably physically fit, at a lean five foot eight inches tall and 129 pounds (we know because he tracked his weight precisely throughout school). His handsome, chiseled features and neat grooming would have been sound foundations for courting eligible women, but he showed little interest in them except as colleagues.

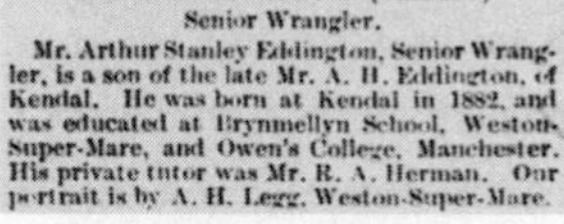

MR. A. S. EDDINGTON (TRINITY)
Senior Wrangler.

Mr. Arthur Stanley Eddington, Senior Wrangler, is a son of the late Mr. A. H. Eddington, of Kendal. He was born at Kendal in 1882, and was educated at Brynmellyn School, Weston-Super-Mare, and Owen's College, Manchester. His private tutor was Mr. R. A. Herman. Our portrait is by A. H. Legg, Weston-Super-Mare.

The young Eddington
(COURTESY OF THE AUTHOR)

Even having gathered a small group of friends, he was still known for being quiet and shy. With one exception. Soon after arriving at Cambridge he met C.J.A. Trimble, a mathematics student with whom Eddington felt an immediate connection. In Trimble's presence he was a transformed man. As his biographer described it: "With this one friend Eddington could throw off all the hesitant diffidence which formed an almost impenetrable barrier to intimacy with others." With Trimble he could be "light-hearted and full of fun." The pair took up hiking and spent much of their free time together exploring the countryside.

We do not know precisely the nature of Eddington and Trimble's connection. Today their relationship would certainly be read as romantic and probably sexual. However, they met in the waning days of the Victorian tradition of romantic friendship in which two men could have an extremely close, intimate, but nonsexual relationship.

Cambridge and Oxford were particularly known for giving rise to these sorts of associations. Same-sex relationships of the past, filtered as they are through social and legal conventions of the time, are notoriously difficult to interpret accurately from the present. Eddington's letters were destroyed at the end of his life, so we lack much of the evidence we use to understand Einstein's romances.

Regardless of whether their relationship was a physically romantic one, Eddington and Trimble remained close throughout their lives. Their different backgrounds occasionally caused friction. Once, they stopped at an inn and Trimble suggested they have ginger wine mixed with a bit of gin—Eddington, still adhering to his strict upbringing, was "quite indignant" and refused. Right around this time he took up the similarly forbidden practice of smoking tobacco, though. He originally started smoking just to ease toothaches or calm nerves before an exam, but it quickly grew into a lifelong habit.

Something soothing was very helpful as the time of the legendary Mathematics Tripos examination approached. This was the climactic moment for a student at Cambridge, a grueling four-day test intended to find the finest scholars at the university. It was a rite of passage for generations of British physicists and had an enormous influence on a young scientist's trajectory. Eddington seized the highest score and was rewarded with the title of "senior wrangler" (the lowest-scoring student was the "wooden spoon"). This was the first time a second-year student had taken the top place, and Eddington was celebrated widely—in particular by the Quakers he had left behind in Weston-super-Mare.

We are used to thinking of Einstein at the end of his career—wizened, pipe held sagely in hand. The Einstein of 1900 was not that Einstein. He was young, vivacious. One friend described him this way: "His short skull seems unusually broad. His complexion is matte light brown. Above his large sensuous mouth is a thin black moustache. The nose is slightly aquiline. His striking brown eyes radiate deeply and softly. His voice is attractive, like the vibrant note of a cello." Despite

The young Einstein
EMILIO SEGRÈ VISUAL ARCHIVES

his preference for old clothes (and a lack of socks), women found him irresistible and he returned the sentiment: "He had the kind of male beauty that, especially at the beginning of the century, caused great commotion." One friend commented that Einstein "acted on women as a magnet acts on iron filings."

He turned his charms on one woman in particular. Mileva Marić was one of the pioneering women studying physics at the ETH. She was from a Serbian family in Hungary and walked with a distinct limp. Einstein was entranced and courted her aggressively. They carried on a torrid premarital affair. Calling her his "beloved witch," Albert's letters to her were a heady mix of physics chatter and promised kisses.

Along with Mileva and Grossmann, Einstein's circle of friends grew to include Michele Besso, an older engineer with whom he loved to perform music. Eventually the group added Maurice Solovine and Conrad Habicht—they grandly called themselves the Olympia Academy. They would gather to talk physics and philosophy over sausage and tea, or play music over fruit and cheese. Albert never much liked alcohol and he declined the beer that was often served. Sometimes they read Henri Poincaré on the nature of time, sometimes *Don Quixote*. The young Einstein was a bohemian. His pursuit of science was part of the same social movement as the artists and writers lining the turn-of-the-century coffee shops. Beauty, truth, and love were of a piece with the conservation of energy and Newtonian dynamics.

The exciting physics of those days dealt with the forces of Einstein's childhood—electricity and magnetism. As those forces were put to work in electrical generators over the nineteenth century, they were studied intently by generations of physicists. Alongside developing applications like the telegraph and the lightbulb, scientists tried to create theories to understand and explain what was happening inside the machines. Both electrical and magnetic forces could be seen to push and pull electrical charges apparently without contact—think of the spinning of the young Einstein's compass needle. This idea that physical effects could be caused without visible physical interaction (a conceptual problem known as "action at a distance") was unsettling to many scientists. To resolve this, the English apprentice bookbinder turned experimental physicist Michael Faraday proposed the notion of "fields." These were invisible entities filling the space around electrical and magnetic sources that carried those forces from place to place.

Even this was not a particularly satisfying solution, and it was eventually accepted that there must be a physical thing that carried or supported these fields: the *ether*. The ether was an invisible, nearly intangible substance that filled all of space and permeated matter.

What we observed as electricity or a magnetic field were actually states of tension or twisting within that subtle substance. This fixed the problem of action at a distance (a magnet twisted the ether, and then the ether twisted the compass needle) at the cost of accepting that the universe was filled with bizarre unseen material. This was not as radical as it might seem. It was already well accepted that there was an optical ether that carried light waves. In an analogy to sound waves, which can only exist in air, scientists concluded that there must be some medium that supported light waves as well. The ether functioned both as an explanation (it tells us how and why certain phenomena happened) and a hypothesis (an idea whose consequences we can extend, predict, and look for).

It was extremely impressive as both. The Scottish physicist James Clerk Maxwell used the concept of the ether to construct an elaborate theory of *electromagnetism*, which connected electricity and magnetism on a deep and profound level. The equations that came out of this theory (now called Maxwell's equations, though he never actually wrote them down in the modern form) were one of the most successful scientific accomplishments of the century, and today underlie everything from your cell phone to your fiber-optic Internet connection.

Maxwell's theory was quite powerful. One of its great benefits was its ability to explain what is perhaps the single most important physical phenomenon for our modern civilization: induction. When you move a magnetic field over a conductor like a piece of wire (or alternatively, move the wire through the magnetic field), an electrical current appears in the wire. This is the principle by which virtually all electricity is generated today. Unless you are in a solar-powered home or outdoors, the light by which you are reading this almost certainly relies on electromagnetic induction. Every electrical generator—coal, gas, nuclear—and electrical motor comes down to a magnet moving past a wire or vice versa. Maxwell was able to not only make sense of induction but also predict a strange new phenomenon from it. His theory said that electrical charges and magnets themselves were not so important; it was actually the electrical and magnetic *fields* (that were thought, remember, to be particular states of the ether) that made induction work. When a magnetic field in a particular spot changed—by moving a magnet close to the spot, say—it generated an electric field. When an electric field changed—by hooking up a battery, say—it generated a magnetic field. So in certain circumstances, the electric and magnetic fields could be separated from their sources and simply generate each other, in a microscopic dance of energy shifting back and forth between electrical and magnetic fields. Maxwell showed mathematically that this shuffling packet of energy would move through the ether in the form of a wave traveling at an astounding 186,292 miles per second.

Maxwell noticed that this was exactly the speed of light, and decided this could not be a coincidence. He concluded that light was nothing but an electromagnetic wave. What we saw with our eyes was merely a special sort of vibration in the ether. There could be many other sorts of vibrations—what we now call radio waves were found in the laboratory by the German physicist Heinrich Hertz in 1888.

So what it meant to do physics around 1900 was to walk in the footsteps of Maxwell. A good theoretical physicist closely examined Maxwell's equations and thought about the physical behavior of the ether in some particular circumstance (say, inside a radio transmitter). Specific equations would be developed for that situation, and then those would be analyzed for a possible consequence that could be seen in the laboratory or on the engineering bench.

This is what filled the science books that the adolescent Einstein pored over while lying on his parents' couch. He thought deeply about the behavior of the ether and wondered about its properties. When he was about sixteen, a particular puzzle occurred to him during these speculations. What would happen if, he pondered, he ran alongside an electromagnetic wave at the speed of light? If you ran at ten miles per hour alongside a train going the same speed, it would essentially appear as though the train was not moving. This was Galileo's great insight (though he used ships instead of trains) and was often called the principle of relativity. It essentially argued that motion was relative—someone sitting on the train appeared to be moving to someone standing on the platform, but the train passenger was perfectly justified in claiming that they were sitting still and the person on the platform was moving.

Einstein wondered how this could apply to waves moving through the ether. Maxwell's equations described a wave of alternating electrical and magnetic fields moving through this universal substance at the speed of light. So if Einstein ran alongside that wave—at the speed of light compared to a stationary ether—the wave should not appear to be moving. Like the person running alongside the train, that very fast runner will see the wave frozen in place. This was what Galileo's thought experiment would seem to suggest anyway. But Einstein realized that such a frozen wave made no sense in Maxwell's theory. There was no equation to represent electromagnetic waves sitting still. It was the very nature of those shifting fields that they *must* move. Galileo's claim—that motion was always relative, that no one observer could declare that they were "really" moving and someone else was not—was not easily applied to the world of electromagnetism. The ether seemed to provide an *absolute* reference. Someone moving against the background of the ether could tell they were *really* moving because of how electromagnetic waves zoomed around them. Einstein realized there was something wrong with how physics was thinking about motion in the ether.

Einstein certainly wasn't alone in this realization, and many of the great minds of physics were occupied with attempted solutions to these problems. Einstein most admired the work of Hendrik Antoon Lorentz, a Dutch physicist who had constructed a masterful theoretical extension of Maxwell's physics, combining those equations with the recently discovered electron. Einstein and his circle of friends eagerly discussed the work of Lorentz and others, carefully studying papers that appeared in journals such as the prestigious *Annalen der Physik* coming out of Germany.

Journals such as the *Annalen* were the lifeblood of science. Whether someone had carried out a new experiment or found a new theoretical explanation, it hardly mattered unless they told someone about it. Sometimes this could be done in person at conferences or similar events. The scientific groups that organized these were often nationally oriented, such as the German Physical Society, the French Academy of Sciences, or the Royal Society in Great Britain. But it was not always possible to attend, and who wanted to wait a year until the next meeting? So those scientific groups often published journals in which scientists reported their latest work. Thus students like Einstein, or someone far removed from the centers of scientific work, could still keep up to date with the latest developments through the mail. The scientific community was defined, in a large sense, by where these journals could travel. A French scientist could feel as though they were participating in German research, reacting to developments and contributing their own ideas.

The papers in these journals provided Einstein the opportunity to argue with Mileva and Michele over equations of electromagnetic optics, the validity of mathematical transformations, and the proper interpretation of experiments with giant electric coils. Many of the papers were incremental additions and adaptations of ether theory. Very few scientists had the virtuosity seen in Lorentz's attempts to overhaul the whole system and grasp the most fundamental principles at work. Ether theory accumulated a number of strange puzzles, though no one truly doubted it. It was the hypothesis that led to the greatest triumphs of physics, from electrification to radio—how could it be wrong? Even Lorentz became frustrated on occasion, though. He wrote down a series of equations—now called the Lorentz transformations—that made some of the puzzles go away if one adopted the ridiculous hypothesis that measurements of time and space could change as one moved through the ether. He reassured his readers that this idea was merely a mathematical convenience to make calculations work out correctly, and should not be taken seriously.

This was the kind of work on which Einstein wanted to spend his days. Einstein even came up with an experiment to measure the Earth's movement through the ether (the so-called ether drift). But in order to do that kind of cutting-edge physics, the young scientist needed to appease certain bourgeois conventions: he had to finish his degree. He and Mileva studied together intently. Einstein was constantly apologizing for wandering off with her physics texts without asking, and his habit of forgetting his apartment keys often complicated getting them back. Come the graduation exams, Einstein received an average of 4.91 out of 6—the lowest grade among those who passed. Mileva received a 4 and was denied her diploma. Discouraged but not defeated, she went back to studying as Albert spent the spring of 1901 trying to figure out what to do with his life.

He had hoped to be able to get a job as an assistant in a physics laboratory. Most graduates of the ETH could. Einstein could not. The chief requirement for getting such a position was a letter of recommendation from the applicant's college professors, usually praising their work ethic and responsibility. Einstein's teachers had vivid recollections of the lazy dog skipping their classes, though, and declined to support his applications. As his mailbox filled with rejection letters, Einstein began to wonder if his chief recommender, Weber, was actively sabotaging his attempts to find employment.

Beyond the inherent anxiety of a new college graduate looking for work, Einstein had some additional pressures. Mileva was pregnant and they were not married. Einstein's mother was deeply traumatized at the thought of him marrying this Serbian—especially while unemployed—and made her feelings known sharply. As Einstein tried to calm his family, Mileva went home to Novi Sad to deliver their baby, a daughter referred to as Lieserl. Einstein sent a steady stream of letters to them, remarking that he had forgotten his nightshirt, toothbrush, comb, and hairbrush while traveling.

Einstein never met the baby. We do not know her fate. She suffered an attack of scarlet fever at one point, and may have died or been given up for adoption or taken in by a family member. Mileva returned to Switzerland alone. She and Albert married in a civil ceremony on January 6, 1903. Returning to their apartment after the wedding that night, he realized he had forgotten his keys. They had to wake the landlord.

Fortunately Einstein had found a way to support his new bride despite his total failure to secure a job in physics. Thanks to his friend Grossmann's father, Einstein began work as a Technical Expert III Class at the Swiss patent office in Bern. The head of the office was somewhat skeptical that this fuzzy-headed theorist could handle the day-to-day practicalities of invention and industry. But Einstein had grown up surrounded by electricity meters and dynamos. He was quite comfortable with machines. He liked standing at his desk and distilling a complicated patent application to its simplest essence: was the machine based on a fundamental principle that would work? Years later he recalled fondly his time at the patent office, remarking that it helped shape his distinctive approach to scientific problems: "It enforced many-sided thinking and also provided important stimuli to physical thought."

Einstein's entry into the federal bureaucracy was just one part of a wider embrace of his new home. He had already taken on Swiss citizenship. And—in an amazing move considering his disgust at the kaiser's army—reported for a physical for military service. The health examination recorded his varicose veins and flat, sweaty feet. He was marked unfit for service and was required to pay a small fee to compensate, which he dutifully sent in for the rest of his life.

He found Bern charming. In a letter to Mileva he sang its praises: "An ancient, exquisitely cozy city, in which one can live exactly as in Zurich. Very old arcades stretch along both sides of the streets, so that one can go from one end of the city to the other in the worst rain without getting noticeably wet. The homes are uncommonly clean." In the summer of 1903 they moved into an apartment at Kramgasse 49, a beautiful old street. The next year their first son, Hans Albert, was born there.

EDDINGTON'S FAMILY WAS perturbed that marriage and children did not seem to be anywhere in his future. After some tense conversations when he apparently declined to marry one Emmeline Yates (her brother Rex was a friendly swimming partner), he focused again on his science. His impressive placement on the Tripos exam led to an invitation from William H. M. Christie, the eighth Astronomer Royal, to become chief assistant at the Royal Greenwich Observatory.

The observatory was the literal heart of the British Empire—it defined the meridian on the globe from which all distances and times were measured. Precision measurements of the motions of the stars made there were the foundation of the navigational tables on which British ships relied. It was the most visible and important site of astronomy anywhere. The importance of astronomy for commercial, political, and military power was undoubted.

The chief assistant position was traditionally filled with one of the top wranglers from Cambridge, in order to put their mathematical skills to work for the nation. Strangely, few of those Cambridge men would have had much experience with *observational* astronomy—that is, actually looking through sophisticated telescopes and making precise measurements of stars and planets. Instead, their education usually focused on *mathematical* astronomy and physics—using the laws of nature to calculate the forces acting on those stars and planets, and predicting their movements.

Those sorts of calculations were, at root, based on the work of Cambridge's own Sir Isaac Newton. In the late seventeenth century Newton revolutionized physics and astronomy with an entirely new set of concepts and methods. The core of the Newtonian world view was his laws of motion and his theory of gravity. The laws of motion gave the basic interactions of forces and matter—objects tend to continue moving in their current way unless acted upon by a force (inertia); larger forces produce greater acceleration; and every force has an equal and opposite reaction. The law of gravity was fairly simple conceptually but rich in consequences: every piece of matter in the universe exerts a force of attraction on every other piece of matter, increasing with the size of the masses and decreasing rapidly with the square of the distance between the masses (that is, doubling the distance quarters the force). Newton also provided new mathematical tools to combine with these ideas, and he was able to explain and predict both the motions of the heavens and Eddington's more terrestrial golf balls. He bound together the universe through gravity.

A basic application of this idea is the planetary orbits. A planet like Earth tries to move in a straight line through space, but the gravity of the sun pulls that line around in a curve, making an ellipse over which the planet travels year by year. High school physics is enough to show why the Earth moves as it does. The calculations become much more complicated once you take into account all the other bodies in the solar system, though. For a solid prediction of the Earth's motion you need to calculate not just the force of the sun, but also that of the moon, Venus, Mercury, Mars, Jupiter, Saturn, and so on, all of which are themselves in motion and also subject to those same forces. These equations are ferociously difficult and require the highest skill. An exact solution to the equations is actually impossible; the best that could be done was increasingly precise approximation.

Newton's own mathematics proved to be largely inadequate for this, and it was more common that the tools of the French astronomer Pierre-Simon Laplace were used for day-to-day work. This detail was often ignored as English scientists proudly claimed ownership of the physics underlying the cosmos. Newton's theory was the most successful of all time and Eddington's new job required an absolute mastery of it. He rose to the challenge and set to applying the equations of gravity to an entirely new realm: the movements of the Milky Way as a whole. He tried to analyze the motions of what he called "star streams"—vast currents of billions of stars spinning through space, held together only by Newton's gravity.

Perhaps Eddington focused on distant oceans of stars to separate himself from developments on the seas closer to home. The Royal Observatory was closely tied to the Royal Navy, which was increasingly worried about a new competitor. As part of Germany's push for imperial status, in 1898 Adm. Alfred von Tirpitz proposed to build a fleet that could challenge the British. It had been British policy for decades to maintain unquestioned naval superiority—without control of the seas, the empire would wither and die. So the German shipbuilding program was seen as a direct existential threat. British industry stepped up not just quantity but quality. The First Sea Lord Sir John Fisher introduced new, powerful designs such as the dreadnought and the battlecruiser. Both Germany and Britain poured resources into an arms race for larger, faster, better-armed ships. Other powers around the globe began to follow suit.

Germany was preparing its military on land as well as sea. Their army was very well equipped and extremely well trained. And despite their size, they expected to be outnumbered in any European conflict.

If Russia and France honored their mutual defense treaties, Germany would find itself fighting on two fronts. And even worse, Russia was gradually modernizing its army and within a generation would be a genuine threat. The German High Command decided that it could not win an extended war and instead gambled on their operational superiority to knock their likely opponents out early in any fighting. In 1905, Gen. Alfred von Schlieffen finalized his invasion plan. This called for a lightning-fast mobilization (made possible by the efficient German railway system) supporting a vast flanking movement through Belgium into northern France. Once the French were outmaneuvered and defeated, forces would then be shifted eastward to confront the slower-mobilizing Russians. This depended on precise timing and a carefully orchestrated logistical plan. A formidable task, but intricate offensive plans were how the Germans won in 1871, and they had only been improving since then.

Having escaped the kaiser's armies himself, Einstein spent little time thinking about these arms races. The diplomatic neutrality of the Swiss meant he could focus on his physics. Some of his attention was spent on what is called statistical mechanics, the analysis of the movement of atoms and molecules on a microscopic level. The hypothesis of molecular motions (it was still not universally accepted that ordinary objects were made up of tiny particles) had been immensely fruitful for understanding the behavior of heat. Einstein found he had some talent at the statistical methods needed for these investigations, and worked steadily at polishing up a paper on molecular motions that would be sufficient for a doctoral thesis.

His scientific passions remained focused on the mysteries of electromagnetism, however. One particular formulation stuck in his head. It was again based on induction, that phenomenon underlying so much of Einstein's world. Imagine the coil of wire and magnet at the heart of every electric dynamo. Place the wire at rest against the unseen ether. Now move the magnet past it. The magnetic field in the ether around the wire will increase as the magnet gets closer. By Maxwell's equations this changing magnetic field creates an electric field that then pushes a current through the wire. But if the magnet is at rest in the ether and the wire is moved past it, Maxwell's equations predict a current without any electrical field. Einstein was frustrated that there was an asymmetry here. The same observable situation—moving wires and magnets past each other—was given a different physical explanation: electric versus magnetic fields. The "what" was the same in both cases, but the "why" was different.

Einstein's use of an *aesthetic* criterion—symmetry—is quite important. The chain of reasoning that would eventually lead him to relativity did not begin with a specific experiment or a mathematical calculation. It began with a sense of the way the universe should be, of what a scientific explanation should look like. It did not seem right that there was this split between what could be seen and the concepts used as explanations. These objections did not spring spontaneously into Einstein's mind. Rather, they were the result of several years of intense thought and consideration. We do know, though, what triggered the avalanche. It was a pair of philosophers.

Einstein loved to read and debate philosophy with his friends. Even though their focus was science, they came from a generation where intellectual training was still broad and technical expertise was expected to be grounded in the liberal arts. Epistemology—the study of how we gain knowledge—seemed to them a basic part of physics. They read Kant along with their Newton. One of the books they read was a classic by David Hume, the eighteenth-century Scottish iconoclast known for his aggressive approach to questioning everything from miracles to causality. Hume located all our ideas in sense impressions and demanded close scrutiny of any entities beyond what could be directly experienced.

Along with the influence of Hume, Einstein had been interested in the work of Ernst Mach for some time. Mach, an impressively bearded Austrian physicist and philosopher, put forward an approach often called positivism. Mach argued that scientific concepts needed to be grounded not only in direct experience but specifically in measurement. One should not talk about "force," for instance, but instead specifically about how one *measured* force (springs, scales, and so on). He warned that scientists often held on to ideas not for good reasons but simply out of tradition and habit. Einstein took away the lesson that it was necessary to critically examine the roots of all of our basic concepts. Once we forget the origin of our ideas, we might try to base our progress on dangerously unstable foundations. Between Mach and Hume, Einstein was primed, as he said to Besso, to "stamp out vermin" hiding within physics.

So Einstein's objection about asymmetry came from this Machian question: how do we measure these electromagnetic phenomena? Or as Hume would demand: what is our actual *experience* of induction? Einstein, the son of a dynamo maker, answered: measuring an electrical current. We do not see a magnetic field. We do not see an electrical field. We do not see the ether. We see a magnet and coil get closer together, and then a needle shift on a device that measures current or a lightbulb start to glow. We can't even really know which of the magnet or the coil is "really" moving—Galileo's principle of relativity demanded that either one could be seen as moving, just like the observers on the train and on the station platform. And if we look only at what can be measured with our electrical equipment, that is true. Either way, we see the same current. Our conclusions about whether the magnet or the coil was moving could not be separated from a careful analysis of the way we measured that movement. Einstein realized he needed a new way of thinking that could make sense of these ambiguities and keep physics close to that which could be directly observed.

This sparked several weeks of what he called his "struggle." The crucial moment came one evening in the middle of May 1905. After a frantic conversation with Besso about these problems, Einstein wandered off in a haze of thought. The next day when he saw Besso he simply said, "Thank you. I've completely solved the problem." He declared that an analysis of time was the key. The historian Peter Galison has richly demonstrated that this moment was not a fluke. To live in Switzerland in 1905 was to be steeped in the technologies of time. Clocks were mounted everywhere on public buildings. Trains were coordinated by electrical time signals. Every day, Einstein passed underneath the most famous clock in Bern as he walked from his apartment to the patent office. At his desk, he examined a steady stream of devices for measuring, marking, and synchronizing time.

So when Einstein decided that time was the key, he meant something very specific. Sitting at the intersection of Hume, Mach, and the patent office, he meant one thing by "time": clocks. Time as an abstract or metaphysical concept was unacceptable. A scientific, positivist notion of time needed to be built on how one measured time, and nothing more.

The heart of Einstein's theory of relativity was one of these clocks, albeit a strange one. You can make a clock out of any kind of repetitive physical process—the motion of the sun, the swinging of a pendulum, or the pulsing of the quartz crystal in a digital watch. We can understand the theory of relativity by thinking about a *light clock*, where the repetitive process is a pulse of light bouncing back and forth between two mirrors. Every time the pulse finishes one cycle, the clocks tick; one second has passed. It is important to note that this is not a real clock (it would be far larger than the Earth). Einstein was proposing a "thought experiment." *Thought* in that it takes place entirely in your head. *Experiment* in the sense that you don't know what the outcome will be before you start. It is a rigorous and careful thinking through of the consequences of some initial idea or postulate, in much the same style as Albert's sacred geometry book. You need no laboratory or equipment, only imagination and mental discipline.

Einstein's scientific paper "On the Electrodynamics of Moving Bodies" dates from June 1905. This was the first appearance of what we now call the theory of special relativity. If Einstein's goal with this paper was to irritate other physicists, he was well on his way. It had no footnotes, no references to other papers, and only a brief note thanking his friend Besso for helpful conversations. Instead of building on an experiment or an existing theory, the paper simply began with his objection to the asymmetry of the magnet-wire situation. He then proposed two postulates that would govern his thought experiments. First, that the laws of nature should be the same for observers in any inertial frame of reference ("inertial frame of reference" is just a technical way of saying sitting still or coasting at a steady velocity). Einstein thought of this as a simple extension of Galileo's principle of relativity—if the observers on the train and on the platform are truly equivalent, this has to be true. Indeed, few scientists of 1905 would have objected to it. It was simply saying that no one person had access to the "correct" laws of physics. The physics jargon is to say that there are no "privileged reference frames." More simply it states that everyone should agree on the basic operations of nature.

The second postulate stated that the speed of light should be the same for all observers in inertial reference frames. This seems in flat contradiction to our ordinary experience—if I throw a ball from a moving train, the speed of the train and my throw are added together, and the ball is going faster than if I threw it from the platform. Similarly, if I turn on a flashlight on the train, surely the speed of the train should be added to the speed of the light beam? This harkens back to Einstein's youthful imagining of running alongside an electromagnetic wave and seeing it frozen in place. If the second postulate is true, the wave would always seem to be moving at 186,292 miles per second regardless of how he was running. Despite this strangeness, Einstein asked for the reader to be patient and follow his reasoning.

Now, armed with our two postulates, we revisit the light clock. We give two of our friends, Alice and Bob, their own light clocks. Being good Swiss citizens, they synchronize their clocks—that is, their clocks always tick at exactly the same moment. They watch the light pulses for a little while to confirm that they are, in fact, synchronized. However, strange things begin to happen if we set Alice on a train moving quickly (the train is coasting, so she is still in an inertial reference frame). As the train passes the platform where Bob is standing, they compare the ticking of their clocks. Bob sees his light pulse tick up and down (the clock on the top). But as he watches Alice's mirror carried along by the train he sees the light pulse *move at an angle* (the clock on the bottom).

This means that the overall distance that Bob sees Alice's light pulse travel is *longer* than the distance he sees his own light pulse travel. The natural solution to this is for him simply to say that the movement of the train is changing the speed the light is traveling at, just as with the thrown ball. That fixes the extra distance, and the clocks still tick at the same time. But suddenly Einstein's second postulate lunges in and reminds Bob that the speed of light *can't* change, even with the motion of the train. Instead, Alice's light pulse has farther to go at the

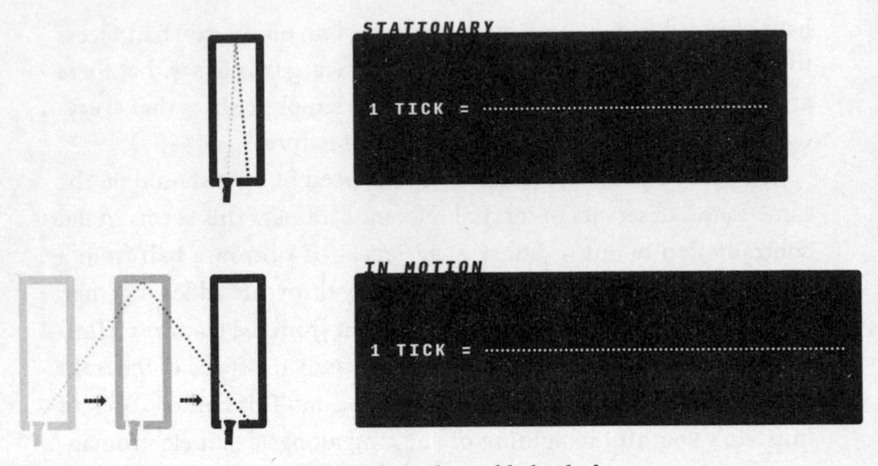

Einstein's hypothetical light clock
ORIGINAL ILLUSTRATION BY JACOB FORD

same speed, and Bob's light pulse finishes its cycle first. Bob's clock ticks before Alice's. Bob is startled to realize that Alice's clock—formerly synchronized with his—is now running slow.

Bob's reflex will be to claim that his clock is right, and Alice's is wrong. He thinks her clock is not running correctly because of her motion. Now Einstein's first postulate enters the picture, which warns that neither Bob nor Alice is allowed to say they are the one that is "really moving." According to Galileo, Alice is perfectly entitled to say that her train is standing still and it is the platform that is moving. So now Alice follows exactly the same chain of reasoning that Bob just did, and arrives at the conclusion that *Bob's* clock is running slow. Now, here is the perverse core of relativity: they are both right. There is no one correct position from which to watch clocks. Any observer will see a moving clock running slow compared to their own. And since time is nothing more than the ticking of a clock, time itself changes with motion.

If time truly changes, then we are in a strange place. People can disagree about whether two events are simultaneous. Identical twins can find themselves to be different ages. Our most basic experience of the world around us—the passage of time—is suddenly relative.

Einstein then performs exactly the same kind of positivist analysis with space instead of time. How does one measure space? With measuring rods—rulers, meter sticks, and so on. Running through the process of measuring length with a rod, just as we did with time and the light clock, results in the conclusion that the rods *shrink* as they go faster and faster. So as time seems to slow down, length seems to be compressed. The former phenomenon is called time dilation, the latter length contraction. One further Machian inquiry, this time about mass, concluded that mass, too, changed with the speed of the observer (an interesting result of this calculation was the simple formula $m = L/c^2$, more often written today as $E = mc^2$).

Einstein's postulates are, on the surface, basic statements of the universality of science. They are guarantees that any scientist, no matter their point of view, will be able to discover Maxwell's equations or Newton's law of gravity. But Einstein's thought experiment with the light clocks shows that there is a cost to this universality. The cost is that, while the laws remain the same, the measurements we make—of time, space, and mass—are all malleable. Motion changes them all. And since motion is relative—thanks, Galileo—time, space, and mass are themselves all relative. These categories, assumed since Newton (and demonstrated by Kant) to be immutable and absolute, were no longer so.

A completely reasonable objection to Einstein's outrageous claims is that you never see any of these things happen. Train stations do not seem to shrink as my subway car passes through them. I will not be able to persuade my boss that time dilation means I should get to go home early. A quick glance at the formulas, though, shows why this is (see page 34). These effects can only be seen when moving very, very close to the speed of light—you need to go about 90 percent the speed of light to see a clock running at half speed. For comparison, the fastest any human being has ever traveled relative to the Earth is about 0.0004 percent of the speed of light (the Apollo 10 lunar module, if you are wondering). Even today, time dilation can be demonstrated only with hyper-precise atomic clocks. Relativity was not amenable to being tested in 1905.

It is important to realize that these are all the results of thought experiments, not anything done in the physical world. Einstein had no tests he could perform, no predictions that could be reliably checked.

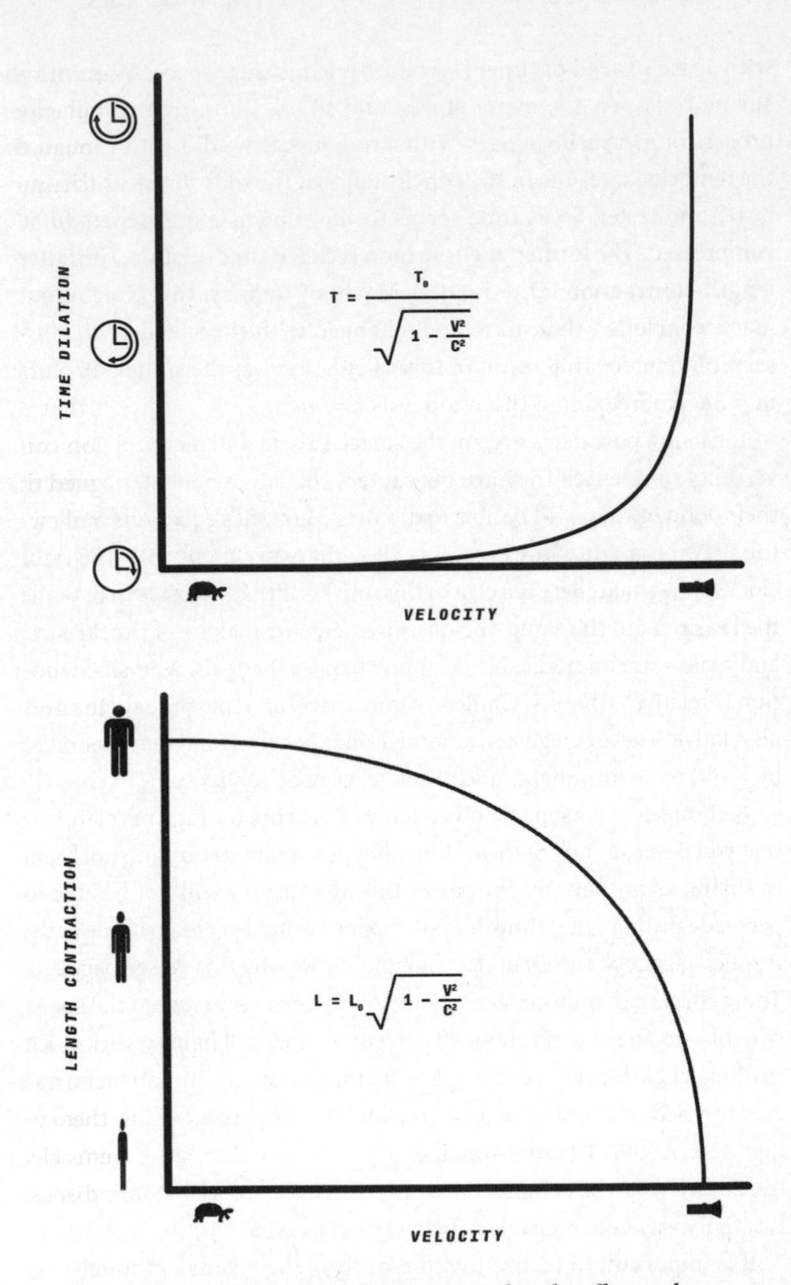

Time dilation and length contraction. Note that the effects only become significant close to the speed of light.

ORIGINAL ILLUSTRATION BY JACOB FORD

The best he could do was to say that his theory was *compatible* with the unusual results of certain experiments. These were the ether-drift experiments intended to measure the movement of the Earth through the electromagnetic ether, the most famous of which today is the Michelson–Morley interferometer. These experiments always returned what is called a "null result"—no apparent motion through the ether. This never made any sense. The Earth's direction of movement through the ether should be constantly shifting as it moved around the sun every year.

The null results were a genuine puzzle. Einstein's new theory had an explanation, though. The ether-drift experiments assumed that the ether was an absolute reference frame against which all motion could be measured, a place truly at rest in the universe. Einstein argued that this was impossible—if we wanted the laws of physics to be truly universal, there could be no such place. Special relativity dictated that any attempt to measure absolute rest would fail. The very concept of the ether was, in the positivist sense, unscientific because it could not be measured. Thus the twenty-six-year-old patent clerk in Bern, Switzerland, never having held an academic position, dismissed the fundamental hypothesis of nineteenth-century physics as "superfluous." Not wrong, not disproven. Superfluous. No longer needed, thanks to Einstein's insistence on the symmetry of the universe.

It is extraordinary that Einstein was able to arrive at these results on his own, and, some say, suspicious. Surely Michele Besso could not have been his only collaborator? What of Mileva, herself a physicist? She must have been a part of the genesis of relativity. Perhaps, even, Albert stole the theory from her and passed it off as his own? The evidence for this claim essentially comes down to a single word in Einstein's letters: "our." On one occasion he wrote to Mileva referring to "our work on relative motion." On others he mentioned "our theory of molecular forces." But there is no evidence beyond this. It seems clear that "our" was used here in the sense of "the theory we were discussing." Mileva never claimed responsibility for any of Albert's work. She was certainly a sounding board for his ideas in the same way that Besso and Grossmann were, and special relativity would no doubt have been different if any of them had not been present—Einstein needed his friends, as we will see over and over again. It would be strange, though, to say that special relativity was created by Besso, even though he was essential to its creation. Similarly with Mileva. There is really no reason to credit her with any of Einstein's work (he had productive decades of work without her, so he was hardly a fraud). There is no question that as a woman in physics circa 1900, Mileva had a nearly impossible road to success. Her social context essentially did not allow her to excel, regardless of her ability. But that does not mean her work was stolen; it simply means that she was one of vast numbers of women denied opportunities by an intensely patriarchal society.

The molecular theory Einstein referred to above is an important reminder that relativity was only one facet of his work at this time. 1905 is sometimes referred to as his "miracle year," an astoundingly productive time for the young man. He published six papers in the prestigious *Annalen der Physik*, all of which changed the world. In March, a paper on the photoelectric effect, which he described to his friends as "very revolutionary," proposed the notion that light can be treated as a particle, a small packet of energy called a quantum. This essentially jump-started the new field of quantum physics. In April, his doctoral thesis provided a new way to estimate the sizes and motions of molecules. Less than two weeks later, and then more completely in December, he put those estimates to work in an analysis of Brownian motion (the observed zigzag motion of tiny particles suspended in liquid). Einstein was able to explain these motions as collisions with molecules, and predict their behavior startlingly well—this was essentially the final nail in the coffin for skeptics of the reality of molecules. In June, his first paper on relativity appeared. September saw his second paper on relativity presenting $E = mc^2$.

As is often the case, the process of publication created the illusion of a sudden wave of discovery, when in fact each of these papers was the culmination of years of hard work. Nonetheless, their appearance en masse attracted serious attention, particularly from the editor responsible for their approval in the *Annalen*: Max Planck. By 1905 Planck was already the most important person in German physics. The wild, unkempt hair of his youth had long since given way to extensive balding, but his thoughtful, hooded gaze remained. His theoretical skill was unquestioned, especially having cracked the fundamental problem of so-called black-body radiation (how hot objects give off light). This theory was the groundwork on which Einstein built his own quantum calculations. It was the beginning of what we now call modern physics.

Planck was the very model of the German professor: morally upstanding, bureaucratically skilled, politically conservative. Colleagues remembered him for his "sense of duty and deliberateness of action." He was a mentor to countless young scientists, and as a three-time president of the Physical Society, essentially ran the field of physics in Germany. The publication of Einstein's papers was, for the most part, Planck's own decision—this was in the days before peer review. Despite their unconventional style, he saw something important in them and called the attention of physicists to this obscure patent clerk. It is sometimes said that Einstein was Planck's second great discovery.

When Planck talked, people listened. Einstein's papers on molecular motions and the light quantum fit well into other work being done at the time, and helped solve long-standing problems in science. Their importance was clear. Less clear was what to do with "On the Electrodynamics of Moving Bodies." There was little initial reaction to Einstein's ideas about time dilation and length contraction. Einstein even sent copies of the paper to physicists who he thought might be interested, including G.F.C. Searle, a physicist in Cambridge. Searle couldn't make heads or tails of it. Those who read the paper usually interpreted it as being a modest contribution to the physics of the ether and not a wholesale restructuring of our basic categories of experience.

But Planck was intrigued by the "absolute, invariant features" of the theory—that is, the way it allowed for the laws of physics to be universal. This may seem strange, given that we call it the theory of *relativity*. Einstein actually never liked that name and didn't use it until 1911. Planck wrote the first paper that dealt with relativity. He presented Einstein's paper at his physics seminar in Berlin that winter, and set his assistant Max von Laue to studying the theory—and to find out who, exactly, this obscure patent clerk was.

CHAPTER 2

Science Across Nations

"This unconstructable and unvisualizable dogmatism."

THE SAME WINTER that Planck was trying to excite his colleagues about relativity, Eddington had a final breakfast with Trimble in Cambridge and moved to 4 Bennett Park in the London neighborhood of Blackheath, just south of the observatory. Although he was hired for his mathematical acuity, his duties were overwhelmingly of the practical and observational variety. He set about learning all the techniques he would need and that Cambridge never taught him. The Astronomer Royal's initial letter had praised Eddington's skill with astronomical instruments, but in fact Eddington spent his last two weeks in Cambridge frantically learning how to use basic tools.

The Royal Observatory's most important project of the time was the new star catalogue, which listed the exact location of more than twelve thousand stars. His first job was to double-check all of those locations. This required long nights at the telescope making minute measurements. There was no room for error or carelessness. In all, he spent more than a thousand hours at the telescope working on the catalogue. Even more challenging were the social duties that came with the position. When he met with the observatory board, he had to wear an old-fashioned frock coat and top hat. When he was taken to dinner at the

Royal Astronomical Society Club, a toast was taken to his health and the teetotaler Eddington felt obligated to drink. Adjusting to life in London was made easier by visits from his mother and sister (they visited all the usual tourist spots) and Trimble (they went to see *The Taming of the Shrew* and swam at the Westminster Baths).

Not all of the measurements needed by His Majesty Edward VII could be done in Greenwich. Part of Eddington's duties was to travel the world to make official determinations of longitude and latitude of various outposts of the British Empire. These were the official map coordinates that commercial vessels used to make their fortunes and that the Royal Navy used to enforce order. One of the most important outposts was Malta, the tiny island absolutely essential to strategic control of the Mediterranean. The two previous measurements of Malta disagreed by one second of longitude. This could conceivably cause danger for some wayward frigate; more to the point, it was unacceptable to not know the exact location of any British possession.

Eddington nervously gathered the equipment he would need—literally a ton of very expensive gear, with which he had been practicing for months. He sailed on the *Japan*, a passenger and mail ship. This was his first long sea voyage, on which he read *Tristram Shandy* and *Don Quixote*. After nine days he arrived at St. George's Bay on Malta. The telegraph station where he was to work overlooked a sandy beach perfect for swimming, but he had an enormous amount of work to do and had to stay focused. He had little opportunity to see the island and declared it an awful place.

His first task was to build the "hut"—a temporary shelter for the equipment. This makeshift observatory contained a telescope, a clock, and elaborate electrical equipment. The process for measuring longitude was fairly straightforward. Eddington would use the telescope to measure the exact time in Malta (determined by the height of a particular star above the horizon). He would then send a signal by electric telegraph back to Greenwich, where, after accounting for transmission time of the signal, they would determine the equivalent time in London by measuring the height of the same celestial body. Because Greenwich was far to the west, it would be earlier than in Malta (the basis of time zones). The difference in time would then be precisely converted into distance.

This was a standard technique for determining longitude anywhere in the world connected to the telegraph network. It was very accurate but required good weather at both observing stations. A snowstorm in southeast England interrupted the work, as did several cloudy nights. Eddington spent about ten hours a day crammed into the hut, assisted by a petty officer from the naval base. After three solid weeks of work he found his goal: Malta lay at 0 hours, 58 minutes, 2.595 seconds east longitude. The precision of his measurement was, as expected, about eleven inches. Eddington became known for the speed with which he could carry out the complex calculations involved in estimating error, once deriving a formula on the back of a dinner menu when someone mentioned the need.

On the journey home Eddington stopped in Tunis, making his first visit to Africa. Tunis was still a French colonial possession, and it was an easy steamer journey to visit Marseilles as well. He deeply enjoyed travel (with the occasional exception such as Malta) and often spent his holidays on the Continent. He took his mother to Norway and his sister to Germany, and usually took advantage of these trips to meet and work with astronomers in those countries. When in Holland on vacation he visited the observatory in Leiden and conferred with the astronomer Jacobus Kapteyn in Groningen. It was Kapteyn's work that got Eddington interested in the problem of the rotation of the Milky Way, and they connected on a personal level as well. The next year Eddington hosted Kapteyn for a visit in London.

International travel became a standard part of his professional life. In 1909 the British Association for the Advancement of Science (known colloquially as "the BA") held its annual meeting in Canada. It was fairly common for British scientific organizations to make this sort of trip to help build imperial solidarity. Eddington gladly went along, departing from Liverpool on the *Empress of Ireland*. The six-day ocean trip gave him many opportunities to get to know other scientists. He didn't care much for Winnipeg in the summer (he declared the plains "not much to look at"), but the Rocky Mountains and Niagara Falls made a lasting impression. On his way home he passed through Boston to meet with the Harvard astronomer Edward Charles Pickering. Scientific journeys would eventually take him to every continent save Antarctica.

The other focus of his travel was the Friends' Guild of Teachers, a Quaker organization dedicated to experiments in education and applying their religious values to the problems of modern society. Their annual meetings were held in various locations across the United Kingdom, and Eddington attended nearly all of them (many with his sister as well). He was strongly committed to education as a powerful tool for improving the world. He remained an active participant in the guild his entire life and served as president for several years.

Eddington often planned travel to those annual meetings to coincide with a holiday with Trimble. They would spend a week or two together hiking. Many stories came out of these trips: being drenched by rain for six days straight; going to the movies but enjoying watching the locals' behavior more than the actual film; Trimble insisting on stopping to investigate every interesting bit of geology. Eddington loved glissading (sliding down hillsides), and on one adventure in Wales tore the seat of his trousers. As they walked to the nearest town, Trimble walked close behind to conceal the opening. Village children gave "shrill cries" at the sight and Eddington remarked, "It was perhaps a mercy that we knew no Welsh." When they reached their destination, they borrowed needle and thread and Eddington insisted that Trimble sew up his pants without taking them off. Trimble said he accomplished it without injury but added, "I believe I sewed him to his shirt."

Even beyond these trips, Eddington made regular trips to Cambridge to see Trimble. Bicycling was a favorite pastime on those visits. Eddington, always an obsessive counter, kept a fantastically detailed journal of his cycling adventures. When mere tallies were not enough (in 1905 he rode 2,669 miles), he began recording his progress with his infamous *n*-number: the number of times *n* he had cycled more than *n* miles (that is, when *n* was 37, he had taken 37 trips of 37 miles or more). Letters to friends would often include an update to *n*. After Trimble finished his degree, he went to London and stayed with Eddington on and off for nearly two years.

When his duties of measurement and observation allowed him time, Eddington continued his theoretical investigations of the Milky Way. At this time, it was not at all clear what size or shape our galaxy was, or even whether it was unique or just one of many similar bodies in the universe. Eddington and Kapteyn's work gave some of the first insight into the structure and movement of nearby stars. Eddington was able to use the measurements he was making at Greenwich to solidify their calculations. The mathematics involved was challenging, though, and he also contacted the German astronomer Karl Schwarzschild for advice. Eddington worked up a full paper on stellar motions for the examination to become a fellow at Cambridge, which also became his first communication to the Royal Astronomical Society. This paper won him the prestigious Smith's Prize. He even received an offer of a professorship of physics at his old college in Manchester. He declined, having decided that he preferred the life of an astronomer. The stars had captured him.

EINSTEIN WAS STILL working at the patent office. While his miracle year had attracted attention, his civil-service job paid more than the open academic positions. Not much changed for his daily life in Bern, save being promoted to Technical Expert II Class. His star within the physics community was certainly rising, though, despite his dismal networking skills (he didn't even attend a physics conference until 1909). The diligent experimentalist Jean Baptiste Perrin in Paris had confirmed his predictions about molecular motions, essentially proving the existence of atoms. The Nobel Prize winners Philipp Lenard and Wilhelm Röntgen wrote to him about his light quantum theory.

Most important to Einstein, though, was the correspondence he began with Hendrik Lorentz. Einstein had admired the elder physicist for many years and even felt that Lorentz had in many ways anticipated his own work. Many of the equations used in special relativity were in fact written down by Lorentz long before 1905, though the physical meaning attributed to them was quite different. The Dutch physicist was unsure exactly what the implications of relativity were, but he nonetheless saw a kindred spirit and took Einstein under his wing. Their relationship quickly became parental, and Einstein wrote, "I admire that man more than anyone else, I might say I love him."

Only a handful of scientists took notice of relativity, though. Planck continued to evangelize for it. Hermann Minkowski, Einstein's old mathematics teacher, was amazed at what his lazy student had accomplished and asked for an offprint. Arnold Sommerfeld, the dean of Munich physics, mixed admiration with some casual anti-Semitism:

> Works of genius though they are, this unconstructable and unvisualizable dogmatism seems to me to contain something almost unhealthy. An Englishman would scarcely have produced this theory; perhaps it reflects . . . the abstract-conceptual character of the Semite.

Nonetheless, he called relativity an "inspired conceptual skeleton" and expressed hope that Lorentz could give it "real physical life."

In September 1907, Johannes Stark, a noted physicist in Hanover, asked Einstein for a summary of relativity theory for a science yearbook he was putting together. This provided Einstein his first opportunity to reflect on the state of his theory and think about how he wanted to extend it further. One issue arose from the problem of how to connect special relativity with the crown jewel of classical physics: Newton's theory of gravity. It had generally been assumed that the force of gravity was instantaneous—if you destroyed the sun, Earth's orbit would be immediately affected. But special relativity set the speed of light as the upper limit for any sort of physical effect, so Earth would not feel the sun's destruction for about eight minutes. This difference was hardly insurmountable. Einstein just had to tweak Newton's equations to match those of relativity.

Unfortunately Einstein was not a fan of tweaking. He distinguished between "constructive" theories (assembled from many small details and facts) and "theories of principle" (guided by universal statements or ideas that applied in all possible cases). The former could be effective and useful, but only the latter could be logically secure and philosophically profound. Slightly changing Newton's law of gravity would yield only a constructive theory. His efforts on the problem so far, even where they looked promising, struck him as "highly suspicious." He needed some new foundational principles on which he could base an entirely original theory of gravity.

Newtonian gravity had inaugurated science as it had been understood and taught for centuries. Any change to it would have enormous consequences. But there were special problems in wedding Newton and relativity. In Einstein's theory, two observers in motion can disagree about the mass of a given object. And since gravitational force is proportional to mass, they will feel different amounts of gravity depending on their motion. This was a profound possibility: could the force of gravity, the very glue holding together the universe, be relative and malleable? Einstein decided that a close consideration of gravity was essential for his theory to grow.

Working with Mach's framework, Einstein's first question was still one about direct sensory experience. How do we experience gravity? How do you know that gravity is pulling on you right now? What does it feel like to have gravity acting on you? Years later Einstein recalled the key moment in this inquiry, what he called "the happiest thought of my life."

> I was sitting in my chair in the patent office in Bern when all of a sudden a thought occurred to me: "If a person falls freely he will not feel his own weight." I was startled. This simple thought made a deep impression on me. It impelled me towards a theory of gravitation.

The detail that he was sitting in his chair is actually important. Right now you are probably sitting. Think carefully for a moment about where you feel the effect of gravity on you. You don't feel the downward pull; you feel your chair pushing upward, keeping you in place. Einstein's realization was that it is actually this stabilizing force that makes you aware of gravity. If your inconsiderate roommate were to suddenly yank your chair out from under you, you would lose the sensation of that force. There would be a brief moment (before you hit the floor) when you would not be aware that gravity was pulling on you. Your perception of gravity would vanish (the same situation astronauts in orbit experience). Perhaps it was the case that gravity could be relative—it was not an absolute truth, but depended on your circumstances and surroundings just as time depended on your movement.

Einstein carefully considered how we perceive gravity and pondered situations where that perception could be changed. Like the light clock, this is a thought experiment. Imagine a physicist who, after a lively night out, awakes groggily in a sealed, windowless room. No scientist would be caught without basic experimental equipment, so she begins studying the chamber. She holds a weight out and releases it, noticing that it moves quickly toward the floor. One explanation for this is that the room is sitting on the surface of a planet like the Earth and the planet's gravity pulls the weight down toward the floor. But she realizes that there could be another explanation. What if the room is in deep space, far from any gravitational sources, but has a rocket strapped to the bottom? The rocket continually fires and accelerates the room in one direction. This time, when she releases the weight, it still moves toward the floor—but because the floor is *accelerating upward toward it*. From the scientist's point of view, though, the two situations are identical. There is no experiment she can perform from inside the room that will let her distinguish these two situations.

Einstein said that this thought experiment shows that gravity and acceleration are *equivalent*. This is again something only a positivist would say. One might object that the two situations are, in fact, physically different. But the positivist says you *can't tell*, so they might as well be the same. You feel your chair pushing on your gluteus maximus right now. That might be gravity, or there might be a rocket strapped to the bottom of the chair. If you can't tell which it is, Einstein said, it doesn't matter which it is. To someone doing physics they are the same, or at least "of the exact same nature." He comes to call this the *equivalence principle*: gravity and acceleration are indistinguishable.

In addition to Einstein's own concerns, this principle would help make sense of an odd feature of physics. An *m* for mass appears both in Newton's equations for motion (called inertial mass, it is a measure

The equivalence principle. The experimenter inside the room cannot distinguish between the two cases.
ORIGINAL ILLUSTRATION BY JACOB FORD

of how hard it is to accelerate something) and his equation for gravity (called gravitational mass, it is a measure of how much gravitational force is created by it). They were generally considered to have the same value (and had actually been measured to be so in the laboratory), even though the equations did not demand it. The mathematics allowed a universe where the two kinds of mass were different, so why did we live in one where they were the same? Einstein finally had an answer: they're the same because you can't tell them apart.

Einstein decided that the equivalence principle would have to be a core principle of any further elaboration of relativity. It not only suggested avenues for solving the problem of gravity, it also hinted at how to fill the other great gap in his theory so far. That was the problem of acceleration. Recall that the "special" part of special relativity is a warning that it only applies in certain situations—when observers are at rest or coasting steadily. Most movement is not like that, though. Usually there are accelerations—changes of speed. Your train has to slowly get up to speed, and has to put on the brakes to stop. When those things happen, you feel it. It's the nudge that spills your coffee or knocks someone into your newspaper. That nudge means special relativity can't apply. Thus, if Einstein wanted his ideas to apply to the world as a whole, he had to move beyond his special theory and make a *general* theory of relativity. The equivalence principle was his first window into how to do so.

Even the basic thought experiment for the equivalence principle could yield important further insights. Consider our imprisoned scientist once more. She now notices a small hole in the wall, at about her shoulder level. A beam of light comes in through the hole. If the box is at rest or coasting in deep space (that is, no rocket acceleration and no gravity), the light beam shoots straight across and hits the far wall at shoulder level. Now put the rocket on the bottom and turn it on. This time, as the light beam moves across the box, the box accelerates upward. In the time it takes for the beam to move across the box, the floor has moved upward, and the beam hits the far wall lower than shoulder height. From the scientist's point of view, the beam has curved downward due to the acceleration. She then invokes the equivalence principle. If acceleration bends light, gravity must do so as well. A sufficiently powerful gravitational field should produce a noticeable effect on the path of light.

Another possible test came out of the equivalence principle. The mathematics suggested that acceleration would cause time dilation much as inertial motion would. Einstein realized that this time dilation would cause another observable effect: a distortion of the light coming off of an accelerated source. The color of light is determined by the frequency with which the electromagnetic waves wiggle, and time dilation would slow that wiggle down. That slowing would be perceived by our eyes as the light becoming redder than it should be. And again, the equivalence principle suggests that whatever happens with acceleration must also be true of gravity. Gravity, then, should make light appear ever so slightly redder, a phenomenon called the *gravitational redshift*. Einstein now had two possible, albeit crude, predictions of his theory.

Finally, there was a third possibility. Astronomers had spent two

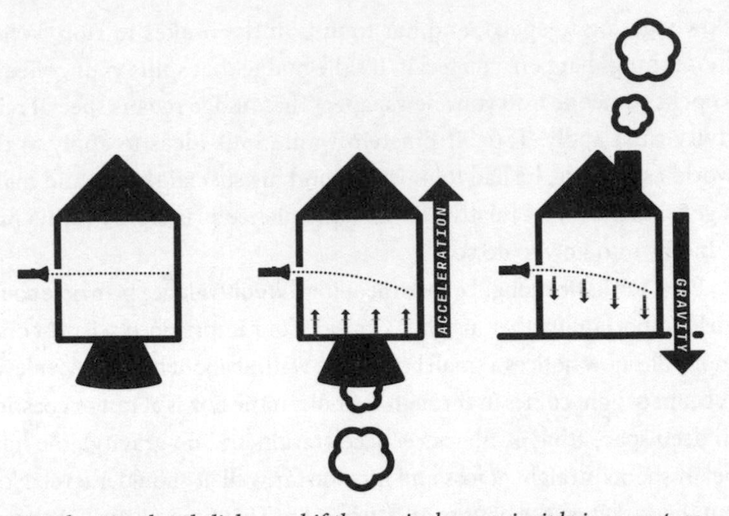

Acceleration bends light, and if the equivalence principle is correct, so must gravity. ORIGINAL ILLUSTRATION BY JACOB FORD

centuries applying Newton's laws to the motion of the planets with fantastic accuracy. Every discrepancy had been accounted for with that Englishman's theory—save one. The orbit of Mercury had a tiny wobble (technically known as the precession of the perihelion). It had been known for decades and resisted easy explanation. Astronomers assumed it was the result of the gravitational pull of an as-yet-undiscovered planet inside Mercury's orbit, referred to as Vulcan. A successful explanation for that anomaly had been the holy grail of anyone proposing an alternative theory to Newton's. Einstein now realized he was in that fantastical place of competing with the greatest mind his civilization had ever produced.

Within a few months of beginning to ponder the relationship of gravity and relativity, he had found three consequences of an extended version of relativity that, at least in principle, could be tested. If he could successfully make relativity into a more general and powerful theory, he could move it from the realm of speculation to that of confirmed science. Abstract ideas could be made tangible—measured and real.

ALL OF THESE were largely background thoughts for Einstein. Most of his work at the time, as encouraged by other physicists, examined molecules and quantum theory. To those distant colleagues, this was work worth supporting, not strange tales of physicists locked in rooms in deep space. And he followed that encouragement, grateful for the first time to have his ideas supported by real, professional physicists. So when a professorship of theoretical physics opened up at the University of Zurich in 1909, Einstein finally had people who would write him letters of recommendation. It was by no means an easy victory, though. One of his primary competitors was an old college friend, Friedrich Adler—ironically, one of the few full converts to relativity. In the end he received the position only because the top candidate suffered an attack of tuberculosis and had to withdraw from the competition. Even when the job was offered, Einstein had to hold firm until they matched his patent office salary. They did, and Einstein finally became a professor of physics. He had mixed feelings about becoming the sort of authority he used to antagonize. To a friend he wrote, "So now I am an official of the guild of whores."

The Einstein family moved to Zurich to take on the more bourgeois lifestyle of a professor. They actually rented an apartment in the same building as the Adlers, and Einstein was happy to have friends nearby with whom he could talk physics and play the violin. Einstein loved playing with his son, building him toys out of string and matchboxes. Despite his father's efforts, though, Hans Albert never took to music. Zurich was a more intellectually exciting place than Bern, and Einstein was expected to move in higher social circles (he ran into Carl Jung once or twice). His grooming habits had not improved and colleagues still remarked on his "somewhat shabby clothes" and "too short trousers."

The end of his first year at Zurich brought two surprises. Another son, Eduard, called "Tete," arrived in July, and the German University in Prague offered Albert a well-paid position as a full professor. The offer was an enticing one, but there were significant obstacles. Prague was part of the Austro-Hungarian Empire, a sprawling state that ruled eleven different major ethnic groups, from Germans to Serbs to Ukrainians. It was the largest political body on the Continent. Austria-Hungary was massively complicated to govern, and its reputation for bureaucracy was well earned. A peculiar moment happened when Einstein had to take an oath of allegiance. In order to swear the oath, he needed to declare a religious denomination. He tried to say "none." That was not an acceptable answer, and for the first time in his life he had to formally identify as a Jew. He liked to joke that it was the emperor that had made him Jewish.

Einstein enjoyed the salon culture in Prague, and this was a critical year for Einstein's professional development. He hosted the physicist Paul Ehrenfest, with whom he formed an immediate, intense friendship. Ehrenfest was an Austrian expat whose clear eyes peered out from wire-framed glasses that seemed to float between his unkempt hair and bushy mustache. He made major contributions to quantum physics, and his frequent smile concealed his severe, and eventually fatal, depression. His manic diary from his Prague trip recorded endless cordial arguments for days straight. With him Einstein found the perfect balance of physics insight, sharp-tongued debating, and musical appreciation.

Einstein made a number of trips around central Europe at that time, including his first visit to Berlin. He met Fritz Haber, the chemist whose nitrogen-fixing process made possible the large-scale production of artificial fertilizer and thus the feeding of billions. Haber, scarred from dueling and fond of wearing military garb, made a strange match for the internationalist Einstein dressed in coats full of holes. But they, too, became fast friends. Their political differences were concealed by deep respect for each other's scientific work. One unusual but crucial encounter was with an obscure Berlin astronomer named Erwin Finlay Freundlich, who had taken an interest in relativity. He and Einstein had been corresponding for a while before meeting in person. Freundlich was interested in the predictions made by Einstein's extensions to relativity: the gravitational deflection of light and the gravitational redshift. These were tasks better suited to an astronomer than a physicist, and Einstein was intrigued by the possibility of testing his new ideas.

The trip to Berlin also brought Einstein together with his cousin Elsa for the first time in their adult lives (she was both his first and second cousin, through both parents). Albert was enchanted upon seeing her again. Three years older than him and recently divorced with two children, Elsa was a lively and intriguing companion. Einstein's marriage with Mileva had become increasingly tense, especially after moving to Prague, where she was almost completely isolated. Mileva resented the way his growing professional success took him away from the family; Elsa was deeply impressed with the young professor's achievements. Elsa and Albert began corresponding regularly and passionately. In one of his first letters to her he declared, "I consider myself a full-fledged male. Perhaps I will sometime have the opportunity to prove it to you."

One of Einstein's journeys took him to the first Solvay Conference, a gathering of two dozen top physicists and chemists. The conference was funded by the Belgian industrialist Ernest Solvay to grapple with the mysteries of quantum theory and what light it might shed on the problems of radiation. It was held in Brussels at a time when hosting a scholarly conference was a mark of a country's development and sophistication. Participation in international intellectual life was crucial for the status of a modern nation and its citizens.

Einstein was the second youngest in attendance, along with luminaries such as Marie Curie, Ernest Rutherford, and—most important to Einstein—Lorentz. Einstein was giddy at the chance to work with his idol. He praised Lorentz, stately and charming with his thinning white hair, as the mentor who "meant more than all the others I have met on my life's journey." He wrote to a friend, "Lorentz is a marvel of intelligence and tact. A living work of art! In my opinion he was the most intelligent among the theoreticians present." To Lorentz himself he wrote a fawning letter, thanking the elder physicist for his "fatherly kindness."

It was quite a coup for Einstein, still fairly new on the European physics scene, to be invited along with this stellar group. His contributions to quantum theory were seen as epochal, even if they were "the strangest thing ever thought up." Curie was quite impressed with him, declaring, "One has every right to build the greatest hopes on him and to see in him one of the leading theoreticians of the future." This was the first chance many of these distinguished scientists had to encounter Einstein in person and not merely through his papers. He earned significant respect at that meeting, even if some dignitaries were taken aback by the discovery that his laughter was like "the barking of a seal."

His increasing status among physicists led to job offers from the University of Utrecht and, again, from Zurich. He returned to Switzerland once the salary was right and tried to focus more on his research. Specifically, he wanted to return to relativity. He hoped to extend the equivalence principle and other insights he had been toying with since 1907. While relativity was still a fairly obscure subject, others had been thinking about it during this time. Planck wrote a few modest papers. More important was the wholesale reconceptualization of relativity from Einstein's old teacher Hermann Minkowski. Minkowski ran a seminar on electrodynamics that included the special relativity papers of 1905. He was a mathematician and focused primarily on the mathematical structures implicit in the theory, rather than its physical meaning.

Minkowski, like Planck, was interested in the less relative parts of relativity. He noticed that while two observers could disagree about their measurements of time and their measurements of space, there were some that they would always agree on. These were measurements of *space-time*, a new mathematical amalgam of three dimensions of space and one of time. While time and space were relative, the *combination* of them was absolute. This is similar to how people facing in opposite directions will give different directions for turning left or right but can both agree about whether you should turn north or south. Minkowski had found a new kind of measurement hidden under the equations of special relativity that would be independent of movement or point of view.

Minkowski was able to write down a new, elegant mathematical formalism based on this idea. For him the foundations of the universe were no longer three dimensions of space plus time, but rather a four-dimensional continuum in which time and space were woven together. Our perceptions of time and space as separate were fundamentally mistaken. We could take different trajectories through these four dimensions, which result in the different experiences of space and time called for in special relativity. He called this new concept "radical" and was not shy about describing the implications: "Henceforward space by itself and time by itself will totally decline into shadows, and only a kind of union of the two shall preserve independence." Like the prisoners in Plato's cave, we see merely the shadows of reality. Only mathematics could reveal the true nature of things. Minkowski had created a new four-dimensional geometry that was supposedly more real than our everyday perceptions of the world.

Einstein was not impressed. The notion of a four-dimensional space-time was exactly the kind of fuzzy-headed metaphysics he had been hoping relativity would dispose of. In fact, he dismissed Minkowski's theory with the same disparagement he brought against the ether. It was "superfluous." The elegant mathematics were no more exciting than were Minkowski's lectures back at the ETH. Einstein reportedly complained, "Since the mathematicians pounced on the relativity theory I no longer understand it myself." Minkowski had reduced Einstein's subtle and powerful theory to mere geometry.

He had not produced much himself, though. In the three years since the happiest thought of his life, he published nothing on relativity or gravity. While in Prague he finally wrote a paper on those. It was a manifesto, a statement of intent to create a general theory of relativity that encompassed gravity. He presented the equivalence principle and its consequences along with some approximate numerical predictions for the gravitational deflection of light and the gravitational redshift. The redshift was absurdly small and seemed well beyond the possibility of measurement.

The deflection of light, however, might be within reach. If an observer on Earth looked at a star that appeared just at the edge of the sun, she would see the sun's gravity bend the light from the star very slightly. This bending would cause the star's image to be slightly pushed away from its true location, like looking through a thick sheet of glass. Of course, seeing a star next to the blazing sun was impossible, so one would have to wait until an eclipse covered the solar disk. Only in that rare moment could one test Einstein's radical idea. He predicted that during an eclipse such a star would appear to be 0.83 arc-seconds away from its real position. This is a tiny amount—about how large a coin appears from a couple miles away. But astronomers regularly measured effects smaller than this. It could be done. Einstein had high hopes but he needed an astronomer to take on the project. He worried that practical-minded astronomers would be uninterested in his theory (with good reason). In recruiting allies he tried to simultaneously assuage those worries and generate excitement: "It is greatly to be desired that astronomers take up the question broached here, even if the considerations here presented may appear insufficiently substantiated or even adventurous."

His only lead was the enthusiastic but somewhat hapless Freundlich. In September 1911, Einstein said he was "extremely pleased" to have him on board. In an impassioned letter, Einstein explained the importance of this measurement. This was the point on which the theory of relativity would live or die: "One thing can, nevertheless, be stated with certainty: If such a deflection does not exist, then the assumptions of the theory are not correct."

Unfortunately total solar eclipses were rare events and Freundlich hardly had the resources necessary to mount an expedition to observe one. Instead, he examined old photographic plates (precision astronomical pictures needed glass plates) taken at eclipses, looking carefully for any stars showing a deflection. Photographs of eclipses were usually intended to capture the corona or spectacular solar prominences, though, which required a setup precisely the opposite of what Freundlich needed. He saw no deflection. Freundlich began contacting astronomers around the world to see if anyone could help in their quest.

BY THIS TIME Eddington had become an expert in maneuvering through exactly those international networks. In 1912 he became

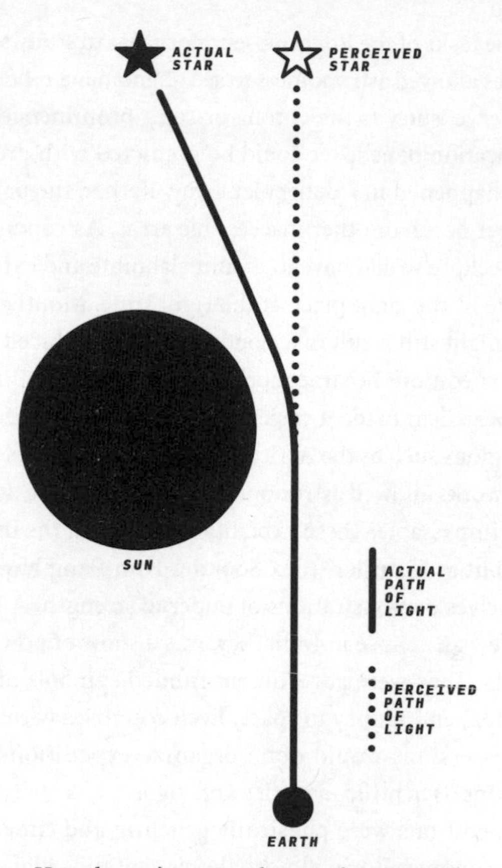

How the sun's gravity changes the perceived position of a star

ORIGINAL ILLUSTRATION BY JACOB FORD

secretary of the Royal Astronomical Society as well as section president of the British Association for the Advancement of Science, and he sat comfortably in the upper echelons of British science. When William Christie had fallen ill, Eddington took over many of the Astronomer Royal's duties at Greenwich. After Christie retired, the post was taken over by Frank Dyson. Dyson had once held Eddington's position of chief assistant and the two became good friends. Beyond their shared passion for precise measurement, Eddington praised Dyson's "complete unselfishness, his sincerity, his sociability." They worked together closely at the observatory, forging bonds that would soon be sorely tried.

One of the tasks of the Royal Observatory was to study solar eclipses. Total eclipses allowed astronomers to see phenomena otherwise impossible to observe, such as the corona or solar prominences. While the time and location of eclipses could be predicted with great accuracy, they rarely happened in a convenient way. Rather, the path of totality might be over ocean or other inaccessible areas. Astronomers wanting to view the eclipse would have to mount elaborate and expensive expeditions to be in the right place at the right time. Months of planning and travel might still result in a cloudy day that produced nothing.

Eclipse expeditions became more feasible in the late nineteenth century as imperialism made it easier and easier for Europeans to access far-flung regions such as the African interior. Naval bases and growing railway networks allowed astronomers to penetrate into areas formerly completely impassable. These expeditions relied on the infrastructure of empire, but as historian Alex Soojung-Kim Pang has shown, they were themselves demonstrations of imperial strength. A British expedition to view an eclipse in India, say, was a show of power over colonial subjects. They were government-funded symbols of superiority, sophistication, and money to spare. Even countries without extensive colonial possessions would often organize expeditions as a way of demonstrating "scientific maturity and vigor."

Such expeditions were physically grueling and time consuming, and the Astronomer Royal almost always sent the chief assistant into the field in his place. When Dyson was the assistant he was a famously lucky observer of solar eclipses, with good weather six out of six times. Now that he was in charge, though, the task of observing the 1912 eclipse was delegated to Eddington. British eclipse projects were organized by the Joint Permanent Eclipse Committee (JPEC) of the Royal Society and Royal Astronomical Society since 1894. JPEC secured government grants, organized personnel, and curated the necessary equipment. Dyson and the JPEC decided that the 1912 eclipse would be best observed from Brazil. Eddington was given the travel details and put in charge of two assistants, Charles Davidson and J. J. Atkinson—generally operating the photographic equipment for an eclipse required three skilled workers. He packed Darwin's *Voyage of the Beagle* to read on the ship and departed Southampton on August 30. Eddington had some experience with preparing and shipping equipment from his time on Malta but there was constant worry that the delicate instruments would be damaged by rough-handed stevedores. This trip had the additional concern of dealing with customs officials, who were often confused by and suspicious of elaborate scientific tools.

The Royal Mail Steam Packet Company carried the group and their equipment free of charge. Their ship passed through Lisbon, which Eddington was delighted to explore. He wrote to his mother about visiting a cathedral where he saw a "mummy of a tiny girl saint." The group landed in Rio on September 16. The Brazilian interior was accessible thanks to new railroads laid down to supply the huge numbers of new plantations that had been springing up. The trains were intended to carry cotton, coffee, and beef, but they handled astronomers and telescopes just as well. The Brazilian government provided them free passage.

It took about six days to get to Passa Quatro, a village about 180 miles from Rio. Keeping track of all their equipment and luggage was hard—at one point Eddington had to sleep on the train platform to make sure they didn't miss their train. The observing site (marked ahead of time by Brazilian astronomers) was a mile from the railway, and the equipment was brought in by oxcart. Local workmen built a brick foundation to support the heavy gear. Three waterproof canvas huts had been brought from Greenwich. The telescopes and spectrograph were set up horizontally in the huts. A clockwork-driven mirror called a coelostat was placed in front of them, reflecting the image of the sun down through the lenses to the photographic plates (see page 58). The sun would move appreciably across the sky during the eclipse, so the coelostat was needed to keep the image steady for the photograph. One of the huts was made light-tight and arranged for developing the photographs. It took nearly two weeks to set everything up, during which the Englishmen complained about the greasy food and the unacceptable tea.

The astronomers had planned an elaborate series of photographs to

An eclipse observation set up in the field. The telescopes lie horizontal under the protective hut. The coelostat mirror can be seen to the left.
SCIENCE AND SOCIETY PICTURE LIBRARY

be taken with four instruments over the few minutes of the eclipse. This would require a finely orchestrated sequence of events executed with military precision: glass plates were slid out of wooden boxes, placed exactly so in the telescope, exposed, removed, and placed into black canvas bags, all while the clocks ticked and the moon crossed the face of the sun. These actions were rehearsed over and over for three days straight until the group was a finely honed machine. One of the coelostats gave Eddington quite a bit of trouble. Its intended smooth turns kept skipping and halting. Eventually they got everything working properly, only to realize that the Brazilians had made a longitude error and the site was some seven miles off the path of totality. But it was too late to move, and a partial eclipse was still valuable.

Crushingly, heavy rain set in the day before the eclipse and continued for a week. The highly disciplined observers waited by the instruments even as they saw nothing but cloud, watching for just a moment of clear sky. There was a sudden onset of darkness at the moment of the eclipse but the sun could not be seen. There was intense disappointment, as the expedition proved entirely unsuccessful after months of planning. Despite the rain, the president and foreign minister of Brazil, as well as several ambassadors, arrived with bands and fireworks. There was a banquet afterward with many speeches in Portuguese, of which Eddington understood not a word.

While a scientific disaster, the trip had other benefits. Eddington was able to get to know Charles Dillon Perrine, an expat American who had been appointed director of the Argentine National Observatory. Like Greenwich, that institution had planned to make several different observations during the eclipse. Most of them would be photographs and spectrographs of the corona, but Perrine also wanted to carry out a test requested by an acquaintance in Berlin—Erwin Freundlich. This was to check a recent prediction that the sun's gravity could bend starlight, displacing stellar images. Eddington surely chatted with Perrine about the project. Perrine would have explained that the prediction had come from a German physicist named Einstein—likely the first time Eddington had heard the name. It sounded like an interesting idea, although not really in Eddington's area of research. And since the observation had failed due to the weather, there wasn't much reason to look more closely at it. Eddington filed it away in the back of his memory.

Brazil had other benefits too. Eddington decided to indulge his glissading obsession and tobogganed some three thousand feet down a mountain. It took about half an hour. He explored the forest on horseback, was charmed by the flying fish in the bay, and watched a battle between two armies of ants. Even more wondrous was seeing fireflies in the rain during a lightning storm: "The scene was like fairyland." All of his local companions gave him the Brazilian farewell ritual, described by Eddington as a hug followed by three pats on the back. The British team sailed for home on October 23 on the *Danube*.

The next April, Eddington and Trimble had their usual Easter walking tour together. They agreed it was one of their best trips despite constant rain. While on that holiday he received some extraordinary news. He had been offered an appointment as Plumian Professor of Astronomy and Experimental Philosophy at the University of Cambridge. This was one of the most distinguished positions in astronomy in the entire country—it dated back to 1704, when its original statutes were drawn up by Isaac Newton. Along with the professorship came the directorship of the Cambridge Observatory. Eddington's time at the Royal Observatory had marked him as one of the leaders in the British scientific community, and this position made that official. He cut short his holiday with Trimble to begin handling the not-insignificant paperwork.

Eddington stayed in London through July, when he moved back to Cambridge. He was delighted to rejoin the Quaker community from his student years. The humble Meeting House on Jesus Lane was unchanged. He was known as an unostentatious worshipper there and refused the honor of sitting in the front of the hall that his status as a professor would have allowed him. For the rest of his life he served on various committees for the Friends and, as a skilled mathematician, handled their accounts.

Without academic responsibilities for the rest of the summer, Eddington went to Bonn for the Solar Union conference and afterward to Hamburg for the meeting of the Astronomische Gesellschaft (the German equivalent of the Royal Astronomical Society). He got to know Karl Schwarzschild quite well on that trip, taking a ten-mile hike with him to Drachenfels Castle. They posed with Dyson and some other friends for a comic photograph with a wooden donkey, a fine composition Eddington titled "Schwarzschild & five mad Englishmen." On his way home he stayed with the family of the astronomer Ejnar Hertzsprung in Copenhagen. He was congratulated widely on his new position and came back to Cambridge exhilarated after a month of intense intellectual engagement.

Upon his return he had to stay in temporary rooms. The Cambridge Observatory had living quarters for the director in the east wing so they would never be far from their work, but he was unable to move in promptly. The observatory directorship was technically a separate position from the Plumian professorship and there was some bureaucratic wrangling before he was able to officially take over. Eddington moved in on March 25, 1914. Cambridge was now his home, though he was regularly in London for scientific meetings and to talk with colleagues.

Typically the director of the Cambridge Observatory would have been married and his wife expected to run the household and act as hostess for scientific dignitaries. The unmarried Eddington brought his sister, Winifred, to live with him and take on those duties. She filled that role for his entire life. As his biographer Alice Vibert Douglas stated, "His interest in women was simply and solely as acquaintances or, in the case of the very few women astronomers in various countries, as friendly colleagues." The Eddingtons brought with them Monty, a cat, and Punch, an Aberdeen terrier "not universally beloved."

The observatory sat behind a modest classical façade about a mile west of the university. Upon entering the building, one turned right into the library. Through a door from there was Eddington's study, the only untidy space in an otherwise fastidious home. Piles of scientific papers on the floor and couch; bookshelves stuffed with both astrophysics and P. G. Wodehouse novels. He would meet with students there, perusing calculations while petting his terrier. Eddington loved the observatory's garden and nearby woods. Visitors were inevitably taken along the walkway and shown Miss Eddington's beehives.

As Eddington settled into his tranquil surroundings in Cambridge, Britain as a whole was more uncertain. The nation had been dealing with episodes of industrial unrest, including a 1911 dock strike and one in coal country the following year. Geopolitical tension had increased steadily if sporadically. Fear of growing German power was quite real, and the *Daily Mail* even ran a series describing a fictionalized Prussian invasion in great detail. Erskine Childers's *Riddle of the Sands* was a bestselling thriller about an imminent German attack. The arms race had a brief respite after diplomatic interventions in 1912, and the kaiser decided that his attention was better spent on the army than the navy (Britain's only real concern).

More likely than a direct confrontation with Germany was Britain being pulled into a difficult situation by their allies France and Russia. Those three countries had formed the Triple Entente in 1907—a loose group intended to counterweight the Triple Alliance of Germany, Austria-Hungary, and Italy. The Entente was a general understanding rather than a formal mutual-defense agreement, so each country was free to make its own foreign policy. The inclusion of despotic Russia with Europe's two great democracies made many uncomfortable.

The division of Europe into these two power blocs was precarious. It was not always certain who was making decisions about foreign relations—Kings? Ministers? Legislatures? When Kaiser Wilhelm threatened the king of Belgium at a dinner party, was that an official statement of policy or merely a drunken monarch known for erratic behavior? Nonetheless the system had already survived various crises; imperial clashes in Africa were generally handled with some level of diplomatic skill. Everyone was fairly cautious. The two alliances were closely matched in military terms, and it was not clear who would prevail in a conflict.

The persistent area of tension between the blocs was the Balkans. Austria-Hungary had been occupying Bosnia and Herzegovina since 1878 and formally annexed them in 1908. Those regions had large populations of Eastern Orthodox Christians isolated by centuries of sequential invasions. The Russians saw themselves as the traditional protectors of those populations and were not at all happy about the Austrian control there. Independence movements spurred by Serbia kept the situation fluid and sabers were rattled constantly. Arguments between Russia and Austria became heated in November 1912, but diplomats were again able to calm the situation.

In a show of imperial control over the region, Archduke Franz Ferdinand, the Austrian heir, planned to visit Sarajevo in the summer of 1914. This seemed a perfect opportunity for a group of Serbian nationalists known as the Black Hand to deal a blow against their oppressors. Three nineteen-year-old men from unhappy households were recruited and given four revolvers and six bombs. As they were smuggled across the border, their drunken escort bragged, "Do you know who these people are? They're going to Sarajevo to throw bombs and kill the Archduke who is going to come there." The sickliest of the group, Gavrilo Princip, nodded and confidently brandished his revolver. He was an excellent shot.

CHAPTER 3

The Wars Begin

"A sin against civilization."

THROUGHOUT HIS LIFE people asked Einstein not just about what his science *meant* but also about how he *did* his science. How did he create the theory of relativity? An often-quoted answer was his claim that doing theoretical physics was merely finding the "simplest conceivable mathematical ideas" and then checking them against experience. "In a certain sense, therefore, I hold it true that pure thought can grasp reality, as the ancients dreamed." This is the image we often have of Einstein—a genius of ultimate rationality, sitting at his desk and smoothly deducing the nature of the universe.

Einstein also warned us, though, not to believe him. In a candid moment he admitted, "If you want to find out anything from the theoretical physicists about the methods they use, I advise you to stick closely to one principle: do not listen to their words, fix your attention on their deeds." Taking his advice, we will follow Einstein down the complicated road he took toward his general theory of relativity. Far from being a serene journey of pure thought, the road wandered from battle to battle. Einstein had to fight through struggles of all kinds—mathematical, conceptual, personal, and, eventually, war.

By the time Einstein was able to bring his focus back to relativity and gravitation he had a lot of catching up to do. Minkowski's reframing of special relativity had attracted a fair bit of attention. Henri Poincaré extended it further, and that pair provided a mathematical foundation for others to explore relativity. Planck's assistant Max von Laue actually wrote the first textbook on relativity in 1911 using Minkowski's, rather than Einstein's, presentation.

Two other physicists, Max Abraham and Gunnar Nordström, also used Minkowski's format to follow up on Einstein's brief 1907 suggestion that relativity might be connected to gravity. They were not the first to propose alternatives to Newton's theory. There were a few options proposed around the turn of the twentieth century, usually trying to explain gravity as a special case of electromagnetism. And electromagnetism at the time meant ether theory—Lorentz commented that ether theory had been so successful with electricity and magnetism that it was "natural" to also use it to understand other forces.

In 1911, Max Abraham, a theorist at the University of Göttingen and former student of Max Planck, started work on his relativistic theory of gravity. Although his theory explicitly built on the equivalence principle, Einstein was generally unhappy with it. It allowed the speed of light to vary and thus broke with the basic postulates of special relativity. It also stuck closely to Minkowski's version, which Einstein thought was mathematically impressive but lacking in physical insight—that is, it told us little about what was happening to real objects in the material world. He also complained that Abraham provided few ways to check his theory observationally. In particular, Abraham did not predict the gravitational bending of light.

Einstein and Abraham fired letters back and forth about these critiques, and their dispute became public in 1912. Einstein saw Abraham's theory as endangering the very foundations of relativity. Abraham's reliance on mathematical elegance as a guide had, Einstein thought, led him astray. "That's what happens when one operates formally, without thinking physically!" He dismissed the theory as "a stately beast that lacks three legs."

The Finnish physicist Gunnar Nordström watched their clash from afar before he presented his own theory. Nordström tried to keep his theory as close to Newton as possible, adopting only the essentials from special relativity. He kept the speed of light constant but did not predict either the anomalous motion of Mercury or the deflection of light. Otherwise, though, his theory was robust and fulfilled most of what someone might have wanted from a Newton alternative at the time.

With competition from Abraham and Nordström, Einstein had to produce something soon. He had been pondering "ceaselessly" on gravity for three years but published nothing. Reluctant to take up Minkowski's mathematical route, he was still trying to understand the physical meaning of his own equivalence principle. What exactly did it mean to say that acceleration and gravity were equivalent? His initial thought experiments involved linear acceleration—that is, being pushed in one direction only. But he started thinking about one special case of acceleration—rotation. Acceleration is any change in the amount or direction of your speed, and if you are spun around in a circle the direction of your travel is constantly changing. So rotation or spinning is considered to be continuous acceleration.

Einstein thought that the equivalence principle, then, demanded some connection between rotation and acceleration. He wasn't quite sure how to apply his own principle, though. He complained to Besso about his lack of progress: "Every step is devilishly difficult." His first results on rotation were published in February 1912 in an uncharacteristically tentative paper.

Einstein still sought physical understanding of what was happening in relativity. When he pondered the bending of light predicted by his thought experiment, he wanted to know what *caused* the bending. *Why* did it happen? Perhaps it was rotation. Circular motion could cause Coriolis forces, which gave a sideways push to something moving on a rotating body (these are what cause hurricanes to spin). Maybe a spinning lab would create Coriolis forces that bent light? That went nowhere.

Einstein then considered an analogy to another phenomenon where light beams were deflected. Light beams were bent as they entered glass or water because they moved more slowly there than they did in air. This was the basic principle of optics that made lenses and eyeglasses work. And in fact, if he was willing to vary the speed of light in relativity he could explain light deflection. But this was precisely the point on which he had attacked his competitors, and Abraham struck back immediately. He gleefully declared that Einstein had delivered a coup de grace to relativity: "Those who, like the author, have repeatedly had to warn against the siren song of this theory, can only greet with satisfaction the fact that its originator had now convinced himself of its untenability."

The key to Einstein extricating himself from this trap was yet another thought experiment. This one involved someone riding a rotating disk, like a carousel at a playground. If the carousel is spinning very, very fast, Alice riding at the outer edge might be moving near the speed of light (the playground has excellent safety equipment). Alice will then experience all the expected effects of special relativity—clocks will run slow, masses will grow, and distances will shrink—compared to what Bob sees while standing on the ground nearby.

Alice and Bob love circles and, as all circle lovers do, decide to measure the circumference (the distance around the edge) and diameter (the distance across) of the carousel. This will let them calculate that beautiful number pi, the ratio of those two numbers. And they will surely get exactly the same number (roughly 3.14159 . . .) that they get every time they calculate that ratio for any circle, ever. The fact that pi is always the same is the very symbol of geometric perfection and has been known since at least the ancient Greeks. Its stability is the guarantee that geometry is universal and (as Einstein liked to say) beyond the merely personal.

So they are not anticipating any difficulties. They will both measure the same diameter: neither of them are moving toward or away from the center, so there are no relativistic effects on that measurement. But they

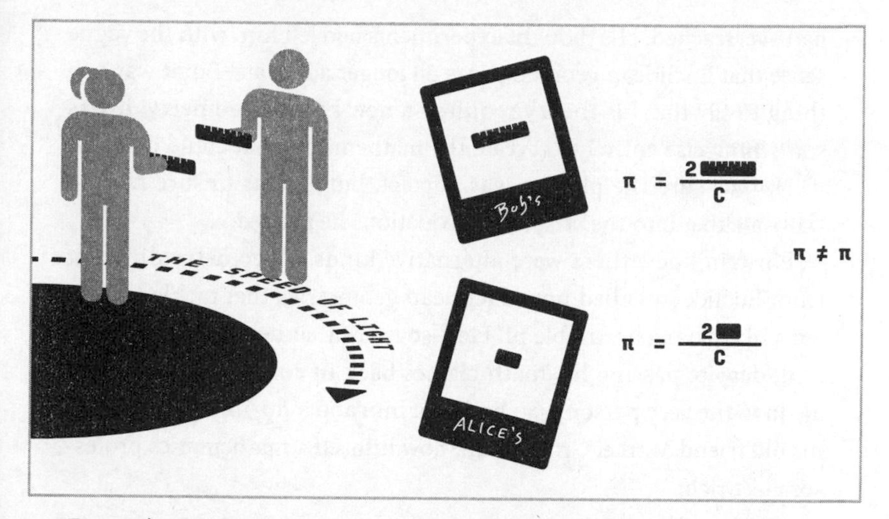

Einstein's rotating-disk thought experiment. Alice and Bob measure the circumference differently because of their relative motion.

ORIGINAL ILLUSTRATION BY JACOB FORD

are moving relative to each other along the edge of the carousel. That movement means that length contraction will occur, and Alice's meter rod will shrink compared to Bob's when they lay them along the circumference. Since they are measuring the circumference with different-size rods, Alice and Bob will come to different conclusions about the circumference of the carousel. When they calculate pi, they hear Pythagoras spinning in his grave: their answers are different.

Relativity had corrupted the soul of geometry. The unshakable foundations of mathematics that so impressed the twelve-year-old Einstein had now been ruptured. Euclid's geometry could no longer hold in a relativistic universe. Moving observers will find that they disagree about the shape of a circle, or how many degrees were in a triangle. Einstein had violated the sacred truths that had set him on this path in the first place. Nonetheless, he had to follow the road he had set himself upon. Even worse than ruining his sacred text was his realization that geometry seemed to be the key to expanding relativity: he grudgingly accepted that perhaps, after all, Minkowski had been onto something important.

As he arrived in Zurich in the summer of 1912, Einstein realized he had overreached. His thought experiments had left him with the vague sense that Euclidean geometry was no longer adequate. But it was one thing to say that his theory required a new kind of geometry; it was something else entirely to create the mathematics that could describe it. He was sure the physics was correct, but he was unsure how to translate that into the crisp, rigid equations he needed.

Einstein knew there were alternative kinds of geometry different from Euclid, so-called non-Euclidean geometries that tackled problems like having a variable pi. He also knew that he wasn't very good at it, despite passing his math classes back in college. So he turned again to the very person who had kept him afloat during those classes: his old friend Marcel Grossmann, now himself a mathematics professor at Zurich.

IMMEDIATELY AFTER MOVING into his new apartment, Einstein rushed over to Marcel's house. Bursting in, he exclaimed, "Grossmann, you must help me or else I'll go crazy!" Marcel had many years of experience calming Einstein down, and began explaining the mathematics he would need. The foundations for what Einstein needed had been laid down by Carl Friedrich Gauss back in the early nineteenth century. Gauss was a fantastically skilled mathematician who invented many of the quantitative methods essential for modern science. Among his other innovations, he pondered what non-Euclidean geometries might look like. For Euclidean geometry, you usually imagine that you are standing on a flat table where you draw your triangles and circles. Gauss wondered what would happen if you were standing on a curved surface, like the top of a hill or the inside of a tube. The answer is that strange things happen—for instance, a triangle drawn on a globe can have 270 degrees instead of the 180 degrees demanded by Euclid.

Gauss, and later his student Bernhard Riemann, developed elaborate mathematical systems that would describe the forms of geometry that would exist on all kinds of surfaces. These geometries relied on the notion of the "curvature" of the surface at a particular point. A flat table, for instance, has no curvature; a water pipe has some curvature; a drinking straw has more intense curvature. The curvature can even change from point to point on a surface—imagine a half-rolled carpet. The triangles you draw on each of those surfaces will follow different rules, and non-Euclidean geometry is all about understanding those new rules.

So if Einstein was going to apply these mathematical tools to Minkowski's four-dimensional space-time, he had to decide what it meant for space-time to have "curvature." If you imagine someone walking along a sheet of space-time, curving or bending that sheet will distort the measurements of space and time they make of the world around them. And distorted measurements are precisely what relativistic effects like time dilation and length contraction consist of—those strange phenomena were just how humans perceived the curving of four-dimensional space-time. Our flawed perceptions like to divide things up into space and time, but the universe sees them as one continuum where stretching time results in squeezing space. As Minkowski said, we see only shadows.

Einstein's spinning-disk thought experiment convinced him that this curvature would be induced by acceleration (since spinning is just a special kind of acceleration). Alice riding the carousel would experience space-time curving around her. And by the equivalence principle, acceleration and gravity have to be indistinguishable. If acceleration curved space-time, then gravity curved space-time. Gravity bends light, then, not because of a pulling force, but because light simply follows the natural curve of space-time.

To talk about things like the path of light through space-time,

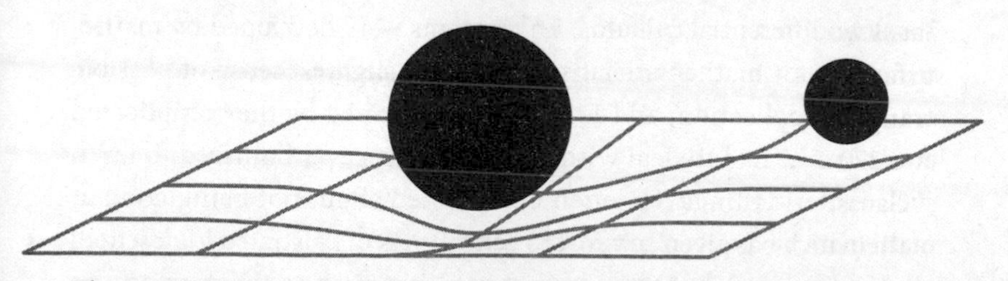

The warping of space-time by gravity. Larger masses create more severe curvatures. ORIGINAL ILLUSTRATION BY JACOB FORD

Einstein needed a way to define distances in four dimensions. Minkowski had found the *interval of space-time*, which worked well for noncurved space. It was invariant, meaning that all inertial observers would always agree on its value. But for curved space-time, Einstein learned that he needed a new kind of mathematical object called a *tensor*. This is a complicated mathematical entity that had a property crucial for Einstein. It is *generally covariant*, meaning that different observers, moving in whatever strange ways, will all agree on its value. This was essentially a promotion of the first postulate of special relativity. Beyond saying there are no privileged inertial reference frames, covariance demands that there be no privileged observers no matter what. Something written as a tensor would be invariant for everyone. Covariance was what would give a truly universal physics independent of the traps of perspective and individuality.

And this universal physics would have to be written in tensors, so it could be recognized no matter one's position or movement. Unfortunately for Einstein, the mathematics of tensors was, to say the least, complex. A single four-dimensional tensor had ten separate elements that one had to keep track of, all of which could be constantly changing. Worse, the tensors he needed were nonlinear, meaning that the result of an equation actually changed the value of the numbers originally put into the equation, creating a kind of feedback loop that made calculating them very difficult.

Grossmann showed Einstein that there were several kinds of tensors: the Riemann tensor, the Ricci tensor, and so on. Pure mathematicians such as Tullio Levi-Civita in Italy had spent decades developing elaborate mathematical systems based on these tensors, known as the absolute differential calculus. The systems were developed by mathematicians for mathematicians without the slightest sense of possible practical application, and Einstein was stunned by the complicated equations he had to deal with. Few physicists ever bothered to learn such esoteric things. He often complained about not being good at math (which has given comfort to generations of frustrated high school algebra students), but that was just in comparison to the company he kept. He was very good at math. If you spent all your time around world-class mathematicians such as Grossmann, though, just being "very good" began to feel inadequate. Einstein wrote, "Never in my life have I tormented myself anything like this, and . . . I have become imbued with great respect for mathematics, the more subtle parts of which I had previously regarded as sheer luxury! Compared to this problem the original relativity theory is child's play." He thanked Grossmann for having "saved me," just like he had back in college.

Einstein asked Grossmann to stay on the project and help him through the mathematics. Grossmann was careful to avoid any responsibility for the physics. By mid-1912 they had a good sense of what they were looking for. In order to make a general theory of relativity, they needed to find equations that fulfilled several criteria:

- They had to be generally covariant. They had to express the laws of nature in a form (probably a tensor) that would make sense to any observer in the universe, no matter how they were moving. This had to include all of the strange features of special relativity as well.

- They had to include the equivalence principle. They had to make gravity and acceleration indistinguishable, and predict things like the gravitational deflection of light and the gravitational redshift.

- They had to contain the great laws of classical physics, the conservation of momentum and energy. To Einstein these were nonnegotiable. Whatever general relativity ended up looking like, it had to preserve the things we already knew were true.

- They had to explain why Newton's law of gravity worked so well. Newton's theory was adequate for almost all circumstances, and Einstein had to show why we spent a couple centuries thinking it was correct. In physics jargon, we say that Newton's law must be a *limit* or *approximation* of Einstein's new law, apparently where gravity was fairly weak. Hopefully general relativity's equations should apply in all cases, but in some of those cases they must look so much like Newton's law that we can barely see the difference. We say that we should be able to *reduce* Einstein's laws to Newton's.

There was a natural split among these criteria that suggested two strategies for finding general relativity. The final three are questions of physics, of how the physical world operates, of the actual processes that make everything go. What kind of law could combine Newton's theory, the conservation of energy, and the equivalence principle? What might nature look like when we mix all of those together?

The first criterion, however, was a question of mathematics. Mathematicians knew a great deal about tensors and non-Euclidean geometry. Could Einstein and Grossmann find what they needed just through manipulating symbols and following abstract principles of mathematical simplicity and elegance? Scholars who have studied Einstein's notebook from this period (notably Michel Janssen, Jürgen Renn, John Norton, John Stachel, and Tilman Sauer) refer to these two options as the physical and mathematical strategies for finding general relativity. Einstein and Grossmann pursued both routes, often switching strategies when they hit a roadblock along one.

They worked ferociously over the winter and spring of 1912–13. Grossmann might suggest a particular tensor, and Einstein would check it against the requirements of physics. This one wasn't compatible with the equivalence principle; that one didn't look enough like Newton's law. Or Einstein would come up with an equation that fulfilled his physics needs and ask Grossmann if there were any tensors that looked like it. Every time, it seemed, there was one piece of the puzzle that wouldn't fit. The tensors were complex objects and the pair often tore them up and rearranged the pieces—what if we use the first half of this one, or just the last chunk of that one? They tinkered, balancing mathematical beauty against the stringent requirements of the physical world.

Einstein's notebook from that winter reads like a mathematical melodrama. We can see his false starts, his notes to himself, his hopeful jottings. The frustration when he simply crossed out an entire page of calculations. One group of equations from late 1912 looked particularly promising, but he couldn't see how it would reduce to the Newtonian limit. The calculations with this set of equations were so tedious and complicated that they weren't worth the time to complete. He complained that his brain was "much too feeble" to handle this. He gave up, and moved on to another possibility.

It did not seem that there were any equations that would fulfill all of Einstein's needs. Something had to give. By May 1913 he decided that he had found the best *field equations* (that is, the equations that describe any gravitational fields) that he could. His equations gave him the equivalence principle, the conservation of momentum and energy, and could be massaged into a Newtonian approximation. But he could not make them generally covariant. They could not provide a truly general relativity—not all observers were equal. This was maddening, as this was essentially the initial premise of the entire theory of relativity. The conclusion of his calculations contradicted his starting point. He complained to Lorentz that without covariance "the theory refutes its own point of origin and therefore hangs in the air."

Could a house be built that destroyed its own foundations? He was both frustrated and proud. These equations were powerful and could do almost everything he asked of them. "Deep down I am now convinced that I have hit on the right solution, but also that a murmur of outrage will run through my colleagues when the paper gets published." This version of the theory came to be called the *Entwurf*, after the label Einstein and Grossmann gave it, meaning something like "outline" or "draft." Einstein admitted that the Entwurf was "more in the nature of a scientific credo than a secure foundation." He could feel it was incomplete but was not sure how to proceed. He tried to express his unease to a friend: "Nature is showing us only the tail of the lion. But I have no doubt that the lion belongs to it, even though, because of its colossal dimensions, it cannot directly reveal itself to the beholder. We see it only like a louse sitting on it."

Nature was not always cooperative. Einstein apologized for the "equations of considerable complexity" required by the Entwurf. One of his colleagues, Max Born, expressed admiration at the stunning abstraction of the theory—gravity, light, time, space, and motion all stripped down and synthesized into unvisualizable, unintuitive symbols on a page. Most physicists were skeptical of that magnitude of abstraction. Wasn't this mere speculation?

Einstein knew he needed to fight back against that critique, to show how his ideas could connect to the real physical world. The best way to do that was with the tests he had already been pondering. He recruited Besso to help with calculating whether the Entwurf could explain the anomalies of Mercury's orbit. An exact calculation was impractical, and even their approximate methods were tedious. They did get a number, though. According to these equations, Mercury's perihelion should advance by 18 arc-seconds per century, less than half of the observed 43 arc-seconds. Not reassuring, but there were still two other tests.

The predicted redshift was still too small to be seen. The deflection of light, then, would have to be decisive (it helped that the deflection would also distinguish relativity from Nordström's competing theory). He pushed Freundlich for the possibility of observing the deflection during daytime, and even wrote to George Ellery Hale of the titanic Mount Wilson Observatory in California. Hale wrote back that it was impossible. Einstein would have to wait for the next solar eclipse.

ZURICH HAD PROVED to be a productive place for Einstein. He was lecturing every week but didn't bother to prepare well, so teaching did not take too much of his time. His office became a central place for faculty interaction—not because of his interpersonal skills, though. The cigar-puffing Einstein made his office the last redoubt against the building's nonsmoking policy, so any colleague needing a quick smoke suddenly found an issue on which they needed his thoughts.

Marie Curie brought her daughters to visit Einstein there and the two families hiked into the Engadin valley in the Swiss Alps. Curie spoke little German, and French had been Einstein's worst subject in school. Nonetheless, they seemed to communicate fine. Curie was most interested in discussing quantum theory, while Einstein was deep into the problems of relativity. One of her daughters recalled: "Einstein, preoccupied, passed alongside the crevasses and toiled up the steep rocks without noticing them. Stopping suddenly, and seizing

Marie's arm, he exclaimed: 'You understand, what I need to know is exactly what happens in a lift when it falls into emptiness.' "

Planck, too, wanted Einstein's insights, and turned his institutional power to bringing him to Berlin. He had plenty of resources to make this happen. Both the German government and industry were interested in making Berlin an intellectual center. As a young empire, Germany was still anxious about showing the world that they were the equal of the other great nations. Patronage of the sciences and intellectual life in general was a way to demonstrate their progressive modernity. Planck put together an irresistible package: a professorship with no teaching duties in which Einstein would become the youngest-ever member of the prestigious Prussian Academy of Sciences. The salary would be 12,000 deutschmarks, the maximum allowed by law, half paid by the Academy and half by the industrialist Leonard Koppel.

Industrial money would also fund a new Kaiser Wilhelm Institute of Theoretical Physics under Einstein's directorship. Much of the rapid expansion of German science in these years was built on the Kaiser Wilhelm Institutes (KWI)—semiautonomous organizations funded largely by private money. Many of the donors were Jewish, since highly visible philanthropy was a good way to demonstrate their German identity. Each KWI focused on a particular scientific field and was intended to conduct cutting-edge research. Einstein's friend Fritz Haber had become head of the chemistry institute in 1912. He used the deep resources available to make Berlin an unrivaled world leader in physical chemistry. When he heard Planck was trying to lure Einstein there, Haber quickly joined in to convince the government to welcome his friend.

Planck laid his personal standing on the line to secure the deal for Einstein, but he still had to convince the young iconoclast to accept it. He and his colleague Walther Nernst went to Zurich to persuade him. Upon their arrival Einstein begged for just a little more time to make a decision. He encouraged them to take a sightseeing trip and return to the train station at the end of the day. In a totally unnecessary bit of theatricality, Einstein said he would be waiting there wearing a white flower if he rejects the offer, red if he accepts. The flower was red.

It was not obvious to his colleagues why Einstein agreed to go to Berlin. He did not have good memories of living in Germany, Zurich was paying him well, and he had many friends and colleagues there. Most of the professors in Berlin were conservative, wealthy, and saw themselves as an integral part of the German Empire. The famous physiologist Emil du Bois-Reymond declared that the university community was "the spiritual troops" of the kaiser's dynasty. These were not people Einstein was likely to get along with.

So why did he decide to go? Not having to teach was certainly a draw; he wrote to Lorentz of his excitement at being able to "indulge wholly in my musings." Planck, Haber, and Nernst promised a particularly exciting intellectual community. But the major reason was almost certainly love—it was an opportunity to continue his affair with Elsa. Their correspondence increased in intensity and he reassured her that Mileva would be no obstacle to their relationship.

The Einsteins moved to Dahlem, a suburb of the German capital, in the spring of 1914. Mileva found her husband distant and uninterested. She was surely not surprised to learn about Elsa (this was not the first time she had suspected Albert of straying). The Einstein household rapidly became an uncomfortable place for everyone. Mileva and the boys moved out just a few months after arriving in Berlin. They stayed in Haber's house (which, they found, was decorated with pictures of himself). Haber's wife, Clara Immerwahr, was a talented chemist in her own right and was not at all happy about having to spend her time cleaning up after Einstein's marital mess.

Haber offered to act as an intermediary between the feuding couple, relaying messages and suggesting compromises. He had been one of Einstein's few confidants during the whole episode (he knew about Elsa before almost anyone else) and looked for opportunities to calm the situation. Einstein refused to continue living with Mileva in any way other than a "loyal business relationship." He said he was only willing to do even this for the benefit of the children. With typically Einsteinian bluntness he composed a patronizing, insulting document listing what he would require of Mileva in order to remain married. She must do his laundry, provide his meals, and keep the bedroom and office tidy. She could expect no social interaction with him, public or private. She must cease talking to him or leave the room if he demanded it.

Mileva agreed, but staying in Berlin was just too painful. She planned on returning to Zurich with the children. Haber had a final three-hour meeting with Albert to hammer out the details. At the end of July the reliable Besso came to Berlin to gather Mileva and the boys and escort them back to Zurich. Haber accompanied Einstein to the rail station to say good-bye. Einstein, crying as the train carrying his children rolled away, leaned on his friend. He wrote a note to Elsa: "Now you have proof that I can make a sacrifice for you."

He decided that he and Elsa had to "act very saintly" while everything was in upheaval, lest they become the focus of gossip, and they did not see each other for a month. Suddenly alone, Einstein found himself able to focus on his work. Berlin's bustle provided some shelter from his family drama: "You get so much outside stimulation that you do not feel your own hollowness so harshly as in a calmer little spot." The intellectual community was remarkable. He reported admiration for the "sheer amount of competence and glowing interest in science one finds here! . . . And the people, you ask? They are basically the same as everywhere. In Zurich they feign republican probity, here military rigidity and discipline."

There was plenty of military display at the Prussian Academy, where Einstein attended weekly meetings. He complained about the "peacocklike" fellows at the Academy and the requirement that he maintain "a certain discipline with regard to clothes and so on." The building sat ostentatiously on Unter den Linden in the center of Berlin. Full of pomp and ritual, it no doubt brought back all of Einstein's unpleasant childhood memories of German formality (it certainly brought back his tendency to laugh at the self-importance of it all). Nonetheless, Planck had worked hard to make everything perfect, to the point of Einstein protesting that he was not a prizewinning hen. And despite everything being in its place, the Academy was at best lukewarm to his work on relativity—he harrumphed that it only got "as much respect as it does suspicion."

It was not until July, as his personal life crumbled around him, that Einstein first addressed his fifty or so colleagues at the Academy. He thanked them for making it possible to devote himself entirely to science. He reminded them that he was a theorist, not an experimentalist. This meant that he was engaged with the difficult task of trying to "worm" general principles out of nature, a task for which there were no blueprints or reliable methods. Einstein said he was looking for no less than a complete reworking of the foundation of physics.

Einstein's address ended on a warning that, even if he found these principles, there was no guarantee that they would be accessible to experiment or observation. This, he said, was where relativity found itself. But while he told the Academy not to expect empirical confirmation anytime soon, he immediately began trying to get precisely that confirmation. Now that he had access to the resources of the Reich he could finally get Freundlich to look for the gravitational deflection of light. There was a perfect opportunity to make this test at the imminent eclipse of August 21, 1914, in the Crimea. Freundlich was eager to organize the expedition and carry out the observation himself. He arranged the use of four astrographic cameras for the eclipse and planned the journey into Russia.

With Planck's help, Einstein was able to secure the serious sum of 5,000 marks to pay for the expedition. The Academy covered 2,000 marks; the remainder came from the Krupp corporation, which had grown rich selling heavy guns to the German Army. On July 19, Freundlich's party embarked on a train in Berlin and headed east.

As the German astronomers entered Russian territory, Eddington was finishing a week of camaraderie in South Africa. The British Association for the Advancement of Science had stopped there on its way to Australia, where a large international scientific meeting was planned. The stated purpose of the meeting was to "celebrate the system of exchange of scientific thought" between the northern and southern hemispheres. The officers of the BA (Eddington had just been made one) left Britain in June, dedicating a full four months to the lengthy round trip.

Eddington was pleased to spend the journey in close contact with his friends Dyson and the geneticist William Bateson. Even with friends, he was not a great conversationalist. Contemporaries noted his "hesitancy, even inarticulateness at time in general conversation." He was not much for small talk, but if one could get him on the right subject—say, mystery novels—he would chat easily, making puns or sly wordplay. Sports was another reliable topic. At one astronomical meeting he cut off discussion of giant star populations so he could go get the cricket scores.

All of the officers of the BA were subjected to the traditional hazing of the Father Neptune ceremony as they crossed the equator. We don't know exactly what rituals they were forced to perform (the ceremony varies from ship to ship), though it was probably more of the embarrassing than physically dangerous variety—perhaps eating uncooked eggs and having to wear their clothes backward. Eddington's letters home to his mother mention his delight in the clear views of the Southern Cross and "the finest stretch" of the Milky Way.

After Cape Town the group made their first stop in Australia, Perth. He and Dyson visited the geological marvel of the Kalgoorlie gold mine and hiked with botanists to see eucalyptus and banksia at the spectacular Lesmurdie Falls. Eddington delivered a public lecture on the movements of the stars, which drew some four hundred people. The rest of the scientists attending the conference joined up with the BA there. The news they brought from Europe gave a somber air to the proceedings.

ARCHDUKE FRANZ FERDINAND and his wife, Sophie, arrived in Sarajevo on their wedding anniversary. They never liked traveling with much security and wanted to have a relaxed trip. The path of the motorcade from the train station to city hall had been well publicized, and the Black Hand waited along the route. One of them threw a bomb, wounding members of the archduke's party and cutting Sophie's cheek. The rest of the assassins either froze or, assuming the sound of the explosion meant the deed was done, waited quietly. Incredibly, the cars continued on to their destination. Franz Ferdinand made a short speech there and then asked to visit one of his men wounded in the bombing. A fantastic combination of poor directions and a car with no reverse gear meant the Austrians had to stop on a crowded street for a moment. A young man on the sidewalk couldn't believe his luck. Gavrilo Princip took aim—he almost hesitated when he saw the duchess—and fired twice. Both Sophie and Franz Ferdinand were shot. Bystanders recall the archduke's ostrich-plumed helmet toppling. The heir to the Austrian throne died about eleven a.m.

Several members of the conspiracy were quickly arrested and it was found that their weapons were of Serbian manufacture. The Austrian authorities knew that this was not definitive evidence of responsibility, but they also noted the excitement about the assassination in the Serbian press. Vienna started a full investigation, though the outcome was never really in doubt. The question was only what the response would be. Austria-Hungary had been vacillating between rough and gentle treatment of Serbia for years. Ironically, Franz Ferdinand himself had always advocated for peaceful measures. His death made that impossible.

Austria-Hungary was well aware that their decision would have more than local consequences. The previous Serbian crisis in 1913 almost led to Russian intervention. So on one hand they knew the stakes could be high; on the other, they had made it through that crisis without disaster. Germany advised their allies that Russia would probably not defend Serbia—they certainly would not want to support attacks on a royal family. Nonetheless, Berlin recommended, if action was to be taken it should be done quickly while public outrage was high.

Unfortunately for the Germans, Austro-Hungarian political culture moved at a snail's pace. There were rules and formalities to follow. They decided they needed to send an ultimatum and receive a reply before anything was done. The ultimatum had ten points, demanding Austro-Hungarian authority over the internal Serbian investigations into the assassinations and suppression of anti-Austrian propaganda. It was designed to be refused (perhaps like Einstein's demands to Mileva). Winston Churchill, then Britain's First Lord of the Admiralty, called it "the most insolent document of its kind ever devised." The Russians expressed support for Serbia. Regardless, Belgrade's reply was as compliant as possible. It was poorly composed—they rushed to meet the forty-eight-hour time limit and their typist was inexperienced. The reply arrived just before the deadline.

The only point they explicitly rejected was Austro-Hungarian control over the investigation and prosecution. Anticipating that their reply would not be well received, Serbia began mobilizing its army and evacuated Belgrade. On the morning of July 28, the emperor of Austria-Hungary, Franz Joseph, declared war. Russia began activating their army to protect their Serbian friends. Germany sent spies disguised as tourists into France and Russia to watch for military movements.

If the Russians were committed, then there was tremendous pressure on Germany to mobilize. The Schlieffen Plan depended on beating their enemies to the punch on two fronts. If they waited too long there would be no chance for victory. Even as his military advisers pressed for immediate action, Kaiser Wilhelm was not sure. He thought the Serbian response to the ultimatum was surely enough to satisfy the Austro-Hungarians. As he usually did in the summer, Wilhelm went on a Scandinavian cruise and put in an appearance at the Kiel Regatta. He socialized with representatives of the Royal Navy, impressing them with his English.

When it became clear that he could not dissuade his allies in Vienna from moving forward, he thought perhaps he could get the Russians to relent. Telegrams went back and forth between him and Tsar Nicholas looking for common ground. It is important to remember that the monarchs of Europe—England's George V, Russia's Nicholas II, Germany's Wilhelm II—were hardly implacable opponents. They were all cousins. Surely Queen Victoria's grandchildren could work something out among themselves?

The moment became the perfect mix of danger and opportunity.

The German military was still the strongest on the Continent, but Russia and France were steadily improving. So a conflict now (as opposed to in a year or two) would be to Germany's advantage. Further, since Russia had already mobilized, Berlin could say they were only acting in defense. Wilhelm was confident that he could keep Britain out of the war too. On July 31 he announced a State of Imminent Danger of War. Mobilization had begun, the Schlieffen Plan was under way. If Germany were to support Austria-Hungary, they had to attack Russia. If they were going to attack Russia, they would have to fight France as well. Carefully crafted tables of troop movements were opened and activated. A small force would go east to keep Russia occupied while a massive attack west hopefully knocked France out of the war.

The French president, Raymond Poincaré, visited Russia, suggesting that they honor their treaty obligations to attack Germany. French diplomats also pressed Britain for a decision. Great Britain was obligated by an 1839 treaty to defend Belgium against attack, which was a likely circumstance. London worked desperately to avoid any formal commitment. This was partially policy—no one wanted to be dragged into a war against their will—and partially because the cabinet simply could not agree among themselves what to do. France hoped that if Britain would make a statement of solidarity, perhaps Germany could be dissuaded. Frustratingly, Britain announced that there could be an "honourable expectation" of intervention, but no "contractual obligation." The prime minister, H. H. Asquith, could not formally mobilize the armed forces. Instead, when Churchill asked about bringing the fleet to war status, he affirmed with "sort of a grunt."

The historian A.J.P. Taylor blamed the start of the war on railway timetables. This is not too far from the truth. Rigidity of planning dominated all the powers' strategies. Germany's hope for victory, the Schlieffen Plan, relied on an intricate plan of armies moving east and west on the nation's preeminent rail network. Every coordinated piece had to be in place, every loading and unloading happening at a precise time. And it was a single unit—removing one part of the plan would ruin everything. It was all or nothing. So at the last moment, when Wilhelm had second thoughts and ordered the attack against France halted, he was told *no*. It was impossible. Once begun, the schedule could not be interrupted.

Similarly, the plan could only work if the German Army passed through Belgium, whose neutrality was officially guaranteed by treaty. But the plan called for it. Military necessity trumped diplomatic niceties. Berlin hoped that the Belgians and the British would understand this and not oppose the advance. The Belgians saw their sovereignty at risk, though, and defended their border ferociously.

As the first German shells dropped on the Belgian fortresses of Liège, British political will stiffened. The cabinet agreed to defend Belgium's neutrality and the British Expeditionary Force (BEF) was readied for deployment to the Continent. After a month of tightening tension, Europe was fully at war. Across the major players there was a sense that something extraordinary had begun. The evening before Britain declared war on Germany, Sir Edward Grey, the foreign secretary, wrote: "The lights are going out all over Europe and we shall not see them lit again in our lifetimes." In Germany, even the more sanguine General von Moltke worried that the war might "destroy civilization in almost all of Europe for decades to come."

EINSTEIN HADN'T BEEN paying much attention. As August began he was anxiously awaiting news from Russia—but news of a solar eclipse, not of troop movements. Freundlich's team had begun setting up their telescopes and cameras in the Crimean countryside, preparing to test Einstein's prediction. The Crimea was a major naval base for the Russians, however, and the authorities began watching them closely as international diplomacy broke down. Once war was declared, they decided that these enemy citizens with sophisticated optical equipment were more likely to be spies than scientists. Freundlich was arrested at gunpoint and the equipment impounded. He was suddenly less concerned with testing general relativity and more worried about surviving the internment camps. He was taken to Odessa to be interrogated.

Eddington received confirmation of war between Germany and Russia on August 3. Four days later, British troops were landing in France. The BA had to decide whether to proceed with their meeting. There were delegates present from all of the warring countries. Surely it was in the very nature of an international scientific conference that it should proceed regardless of nattering politicians back home? It was decided that the meeting program would continue as planned. At the plenary dinner, Sir Oliver Lodge, the outgoing president of the BA, stood to declare that science knew no politics. He toasted the health of the Germans present, and his words were received with brief applause. In an effort to make the German and Austrian attendees feel welcome, the Australian hosts carried out planned tours to scenic spots such as sugar plantations. Trying to put a happy spin on the fighting in Europe, some wag rewrote the military anthem "The British Grenadiers" to honor the valor of scientists instead of soldiers.

As the BA boarded the ship that would take them home, some of the new realities began to set in. All the deck lights were covered and all cabin windows boarded over to present less of a target to enemy ships. The mood on board was increasingly patriotic as the ship approached home. The chemist William Herdman conducted experiments making knockout gas for use in battle. Eddington glimpsed the volcanic peak of Bali as they diverted to Singapore, where the ship took on soldiers heading to the front. He noted grimly that they were now a military vessel and would be a "fine prize" for German raider ships like the *Emden*. This was a real possibility—the *Emden* captured two dozen ships in the Indian Ocean, and at one point a false report was circulating in London that the BA ship had been sunk with all scientific papers lost.

They passed through the well-fortified Suez Canal, skirted the site of Eddington's youthful work on Malta, and returned to an England three months into the war. The country had been transformed. Through July, as the crisis had worsened, the British public and press had been generally skeptical that the nation should become involved. It was yet another Balkan crisis of no particular concern—everyone was much more worried about the possibility of civil war in Ireland.

As involvement became more and more likely, though, liberals and conservatives alike warned against being drawn in. The *Manchester Guardian* ran antiwar campaigns; the *Yorkshire Post* called for isolation; the *Cambridge Daily News* called British interests in the situation "quite negligible." Memories of the Boer War, where the overconfident British Army found itself dragged through three years of brutal guerilla fighting, were still relatively fresh. There was much worry about the economic consequences as well—the modern international trade networks would all be disrupted, with unpredictable consequences. Suffragettes and socialists held peace rallies. Two of the most important scientists in the country, William Ramsay and J. J. Thomson, published a letter in the *Times* arguing that Britain must stay out of the war because fighting against Germany, that font of art and science, would be "a sin against civilization."

The last few days before Britain declared war coincided with a bank holiday. Most people tried to ignore politics and enjoy their time off. The politicians continued to debate in Whitehall, which provided some entertainment for the holidaymakers. Groups strained to catch "a glimpse of Ministers as they arrived. Quiet and orderly, this typical English crowd . . . There was no feverish excitement." The police remarked on how calm they were.

The mood shifted almost immediately upon declaration of war. Thirty thousand people gathered around Buckingham Palace to sing "God Save the King." The windows of the German embassy were quickly smashed. The Church of England called for men to join the army: "The country calls for the service of its sons. I envy the man who is able to meet the call; I pity the man who at such a time makes the great refusal." Belgium was compared to biblical Israel, a tiny nation struggling against overwhelming enemies. Anglicans, Baptists, and Catholics competed to see who could summon more recruits.

Crowds became more exuberant as military preparations advanced, surely at least in part to send the troops off with high morale. It had been tradition for the United Kingdom to maintain only a small standing army, relying on the Royal Navy to protect it. So when the British Expeditionary Force was deployed to help stem the German advance, they numbered only 250,000. This was one-eighth the size of the German Army, then the largest in history. Rumors spread that the kaiser ordered his soldiers to crush the "contemptible little army." The BEF happily embraced the insult and began calling themselves the "Old Contemptibles."

EINSTEIN STILL WASN'T paying the slightest bit of attention. As the war erupted he was focused completely on the disintegration of his marriage: sending money to Mileva, deciding who gets to keep the blue sofa. The first mention of the war in his letters was two weeks after the fighting began, when he worried that "My good old astronomer Freundlich will experience captivity instead of the solar eclipse in Russia. I am concerned about him." To his friend Paul Ehrenfest in the neutral Netherlands he lamented that the war revealed "what [a] deplorable breed of brutes" Europeans were. He had isolated himself, though, and tried to keep his attention on his calculations. "I am musing serenely along in my peaceful meditations and feel only a mixture of pity and disgust."

Max Planck felt serene too, but for very different reasons. He was delighted, not by the war per se, but by how it had brought unity to the country: "The German people has found itself again." The war had created a common will that plastered over domestic disagreements. The beginning of the war happened to coincide with the University of Berlin's Founders Day festivities, where Planck was delivering a lecture. His speech, nominally scientific, began with a celebration "that all the moral and physical powers of the country are being fused into a single whole, bursting to heaven in a flame of sacred rage."

His two sons were of military age and headed to the front lines; his daughters went to work in field hospitals. Planck was very proud. Most scientists in Berlin felt similar enthusiasm, though not many followed Walther Nernst's example of enlisting in the army at age fifty (both of his sons had already joined). Einstein's friend Haber applied for an officer's commission and began reorganizing his institute for military projects.

Einstein's self-enforced isolation couldn't hide the surging patriotism around him. A state of siege was declared in Berlin and militarism was everywhere. The streets were full of people waiting for news of the war. Excitement gripped the crowds. One journalist wrote: "The individual disappeared. The intellectual, nervous, distinguishing man of culture lost control of his feelings and belonged to the masses." As Einstein saw it, this was precisely what was happening to his colleagues around him. The most refined, learned people he knew were suddenly rallying for "senseless butchery." Rationalism had been displaced by nationalism. It was "horrific," he said. "Nowhere is there an island of culture where people have retained human feeling. Nothing but hate and a lust for power!"

EVEN HIS COMMUTE from Dahlem to Berlin was disrupted by the mobilization of the army and by the thousands who gathered in the stations to watch the soldiers depart. For a while he simply worked from home. The trains were again disrupted on August 15, when the first trains carrying wounded arrived in the city. A week later the stations swelled with thousands of refugees from the fighting in East Prussia.

For weeks those trains darted back and forth across Germany, depositing soldiers and equipment at their embarkation points. For all the technical sophistication of the army, once they left the railway sidings they were largely powered by muscle. Every soldier moved forward on their own boots; heavy equipment was pulled by the largest gathering of horses in human history. The lightning advance envisioned by Schlieffen moved at the speed of footsteps.

The first obstacle of that advance was the heavily fortified Belgian border. The city of Liège, a major industrial and cultural center, sat astride the path Moltke needed to invade France. Four hundred heavy guns and forty thousand troops held the walls of a dozen concrete fortresses. The Germans had to not just take these forts but take them quickly, as the whole advance depended on swift movement at the start.

Krupp, the same manufacturer that funded Freundlich's expedition, had built special howitzers to penetrate these specific defenses. It took a few days to set up the massive cannons, which could hit targets from four kilometers away. The guns were so powerful that their crew had to retreat three hundred meters before activating them, lest the pressure wave be fatal. Their 2,000-pound shells were unlike anything imagined by the architects of the Belgian citadels, and the defenders were quickly hammered into submission.

By August 20, the Germans had occupied Brussels. Their advance through Belgium had been fast but cruel. Any village that resisted was burned. To discourage partisans or other guerilla resistance the attackers took many civilian hostages, who were often killed. The Germans were obsessed with maintaining their schedule and allowed nothing to impede them. Even Moltke described the advance as "certainly brutal."

As the kaiser's troops approached the Belgian–French border, the British raced to meet up with their French compatriots. The BEF would make up the far left of the French line and would be crucial to preventing the vast encirclement Schlieffen had envisioned. On August 23, they were ordered to hold the Mons-Condé Canal, an industrial mining area excellent for defense. The British force was tiny but well experienced from colonial fighting. There was one tactic in particular they learned during the Boer War that made its first appearance at Mons: entrenchment. Neither the French nor the Belgians had dug trenches, and the Germans were startled at the volume of fire they received from improvised positions. Mons became a snapshot of the future of the war as precise rifle fire, machine guns, and well-aimed artillery savaged advancing troops. The British superiority in musketry (as they still called it) was limited by slim ammunition supplies and a lack of coherent command. The Germans turned their right flank and forced a retreat. In the resulting confusion Lt. Kinglake Tower of the 4th Royal Fusiliers went looking for his friend Lt. Maurice Dease, who had been manning a machine gun: "I went along under cover and saw his body.

He had been hit about a dozen times. I remember lying there wondering what it would be like to have a bayonet stuck into me and I admit that I have never felt so frightened in my life." During the battle about 1,600 British troops were killed and wounded; German losses were about triple that.

Despite some local successes both British and French forces had to continue falling back before the German juggernaut. Both advance and retreat were exhausting, infantry having to walk fifteen to twenty miles a day in the summer heat. The speed of the German advance seemed to guarantee their victory; the kaiser declared that the war would be over "before the leaves fall." The French commander Joseph-Jacques-Césaire Joffre began desperately planning for a counterattack before Moltke could envelop Paris. Fortunately retreat meant his supply lines were shortening, making logistics easier, while the German ones were stretching daily.

Fighting continued in Belgium as the Germans sought to secure their flanks. They were still obsessed with the dangers of attacks by individuals or irregular forces. They felt that civilian resistance was entirely against the proper conduct of war. During the Franco-Prussian War the so-called *francs-tireurs* (free shooters) had caused endless problems, and they were determined to prevent that from happening again. This generally meant intense, immediate reprisals against civilians.

One of these dramas played out over the night of August 25. The town of Louvain (often written Leuven today) was a beautiful medieval settlement, home to the oldest university in that part of Europe. It was occupied by 10,000 troops who became confused about night movements of their own units and began shooting. In an effort to flush out suspected snipers the occupiers burned buildings and drove people from their homes. During the chaos the university library was set aflame. Nearly a quarter-million books were destroyed, among them irreplaceable medieval manuscripts and incunabula. In all, more than 1,000 buildings were burned, nearly 300 civilians killed, and some 42,000 people forced out. Among the troops involved in the disaster was an infantry platoon of 85 men led by the chemist Otto Hahn, who had immediately left his position at the University of Berlin when the war started. Hahn would go on to win the 1944 Nobel Prize for his discovery of nuclear fission. For this war, at least, that particular terror remained unknown.

At the end of August a German airplane appeared in the skies above Paris. The bombs it dropped did little damage but indicated how close the invading armies were to the capital. On September 2 the French government evacuated to Bordeaux and plans were laid for the demolition of the Eiffel Tower and its powerful radio transmitter. Three days later the Germans were within ten miles of Paris. Advancing with the army was Walther Nernst, the scientist who helped recruit Einstein to Berlin, serving as a driver. They were so close to the capital that he said he could see the glow of the city's famous lights. The same day the HMS *Pathfinder* was sunk, the first British warship destroyed by a submarine. As clashes continued on land and at sea, Britain, France, and Russia signed the Pact of London agreeing that there would be no separate peace. The war would continue until victory or defeat.

French and British forces set their final defense of Paris along the river Marne. Reinforcements were rushed to the site of the decisive battle (including the legendary troops delivered to the front by taxi). The German forces, exhausted by their march, were spread out, inviting counterattack at what should have been the moment of their decisive victory. Joffre rallied his troops and ordered that every soldier "be killed on the spot rather than retreat." After a few days of intense fighting Moltke decided his position was untenable and ordered a retreat back to the more easily defended Aisne River. With that withdrawal, all hopes for the Schlieffen Plan and a quick victory dissolved. Moltke was relieved of command—his final order to his armies was to begin digging trenches and developing fortifications.

Thus began the so-called Race to the Sea, as each side pushed north to try to gain one last flanking action around the growing trench line. Soon, the only remaining opportunity for a breakthrough was in the flat, muddy region of Flanders (meaning "the flooded place"). What remained of the British Expeditionary Force arrived there to plug the gap, crucially reinforced by the first colonial troops—the Indian

Expeditionary Force. The Indians were poorly equipped (many had never received rifles until they landed in France) and were thrown into combat with little support.

This clash, with the Germans desperately trying for a breakthrough and the British forces desperately trying to stop them, became known as the First Battle of Ypres. Both sides were still learning the realities of the new trench warfare and casualties mounted. The battle essentially saw the destruction of everything that was left of the prewar British Army. Fifty thousand Germans died there. Those losses happened to fall hard on the German volunteer corps, which drew heavily on young men who had been students at university when the war broke out. The death of so many students came to be called the Kindermord bei Ypres, the Massacre of the Innocents of Ypres. By the time the battle ended later in the fall, the Germans had the high ground all along the line. An unbroken line of trenches scarred the European countryside for 475 miles, encrusted with the finest tools of death available. By the time the trench lines had solidified, 300,000 Frenchmen and 241,000 Germans had died.

An amazing exception from those terrifying numbers was Einstein's friend Erwin Freundlich. After several weeks in an internment camp, he and his crew were exchanged for Russian officers in one of the first prisoner-of-war swaps of the conflict. Their astronomical equipment was confiscated, making it impossible to even attempt another observation of Einstein's prediction (even the astronomers from America, still neutral in the war, had their equipment impounded). Against all the awful odds of the World War, by the end of September, Freundlich was back safely in Berlin. But Einstein despaired of ever testing his theory: "The observations of the solar eclipse have surely been suppressed by Russia's floggers, so I won't live to see the decisive results about the most important finding of my scientific wrestling."

His despair grew from the purely scientific to the broadly humanistic. He was increasingly sickened by the growing militarism around him: "heroism on command, senseless violence, and all the loathsome nonsense that goes by the name of patriotism—how passionately I hate them." Few of his colleagues agreed. Instead, he had to look abroad to his friends in neutral countries. Writing to Ehrenfest, he wondered if there was anywhere that would feel like home to him:

> The international catastrophe weighs heavily on me as an international person. Living in this "great age" it is hard to understand that we belong to this mad, degenerate species which imputes free will to itself. If only there were somewhere an island for the benevolent and the prudent! Then also I would want to be an ardent patriot.

Without that island of sensibility, the best he could do was isolate himself within the heart of the Reich. He focused on his calculations, and searched for what his equations might still hold.

CHAPTER 4

Increasing Isolation

"Until the truth and German honor have finally been recognized by the entire world."

EINSTEIN'S HOPES OF seclusion did not last long. The still-warm ashes of a medieval library would ignite a firestorm that would consume the world of science, and Einstein was not able to escape. The humble town of Louvain, one of the many shattered by the fighting, held a special significance. The destruction of the ancient university became a symbol of German Kultur run amok. And a critical pillar of that Kultur was science.

About one-sixth of the town had been wrecked by soldiers chasing down imagined civilian snipers. The university library was utterly ruined. Irreplaceable cultural treasures were destroyed. When word of the devastation spread to Britain, it was seized on to mobilize support for the war. London's *Daily Mail* blared, HOLOCAUST OF LOUVAIN—TERRIBLE TALES OF MASSACRE. Burned buildings and bayoneted corpses could be found across Europe by this time, but only in Louvain did there seem to be an assault on the shared European heritage itself. Books and scrolls that should have belonged to scholars of all nations were now gone forever.

Germany had spent decades making itself an intellectual center—remember the Kaiser Wilhelm Institutes used to draw Einstein to

Berlin—and it now appeared to have sacrificed all of that for naked military advantage. Intellectuals and artists around the world accused Germany of reverting to savagery. Britain's Prime Minister H. H. Asquith declared that Louvain could only be considered to be a "blind, barbarian" act. The town's name quickly became shorthand for atrocities against culture and learning.

Scholars in Germany were deeply offended by their country's portrayal as a brute enemy of knowledge and art. In October, Ludwig Fulda, a well-known Frankfurt playwright, decided to defend German honor on the world stage. He drafted a powerful, indignant statement titled "The Manifesto to the Civilized World." It circulated among the German intellectual and cultural elite for signatures of support; ninety-three of them attached their names. Once complete it was translated into ten languages, published in all the major German newspapers, and sent abroad in thousands of letters.

The document was a fiery defense of both the actions of the German Army and the German love of culture. Denying all accusations of wrongdoing, it read:

> IT IS NOT TRUE THAT OUR TROOPS BEHAVED BRUTALLY IN REGARD TO LOUVAIN. They were forced to exercise reprisals with a heavy heart on the furious population, which treacherously attacked them in their quarters, by firing upon a portion of the town. The greater proportion of Louvain is still standing, and the famous town hall is still uninjured. It was saved from the flames owing to the self-sacrifice of our soldiers. Every German would regret works of art having been destroyed in this war or their being destroyed in the future. But just as we decline to admit that any one loves art more than we do, even so do we refuse no less decidedly to pay the price of a German defeat for the preservation of a work of art.

Many of the critics of the destruction in Louvain had blamed "militarism"—the kaiser's pursuit of strength of arms and imperial power—rather than Germany as a whole. The manifesto went out of its way to reject this distinction:

> IT IS NOT TRUE THAT FIGHTING OUR SO-CALLED MILITARISM IS NOT FIGHTING AGAINST OUR CIVILIZATION, AS OUR ENEMIES HYPOCRITICALLY ALLEGE. Without German militarism German civilization would be wiped off the face of the earth. The former arose out of and for the protection of the latter in a country which for centuries had suffered from invasion as no other has done. The German Army and the German people are one . . .

These intellectuals bound themselves to the deeds of the military. It was impossible, they said, for them to commit a crime against art and science. So, then, was it impossible for the army to have done so. The document denied any distinction between their scholarship and their country's actions. National solidarity was more fundamental than disciplinary ties.

German science was well represented among the signers of the "Manifesto of 93," as it came to be known. Six Nobel Prize winners signed: Wilhelm Röntgen, Philipp Lenard, Wilhelm Wien, Adolf von Baeyer, Emil Fischer, and Wilhelm Ostwald. A dozen other scientists attached their names. Einstein's friends Haber, Klein, and Nernst—and, most painfully, Planck. Planck and Klein later claimed to have not seen the details of the manifesto and to have only signed based on the reputations of those who had signed before them. Not two weeks later, though, Planck affirmed that "as members of our university, we unite as one—with all that we morally feel and represent scientifically—until the truth and German honor have finally been recognized by the entire world, despite all the defamation by its enemies."

The Manifesto of 93 seemed to represent the entire German intellectual elite. One observer commented that the signatories essentially included every German thinker "of real celebrity." The document was aimed at neutral countries not yet involved in the fighting. The intent was to persuade them of Germany's moral superiority in the war, particularly against accusations stemming from Louvain. It backfired in every important way.

It was supposed to use the moral authority of German intellectuals to lend respectability to the army. Instead, it dragged those intellectuals down to the level of atrocities. Scholars in other countries were shocked to see their German colleagues deliberately associating themselves with the terrible acts in Louvain. Many of the signatories followed up the manifesto with personal letters to neutral countries. Wilhelm Wien wrote to Lorentz further denying the accusations against the army's actions in Belgium. Lorentz was deeply offended by both the public and private claims. He complained that Wien and the other signatories were not acting like scientists—they were claiming full knowledge of events that they had not witnessed themselves. Instead of saying "it is not true" they should say "we do not believe this." Even worse, they seemed to be celebrating the same events whose existence they denied.

A formal response to the manifesto appeared in the *Times* of London at the end of October. The names of 117 British scholars were attached, including distinguished scientists such as J. J. Thomson, Oliver Lodge, and William Crookes, president of the Royal Society. They expressed sadness that their former colleagues had been dragged down by the kaiser: "We grieve profoundly that, under the baleful influence of a military system and its lawless dreams of conquest, she whom we once honoured now stands revealed as the common enemy of Europe and of all people which respect the law of nations." They declared their solidarity with the British government and their determination to carry on the war. A critical mass of British scientists had explicitly broken with their German colleagues and renounced the international basis of their disciplines. Eddington was no doubt distraught to see two names in particular attached to the letter: Arthur Schuster, his former professor at Manchester, and Frank Dyson, the Astronomer Royal and his close friend.

The published response was no spontaneous gathering of scholarly opinion. The reply had been organized by Wellington House, the British national propaganda headquarters (both Rudyard Kipling and H. G. Wells wrote for them). The Manifesto of 93 had made Germany's intellectual leadership into a weakness rather than a strength, and the

British government took advantage. This is not to say the document did not represent the scientists' views—they were genuinely offended by the manifesto and were happy to express their anger at their German counterparts.

EINSTEIN WAS AS horrified by the manifesto as any scientist in London. His thoughtful friends had suddenly become patriotic sheep. Shocked, he protested to Lorentz that he was surrounded by "collective insanity." The whole world seemed to be "like a madhouse." This "mass psychosis," he said, was particularly powerful in Germany, though it was affecting all the combatant nations. He wondered whether the psychosis was somehow related to the "sexual character of the male."

He commiserated with his friend Georg Friedrich Nicolai, a Berlin doctor and professor of physiology. Nicolai had been a friend of Elsa's before Albert even moved to Berlin, and they had bonded over shared politics. He was an odd character: a resolute pacifist who carried scars from dueling and married the daughter of an arms manufacturer. Handsome, he sported a fashionable monocle. He was actually in France when the war broke out and was nearly shot as a spy. His close connections with French professors got him released.

Nicolai was outraged by the Manifesto of 93, particularly the distinguished scientists who had signed it. Science was supposed to be above patriotism. It made him "wonder for the first time whether the edifice of German science, so splendidly solid in its outward appearance, might not be inwardly rotten."

To battle this, Nicolai drafted a countermanifesto, "An Appeal to Europeans." It called out the disruption of international scholarly networks caused by the war and summoned intellectuals to fight for peace:

> Never has any previous war caused so complete an interruption of that cooperation that exists between civilized nations. . . . Educated

men in all countries not only should, but absolutely must, exert all their influence to prevent the conditions of peace being the source of future wars.

He circulated it around Berlin, hoping he would find supporters who, like him, felt alienated by the Manifesto of 93. The silence was deafening. Einstein enthusiastically signed on: "Although I am convinced that the voice of a handful of the informed carries little weight against the lust for power of the mighty and the fanaticism of the many, I still welcome your manifesto with pleasure and am pleased to be permitted to add my name to it." Only two others signed.

This was a disaster. Worse than being controversial, the counter-manifesto was almost completely ignored. It circulated just widely enough to ruin Nicolai's academic career. He then volunteered to run a cardiac clinic for the army, which, at least in the short term, soothed the authorities enough to allow him to keep working.

Einstein was extremely disappointed. This was his first public statement on politics. The Manifesto of 93 had galvanized him into action, and that action turned out to be completely useless. His colleagues did notice his willingness to buck the nationalistic tide, and he gained a reputation as a political oddball. He was already known to be somewhat eccentric, so no one was particularly concerned.

The Berlin scientific community was in a wartime frenzy. The famed zoologist Ernst Haeckel—Darwin's greatest apostle in Germany—publicly rejected all of his English honorary degrees. Wilhelm Wien composed an avalanche of furious individual replies to the statement of British scientists. He had them delivered via the American consulate. In those letters he announced that he no longer thought the rift between German and British science could "be healed in the foreseeable future."

Einstein saw blinding patriotism everywhere he looked. Schoolteachers forbade the use of loan words from English and French such as *intéressant(e)* ("interesting"). Students studied arithmetic through *Kriegsrechnen*—war calculations. Even after the failure of Nicolai's appeal, Einstein started looking for opportunities to push back against the madness around him. Searching for like-minded intellectuals, he heard about the November 1914 founding of the New Fatherland League (abbreviated BNV in German). This was an eclectic group of people all opposed to the war—liberals, conservatives, socialists, internationalists. They were a small gathering of elites, never having more than two hundred members. Meetings were Monday evenings. Their goals were an early peace and an eventual Europe-wide organization to make future wars impossible. Einstein was member number 29.

IT IS SURPRISING that Einstein did not seem to consider leaving Berlin. He had great confidence in his ability to compartmentalize and stay focused: "Why should not one, like a servant in an asylum, be able to live cheerfully? . . . Up to a point one can choose one's madhouse." He continued working on relativity as best he could. He lectured to the Prussian Academy on October 19 and 29, informing them that the final aspects of his theory were near at hand. Also on October 29 the List Regiment joined the battle at Ypres, among them a young soldier named Adolf Hitler.

Einstein's lectures focused on his Entwurf theory, which he was pretty sure was correct. It still had that enormous flaw, though, which was that it was not generally covariant—the equations did not allow all observers to be equal. He had to decide whether this flaw was fatal. He could say covariance was too important to give up (in a sense, it was the core principle of relativity). But then he would have to start over. Or he could accept the loss of covariance, and perhaps even explain how it wasn't really a problem. He chose the latter.

Einstein began with a thought experiment proposed by his intellectual godfather, Ernst Mach, which in turn referred back to a thought experiment proposed by everyone's intellectual godfather, Isaac Newton. Newton thought he could demonstrate the reality of absolute motion in the following way. He admitted that Galileo's relativity principle sometimes made it difficult to tell who was moving: you or your friend. But now get off your train and fill a bucket with water. When you spin the bucket, you see the outside of the water creep up toward the edge. That sloshing was a result of centripetal forces from the spin acting on the inertia of the water. This, Newton said, was evidence that you were *really* spinning. No combination of relative motion could create that illusion. Ergo, absolute motion against an absolute space must exist.

Not so fast, said Mach. He warned us that we were thinking too small. The creeping rise of the water was indeed relative to something—it was relative to *everything in the entire universe*. The whole universe, which we see as the distant stars, was just one more frame of reference. A big one, to be sure, but still just another frame. Mach's critique was that the inertia that caused the sloshing, then, was actually only determined relative to those distant stars. The bucket of water had no *inherent* inertia, only this effect that was dependent on the location of the observer. Einstein called this relativity of inertia "Mach's principle."

Einstein was determined that general relativity include this Machian perspective. This led to what he called the "hole argument," which helped him feel better about his problem with covariance. Imagine a universe full of stars, except for one empty area (the hole). We perform our bucket experiment in the center of this area, and get certain measurements of space and time that convince us that inertia works a particular way. Our measurements of space and time are Machian—ticking clocks and rulers laid end to end. They are the results of clear, physical methods of measurement, and they result in a series of numbers.

Then we move to an observer waiting at another place within the hole and take those same measurements of space and time again. Once more, clocks and rulers. Our movement to a new place has put us into a different frame of reference—physicists call this shift a *coordinate transformation*. This coordinate transformation has changed the measurements we make (for example, walking across the room might change my measurement of the distance to a chair from one foot to ten feet). Now, since the numbers we get from our measurements of space and time are different, we will get a different understanding of inertia.

Einstein then reminds us of the equivalence principle, which tells us that inertia is the key to understanding gravity. So our coordinate transformation—moving around in the hole—has changed the way we see gravity. The change in numbers must have a physical meaning. Fine—this is expected in Mach's universe. Unfortunately this was not compatible with Einstein's great hopes for general covariance. In a universe with general covariance, everyone should agree on the nature of gravity. It should be independent of the way we move around. But the hole argument has shown that this is impossible—the change of measurements that comes with a coordinate transformation prevents covariance.

Mach's principle of the relativity of inertia, then, was incompatible with covariance. General relativity had to give up one or the other. And since Einstein had already failed to find covariance in his Entwurf equations, that was actually a relief. The hole argument, and Mach's principle, apparently demanded that there be *no* general covariance. So what seemed to be a major failing of the Entwurf was neatly explained away; he could continue using these equations with a clean conscience.

ONE OF THE journals in which Einstein might publish his hole argument was the *Physikalische Zeitschrift*. If Einstein picked up a wartime issue he would have noticed some lists of names: physicists fighting at the front; physicists who had been decorated for valor; physicists who had been wounded or killed. The editor, Max Born, announced that this had been done to show that "physics too is at one with the Fatherland in this time of peril and danger." Johannes Stark, an editor of another important science journal, decided to go further and remove the names of English scientists from the pages of his journal. He was then persuaded that all "enemy foreigners" should be treated equally, and Marie Curie disappeared along with Ernest Rutherford.

The Nobel Prize winner Wilhelm Wien thought that even this was not quite enough. To truly demonstrate how physics had committed itself to the war, Wien said, the language of science itself must change.

He proposed that German physicists not refer to the work of English physicists unless absolutely necessary. No citations, no mentions of their papers. Foreign terms that were usually used in their original language (for example, "equipartition theorem") should be replaced with German equivalents. Even translations of English books should no longer be published.

Wien circulated his proposal in December 1914, looking to convince his colleagues of the danger that English science posed:

> Unjustified English influence, which has infiltrated German physics, will be set aside. Naturally, this does not mean rejecting English scientific ideas and suggestions, but the [adoption of foreign habits] has now ominously appeared in our science. Among significant examples, scientific achievements are very often attributed to Englishmen in our physics publications, though, in reality, they originate from our countrymen.

This last concern, that enemy scientists had been claiming German discoveries as their own, had also been voiced by Philipp Lenard. Because of this, Lenard said, war against Britain was "a crusade for the recognition of honesty on earth." The lifeblood of physics—the papers and journals that communicated ideas—had become explicitly patriotic.

The everyday work of physics was hobbled too. Scientific institutes were drained as both professors and students went to the front. Germany's universal service system meant that essentially every able-bodied male was well trained and ready to fight (Einstein could avoid it due to his Swiss citizenship). There was a basic social expectation of serving in the military, and an enormous bureaucratic apparatus was already in place to funnel men into the battalions and brigades.

In Britain the situation was quite different. The Continental model of universal service was seen as a gross violation of individual liberty, and it had never been adopted. The Victorian age had been built on the idea that the state should be barely seen, which was credited with the great economic success of the period. A government that was reluctant to forbid children from working in coal mines was hardly going to compel adults to wear a military uniform.

So in August 1914 it was widely expected that Britain would fight the entire war with its initial tiny standing army. But the stunning lists of dead and wounded produced by the first few months made it clear that this was unsustainable. For the first time, the country would have to mirror France's and Germany's multimillion-soldier armies.

The obvious move was to institute mass conscription just as those countries had. That was essentially politically impossible. Critics argued that adopting Prussian military policies in order to defeat Prussian militarism was unacceptable. Lord Kitchener, the secretary of state for war, was one of the few members of the government who expected a long conflict, and he immediately set out to raise troops by voluntary enlistment. At his request Parliament approved the induction of 100,000 men just a few days after the war started.

Kitchener's stern gaze and sharp cheekbones stood out from ubiquitous recruitment posters. There was a huge rush to enlist before the war was over, and initially the army had no trouble filling its ranks. A toy company even made a board game about enlistment called "Recruiting for Kitchener's Army," in which players try to pass the medical exam so they can serve. The military bureaucracy was completely overwhelmed by volunteers. Recruiting stations ran out of application forms. There were not enough doctors to conduct medical exams—one said he inspected four hundred recruits a day for ten days straight. Officers were in particular demand and a commission was given to almost anyone with a boarding school or university background. Cambridge and Oxford were rapidly depleted of undergraduates. Both of Eddington's assistants at the observatory enlisted. Neither would ever return.

Despite the enthusiasm, no voluntary program could keep up with the army's needs. At the end of August, Parliament approved another 100,000 troops. Two weeks later, another half million. The increasing pressure for new bodies appeared in subtle ways. At the beginning of the war a volunteer needed a minimum height of 5'8". By October it was 5'5", by November, 5'3".

By 1915 enlistment numbers had dropped enough to require action. Lord Derby, the director general of recruiting, proposed a novel alternative to conscription. This asked men to "attest" that they would serve if called on. Those who attested were given an armband to show that they were not shirking their duty. This was combined with less formal efforts—enlistment was supposed to be a choice, but men were actually subject to intense social pressure. Adm. Charles Penrose Fitzgerald deputized women to hand out white feathers to men not in uniform to shame them into enlisting.

Recruitment was also fueled by increasingly intense hatred of all things German. Performances of German plays were canceled. The author Graham Greene recalled seeing dachshunds stoned to death in the street. Gresham College received an angry letter asking why German continued to be taught (it was justified on the grounds of knowing one's enemy). One of the highest-ranking naval officials, Prince Louis of Battenberg, was forced out because of his Teutonic name.

The Derby Scheme, as the voluntary recruitment plan came to be known, was of limited effectiveness. Fewer than half the 2.2 million men asked to attest did so. Of those, only a quarter were fit for duty. The scheme also formed the first exemption tribunals, which decided whether certain important workers could be exempt from the system. So the owner of a munitions factory might want to retain a highly skilled worker, though they were pressured toward providing soldiers. The whole plan was voluntary, though, so the tribunals had little to do.

Eddington was exactly the sort of man Lord Derby wanted: thirty-one years old at the start of the war, unmarried, and in excellent physical condition. But as a Quaker, he was exactly the sort of person who would refuse. The Quakers' best-known principle was their pacifism. Their justification for rejecting violence was called the Testimony Against War, or the Peace Testimony:

> It springs from our belief of the potentiality of the divine in all men—the Inner Light, as we call it, which is in every man, no matter how hidden or darkened it may be. . . . Hatred and violence only

> feed the flame of evil. . . . If this be true of personal relations, we believe it to be true equally of civic and international ones.

Almost all the Quakers in Britain took this as guidance not to join the military as soldiers. There was some uncertainty about what other actions it forbade or allowed. There was the Friends Ambulance Unit, a volunteer medical corps that sought to alleviate suffering on the battlefield without carrying any weapons themselves. This proved a popular option for many Quakers—using their courage and energy to protect the divine spark within every combatant rather than kill and maim.

Surrounded by the anti-German violence at home and the organized violence across the Channel, they felt clear calls to duty. They mobilized to testify for peace—to show how their religion demanded they take action. Modern Friends like Eddington not only refused to fight, they wanted to take positive action to end violence everywhere. The Quakers in Britain held a conference at the opening of the war to discuss their options, out of which came a public message. It read, in part:

> We find ourselves today in the midst of what may prove to be the fiercest conflict in the history of the human race. [We reaffirm that] the method of force is no solution of any question [and] that the fundamental unity of men in the family of God is the one enduring reality. Our duty is clear to be courageous in the cause of love and in the hate of hate.

Their pacifism was based on the idea that the war was secondary to the dangers of nationalism: "Whatever may be the guilt of the individual countries concerned, it is the system which is much more at fault." Despite its militarism, Germany was not the enemy; the true opponent for the Society of Friends was the human misery that came out of any war. Eddington was an astronomer; the subjects of his study were vast distances away. How could he put his skills to work resisting the war? He wondered how he could protest militarism with equations and star charts. He would have to watch for opportunities.

Many Quakers devoted themselves to helping the refugees from the Continent that were streaming into Britain. About 200,000 Belgian refugees came into the country, for whom the government planned nothing. Most simply arrived at Charing Cross or Liverpool Street railway stations, not knowing what to do next. Religious groups were at the forefront of finding them shelter, clothes, and food. Those philanthropists often splintered into redundant or strangely focused groups such as the Duchess of Somerset's Homes for Better-Class Belgian Refugees.

Eddington met one of these refugees at the RAS in London in November. Robert Jonckhèere, an astronomer from Lille Observatory in northern France, had fled German shelling on foot. Jonckhèere walked only in stockings, as his shoes had worn through. He arrived in London after five days of travel. He did not know if the observatory itself had survived the fighting. His appearance at the Royal Astronomical Society, behind a podium usually used for discussions of stellar aberration and planetary precession, transformed that scientific space into a political one. This would not be the last time.

Stories such as Jonckhèere's helped paint the Germans as savage occupiers ignoring all conventions of war, and encouraged agitation in England. These tales were amplified by press coverage (particularly the *Daily Mail*) and an official British investigation into atrocities. The atrocity report was lurid and provocative and helped bring the remaining liberal objectors around to supporting the war. When the Germans learned of these reports, they started their own investigations into war crimes—focusing on civilian attacks on German soldiers. Louvain was again recast as an illegal civilian uprising. Not many observers from neutral countries were convinced.

Part of the British fascination with atrocities early in the war was surely that, unlike the French and Belgians, they did not see fighting in their own lands. However, the fighting was strangely close—"home front" was a term invented for this war. The economic needs of mobilizing for total war meant that almost everyone had some connection to war production. The effects of siphoning men into the trenches could be seen everywhere. Half of the railway workers went into the armed forces, and they were replaced by women. The female railway worker became a standard image of the home front. They were uniformed just as soldiers were, reminding everyone that women were a critical part of the war effort. Everyone recognized the tremendous importance of women to the war, which eventually allowed them to secure the vote. Suffragette organizations saw this connection quite clearly and became enthusiastic supporters of the war.

There was no way for Eddington to escape reminders of the conflict. In Cambridge, his students were gone. In London, a relaxing walk in Kensington Gardens brought him to the exhibition trenches dug there. The green lawns were scarred by the orderly earthworks intended to reassure city dwellers that their sons were being well taken care of.

WHILE BRITAIN STRUGGLED to fill its ranks in Flanders, it was quite confident with its domination of the seas. One of its first wartime actions was to set up a naval blockade of Germany and its allies. Land borders were mostly locked down by the trenches, so the Royal Navy was essentially able to cut Germany off from world trade.

By 1914 the economies of the major European powers were intensely globalized. Germany's industrial success meant it could buy about a fifth of its food from abroad. Further, its domestic agriculture depended heavily on imported materials for fertilizer (and farm labor siphoned off for the front did not help production). The impact of the blockade was felt immediately. Bread became scarce even before the trench lines had settled. Angry Berliners were told by the government that they had "grown used to the luxury of overeating."

Even beyond what the blockade stopped, the government quickly gained a reputation for terrible management of the remaining supplies. In an effort to save grain being used as animal feed, virtually every pig in Germany was ordered to be slaughtered—the so-called *Schweinemord*. This, of course, led to a glut of pork products immediately followed by a huge meat shortage. The black market quickly became essential for finding food.

The kaiser's government worried that food deprivations were overshadowing their battlefield victories, and with good reason. As early as February 1915, markets were being stormed for a few pounds of potatoes. Berlin police warned of "butter riots." They were given orders to fire warning shots at signs of unrest.

The fertilizer shortage pointed to another critical consequence of the blockade. Fertilizer production depended on nitrates imported from South America (for many years the best source of nitrates was bat guano scraped from cave walls). At the outbreak of war, 80 percent of the world's nitrates came from Chile. And nitrates were not only used in fertilizer, they were absolutely necessary for the production of explosives. Every combatant relied on these imports, and it was immediately realized that control of the nitrate supply would be crucial to victory. The German Navy's last chance to control access to South America ended at the Falkland Islands; *Admiral Graf Spee*'s squadron was ambushed and destroyed by British battle cruisers. The Central Powers were effectively cut off from Chile for the remainder of the war.

As armament production ramped up, the blockade of nitrate supplies meant that Germany could not continue fighting for long. Every cannon shell was the end of an irreplaceable supply line stretching eight thousand miles. Domestic reserves would be used up in just a few months. Fortunately for them, Berlin was home to the one person in the world who could solve this problem: Einstein's friend Fritz Haber.

Haber's great scientific achievement was developing an artificial method to synthesize ammonia. Linking up four hydrogen atoms to one nitrogen might not seem too exciting, but this meant that humans were able for the first time to create the chemical bond that was essential for making compounds like agricultural fertilizer. Before this, humanity had needed to rely on slow and temperamental natural processes to create these; now they could do it at will.

And if they could create fertilizer, they could create explosives. Haber had been worrying about German nitrate supplies for years and developed a plan for industrial-scale production of the needed chemicals. He was largely self-directed on this. He realized the nitrate crisis before anyone in the government and focused his own Kaiser Wilhelm Institute on cracking the problem. It helped that Haber was personal friends with the ministers in charge of procuring raw materials for the war. When Walther Nernst returned from the front, he set to work on explosives as well. Synthetic nitrate production increased from zero to tens of thousands of tons per month.

This effort took over Haber's institute—he proved to be an excellent organizer and kept his subordinates working relentlessly. He even organized child care for his employees. Einstein was supposed to have his own Kaiser Wilhelm Institute, but until that building was complete, his office was actually in Haber's building. We don't know how often Einstein worked there (as opposed to his home office). When he did, he would have been surrounded by scientists focusing all of their energy on new and better ways to destroy the enemy. He was the only person in the building not trying to win the war.

These Berlin chemists had essentially solved the problem of explosives production almost before the military was aware of it. The government came to view scientists as incredible problem solvers, and the war as a series of technical challenges that could be overcome by straightforward solutions. Haber let them know that, with the nitrate problem solved, his laboratory was working on new tools for the battlefield. On December 17, 1914, there was an explosion in the lab (perhaps it knocked over the pencils on Einstein's desk). It just missed Haber but killed Otto Sackur, a brilliant young scientist. Sackur had been a personal friend of both Fritz and his wife, Clara. They were both shocked. Fritz soothed the tragedy by reminding himself of the greater cause: "He died as a soldier on the battlefield in the attempt to improve the technical means of warfare with the help of our discipline."

SCIENTIFIC INSTITUTES SUCH as Haber's had helped make German science the envy of the world by 1914. One of the effects of this leadership was that the institutes became home to many international students and young scientists eager to participate in the research being done there. This became a problem once the war began: "enemy aliens" could not be allowed to roam freely. The faculty of the Kaiser Wilhelm University compiled a list, and 7 professors and 568 students were expelled.

James Chadwick, a young physicist from Manchester who would go on to discover the neutron, had been studying with Hans Geiger in Berlin. Geiger left his lab to serve in the artillery and Chadwick was arrested. He, along with every British subject between the ages of seventeen and fifty-five in Germany, was interned in a prisoner camp built at the Ruhleben racecourse. He slept in a horse stall with five other men. Food was scarce and poor; disease was rampant. When word of Chadwick's detention got out, Geiger was asked about allowing his former student to be held in such conditions. He replied that Chadwick was atoning for the sins of the English, and that in any case he was better off than many Germans living under the blockade.

On the other side of the Channel, some London academics were congratulating themselves on their tolerance. King's College and University College London allowed their foreign students to stay. The British government had other plans, though. Immediately after the war began, Parliament passed the Aliens Registration Act. Elaborate rules were set for citizens of Germany or Austria-Hungary in the country, controlling their residences, work, and movements. They had to register at the nearest police station and could not go more than five miles from it. They could not live near areas of military importance, which included the entire east coast. German newspapers were banned and communication devices were forbidden. Immigrants were no longer allowed to change their names—it would be too easy for them to hide. Postal censorship was introduced to catch German spies but was quickly used to build a list of anyone with antiwar views.

Internment of suspicious persons began as well. The German physicist Peter Pringsheim had been at the BA meeting in Australia and heard Oliver Lodge's declarations in support of international science. He spent the length of the war held in a British camp. Albrecht Penck, a geologist who had also been at the meeting, narrowly escaped internment in England but was then arrested as a spy.

ISOLATION WAS A double-edged sword. Einstein had hoped isolation would let him ignore the madhouse and focus on his work. Instead, as his colleagues rallied around the flag he realized that solitude could be a curse. How could he talk to Haber and Planck about scientific matters when they proclaimed that science was united with the Reich? The militaristic trappings of the Prussian Academy had unexpectedly become more real.

Eddington, too, was shocked to find his friends declaring that British science was, indeed, British and not international. Dyson had decided to embrace the "Royal" part of his title, and astronomy was now supposed to be explicitly in service of the Crown. Eddington was sharply reminded that as a Quaker he was an outsider. His professional success in moving into the inner circles of the scientific establishment had suddenly been compromised. As a pacifist in a newly militaristic nation, he was increasingly lonely.

The destruction of Louvain had set in motion a chain of events that cut both Einstein and Eddington off from their intellectual communities. This separation would only grow as the war increased its terrifying scale. Both had begun searching for new routes forward. Einstein's first attempt with Nicolai had failed, but perhaps the BNV showed some promise. Eddington's Society of Friends had begun mobilizing for peace, though it was not yet clear how they could best work for their cause. Pacifists in Britain and Germany had begun organizing, but would they be strong enough to resist the militarism that had enveloped Europe? And would science place itself on the side of war or peace?

CHAPTER 5

The Collapse of International Science

"No Compromise with a Race of Savages."

On August 10, 1915, a twenty-seven-year-old signals officer was shot through the head by a Turkish sniper. He was in the Dardanelles as part of Winston Churchill's daring, perhaps foolhardy, plan to open a new front in the war. The Ottoman Empire had joined Germany and Austria-Hungary months before, widening the conflict even further. What became known as the Battle of Gallipoli was a disaster, failing to achieve any of its strategic goals. It claimed hundreds of thousands of lives, including that of the young English signals officer—Henry Moseley, who had just been nominated for the Nobel Prize for physics.

Moseley had published only eight papers in his short life, but he was on the verge of revolutionizing physics and chemistry. He had developed a method to use X-rays to probe the inner structure of the atom in ways that seemed almost fantastical. He had been about to receive a prestigious professorship during the summer of 1914 when the war broke out. In Australia with Eddington, he cut the trip short to return to England and volunteer for the army.

Scientists around the world reacted with shock at the news of his death. The physicist Robert Millikan in America lamented, "Had the European War had no other result than the snuffing out of this young life, that alone would make it one of the most hideous and most irreparable crimes in history." Even the Germans called it "a heavy loss for science."

Moseley's mentor, the famous experimenter Ernest Rutherford, wrote an obituary for the flagship journal *Nature*. It was a tribute to genius cut short far too soon. He described the mixture of emotions that he and his colleagues felt about the enlistment of talents like Moseley: "pride for their ready and ungrudging response to their country's call, and with apprehension of irreplaceable losses to science." Rutherford, himself one of the greatest experimenters of all time, saw in Moseley all the future of physics. His frustration came through clearly. He called it a national tragedy that the government regarded that young man only as a combatant: "we cannot but recognise that his services would have been far more useful to his country in one of the numerous fields of scientific inquiry rendered necessary by the war than by exposure to the chances of a Turkish bullet."

Other British scientists were lost at Gallipoli: Charles Martin, a Glasgow zoologist; Keith Lucas, a Cambridge physiologist. A single lab in Oxford lost five men. It seemed that the next generation of scientists were being scythed down by the war. An article in the *Times* complained about this "waste of brains." Imagine, it said, if the war had taken place eighty years before. There might be a burial plaque that read "Killed in Flanders, Charles Darwin."

These mournful warnings showed one of the fundamental tensions of science during the Great War. Even as most scientists on all sides supported the war, it was not at all clear what that meant. Should they support it with their intellect and their scientific skills? Or should they support it with their body and their skill at arms?

IN LONDON, THESE questions were bound up with the expanding sense of the home front. German zeppelins dropped bombs in January. While these early air raids did little damage, the psychological impact of the war literally landing on one's doorstep was immeasurable. Soon after this, the Germans aimed another new weapon at the civilian population. In retaliation for the British blockade of their ports, their submarines began attacking any ship headed to the British Isles. These new craft could evade the Royal Navy almost at will, striking and vanishing before overwhelming power could be brought against them. Like Germany, Britain's industrial might relied on a worldwide network of raw materials and foodstuffs to support it. They were just as vulnerable. Between fifty and a hundred merchant ships were sunk each month by U-boats. Sugar started disappearing from market shelves in Cambridge and York. Two imperial giants, unable to advance on the battlefield, began slowly strangling each other.

A decisive moment came in May, when the British liner *Lusitania* was finishing its 202nd transatlantic crossing, and passed in front of the lurking submarine U-20. It sunk after a single torpedo, taking 1,201 passengers and its cargo of ammunition to the depths. The huge casualty list sparked international outrage. Of those killed, 128 were Americans, which nearly brought the sleeping colossus into the war. Skillful German diplomacy kept that at bay for the time being. King George V finally removed the banner of the kaiser—his cousin—from St. George's Chapel at Windsor Castle. He had resisted public pressure to do so since the declaration of war. No longer.

The sinking triggered anti-German riots across the UK. Shops were ransacked and people dragged from their homes. A piano belonging to a German immigrant family was hauled into the street, where it was used to play patriotic songs. Germans who had not yet been interned were quickly arrested. Some 32,000 were placed in camps by that fall. Mob actions continued to be spurred by the press. The hugely popular magazine *John Bull* ran an editorial calling for a "vendetta" against all Germans. Headlines such as "No Compromise with a Race of Savages" were common.

A MEMBER OF that race of savages was having some trouble with his calculations. Einstein was German by birth and residence, but he was quick to correct anyone who described him that way. He clung to his

Swiss citizenship like a talisman. It not only prevented the kaiser from trying to conscript him, it provided some mental solace—a way of keeping himself apart from the patriotic madness around him. Even so, his scientific correspondence often seemed to bring him face-to-face with his isolation:

> I love science twice as much in these times when I feel so painfully for almost all of my fellows about their emotional misjudgments and the sad consequences. . . . We scientists in particular, must foster international relations all the more and must distance ourselves from the coarse emotions of the mob; unfortunately we have had to suffer serious disappointments even among scientists in this regard.

He found himself relying more and more on those few colleagues with whom he could speak freely. The war had made it obvious, he wrote, that "the only thing really worth striving for in this world is the friendship of exceptional and independent persons."

Paramount among those remained Lorentz. No one had his perfect balance of scientific expertise and diplomatic presentation. Einstein could sometimes be prickly when criticized, but he always welcomed Lorentz's amendments. After one such correction he wrote back: "The theoretician is led astray in two sorts of ways: 1) the devil leads him up the garden path with a false assumption. (For this he deserves pity.) 2) He argues inaccurately and sloppily. (For this he deserves a thrashing.)." He humbly accepted his Dutch mentor's thrashing.

Lorentz was one of Einstein's main confidants about the great weakness of the Entwurf theory—the lack of general covariance. Einstein wavered about whether this was a real problem. Had his hole argument put covariance to rest? Or was covariance still hiding somewhere within the equations? Lorentz thought it wasn't necessary. He was still confident in the ether, whose absolute reference frame would make covariance impossible. He gently suggested to Einstein that, perhaps, his attachment to the principle of relativity was merely a "personal view" and not "self-evident" as the younger man had hoped.

Generations of scientists had worked with the ether—perchance it could be brought back into Einstein's gravitational theory?

Einstein wasn't willing to go that far. The lack of covariance was a flaw, but he could live with imperfection. And it did not shut down further work. He had two avenues of research into general relativity—the physical and the mathematical, what Einstein called a "kaleidoscopic mixture." Covariance was a physical problem, so he decided to focus on the mathematical approach for a time. He began corresponding with Tullio Levi-Civita in Italy, a mathematician who had helped establish the absolute differential calculus Einstein had built his theory with. Their conversations were fruitful—only to be cut off by Italy joining the Allied powers.

He checked in regularly with Freundlich about possible ways to test his theory. Einstein was still trying to get him time off to work on relativity. Freundlich's supervisor, Otto Struve, continued to say no—this time because the young astronomer was already notoriously unreliable about getting his assigned tasks done.

Without a handy eclipse, that left the wobble of Mercury's orbit as Einstein's best bet. This wobble (technically a "precession of the perihelion") had been known since the 1850s and had resisted all attempts to explain it with Newtonian gravity. If the sun was a little fat around the waistline, that would explain it (but it wasn't). A planet inside Mercury's orbit, code-named Vulcan, could explain it (but neither it nor its effects had been seen, and astronomers had been looking hard). Ad hoc adjustments to Newton's equations would fix it (but who was willing to alter those sacred texts?). The standing solution was to assume diffuse bands of gas and dust floating near Mercury—conveniently just thick enough to cause the precession but not thick enough to be seen. Einstein had gambled that the Entwurf would provide a clear mathematical prediction of the precession and thus convert disbelievers to his side. It didn't—the calculation it offered was way off.

Despite the problems, he felt pretty optimistic. The mathematical techniques of coordinate transformations were working fairly well, the equivalence principle was holding up against all challenges, and it was firmly grounded in conservation laws. The struggle had been painfully difficult, though. He complained to Ehrenfest that his "work on gravitation progresses, but at the cost of extraordinary efforts; gravitation is coy and unyielding! . . . What has been found is simple, but the search is hell!"

In January 1915 he similarly wrote to a friend that general relativity was about to be "brought, in a sense, to a close." In another letter he sounded satisfied: "I am working tranquilly in my booth in spite of the distressing, abhorrent war. The general theory of relativity has now been relieved of most of its obscurities." He complained that it was "a pity" that the mathematics of the theory made it so difficult to study. He consoled himself with the memory that this was once true of Maxwell's theory before it was simplified. The letter took a sudden turn once he had given the relativity news, though:

> Why do you write to me, "God chastise the English"? Neither to the former nor to the latter do I have any close relations. I just see with great dismay that God punishes so many of His children for their ample folly, for which obviously only He Himself can be held responsible; I think, His nonexistence alone can excuse Him.

Even in nonpolitical communications, the war couldn't be avoided.

Support from other scientists for relativity waxed and waned. One week he felt that Planck, previously a determined opponent, was finally coming around. Another week he complained that interest in relativity was "exceedingly modest for the time being." He had essentially run out of friends in Berlin. And with his letters and telegrams constrained by the Royal Navy's vigilance, he did not know where he could find new recruits for his cause.

THE EFFECT OF the blockade was two-sided. As war was first breaking out in August 1914 and Eddington was still returning from Australia, British astronomers noted something odd. The most recent issue anyone had of the *Astronomische Nachrichten*—the main journal for

German astronomy—was dated July 22. Surely there was a more recent volume? There was not. The trenches cut off scientific communication as efficiently as they stopped commerce. The British blockade kept out scientific journals along with Prussian propaganda and personal letters.

Print was not the only medium the war disrupted. For years international astronomers like Eddington had relied on an elaborate telegraph network to spread the details of observations and discoveries. This "Science Observer" system used a special code to compress large amounts of astronomical data into an efficient form. This allowed for both speed and more reliable claims of scientific priority. The coded telegrams were sent to a hub in Kiel, Germany, and then disseminated to observatories around the world.

Once the fighting began, though, the telegraph lines were cut. The astronomers' well-oiled system had been decapitated. A Japanese observer wanting to get credit for discovering a new comet, or Eddington requesting confirmation of some unusual stellar motions, had no way to do so. Because the system ran through Germany, even communications between non-belligerent nations were disrupted. Edward Charles Pickering, the head of the Harvard Observatory in the neutral United States, complained to the Astronomer Royal, Frank Dyson, that the global project of astronomy had ground to a halt. "Owing to the cutting of the cables it is now impossible to communicate with the Centralstelle at Kiel. There is, therefore, no official means of intercommunication of astronomical discoveries."

The two scrambled to put a stopgap system in place. American astronomers would send telegrams to Pickering, who would send them to Dyson, who would then send them individually to Allied and neutral observatories in Europe. This situation was extremely confusing. Just as Dyson was setting up his procedures, Elis Strömgren, the leading astronomer in neutral Denmark, was trying to do the same thing. Strömgren thought it made more sense for his institution to replace the cut-off German hub. The astronomers within the Central Powers knew and trusted him to act in a genuinely international fashion. Strömgren assured astronomers on both sides of the fighting that he would take care of the system for however long the war lasted.

British censors, however, were not so sanguine. To the government it appeared simply that streams of coded telegrams were passing back and forth from enemy countries. The postmaster general objected and announced this would not be allowed. The Astronomer Royal appealed personally for an exception, providing the key to the technical code. It was refused again with no further explanation. Early in the war these matters were handled by the post office, which had little experience with censorship, and many of the astronomical messages made it to Copenhagen anyway. This changed abruptly when the Defence of the Realm Act was passed and the War Office took over censorship. Individual telegrams were examined and deciphered versions demanded.

Even when the messages were decoded, scientific communication with the enemy was still seen as deeply suspicious. W. E. Plummer's transmissions from the Liverpool Observatory marked him as possibly being in league with the enemy. The board of directors of the observatory grew suspicious and demanded that he cease all communications. He explained the telegrams' purpose; nonetheless, they had "the appearance of assisting the enemy." R.T.A. Innes did not think it was a good idea to allow neutral countries access to imperial data. He confessed that he "would prefer to send such messages to a British institution."

Suspicion grew about possible German agents within British science. Names were an easy target. Alexander Siemens, the famous electrical engineer and naturalized British citizen since 1878, was pressured into making a public statement of loyalty. Hugo Müller, resident of England since 1854 and former president of the Chemical Society, was pushed to withdraw from that organization. He was praised as a "considerate German" for doing so.

It was noted that the Royal Institution had recently dismissed two scientific officials of English origin and replaced them with Germans. Scotland Yard inspected the premises and found nothing amiss. A. B. Basset, a Fellow of the Royal Society, warned that the Society had already been compromised: one of the secretaries was a German by birth; another had three German names and spoke English with a pronounced accent.

Basset's target was Arthur Schuster, who had spent forty years teaching physics at Manchester University (he was one of Eddington's mentors there). Schuster, known for his kindly gaze and easy smile, was born in Germany but had been a bedrock of British science for decades. As part of his research he had a radio set, which he used to receive meteorological and solar observations from scientists on the Continent. When the police heard about this he came under immediate suspicion and they confiscated the device. At the 1915 BA meeting he was elected president, which ignited a storm of protest. Newspaper campaigns tried to get him to step down and there were rumors that scientists would boycott the meeting. In the end he accepted the office just as he received word that his son had been wounded in the Dardanelles.

The campaign to oust him from his position only increased in intensity. Henry Armstrong, a chemist who held the honor of being the oldest Fellow of the Royal Society, kept up a nonstop barrage of letters to newspapers. He worried about British science being in Schuster's "unimaginative German hands." The Oxford zoologist Sir Edwin Ray Lankester called for Schuster to resign "out of consideration for his colleagues in the Society, to remove the ill feeling, which his presence evokes and the possible injury or at any rate arrogance to the Society." The Oxford astronomer H. H. Turner asked for his resignation in a "friendly" way. Nonetheless, he remained as head of the BA for much of the war. He declined to stand for reelection in 1918.

The attacks on Schuster were not only worries about espionage. There was a widespread sense that the war had revealed that Germans *simply could not do science.* Pierre Duhem, the physicist/historian/philosopher, described German science as "abstract, heavy, obscure" and dismissed Germany's scientists as working "under the orders of an arbitrary and insane algebraic imperialism." Sir William Ramsay—the Scottish chemist who, confusingly, received the Nobel Prize for discovering the noble gases—declared that "German ideals are infinitely far removed from the conception of the true man of science." He supposed that the previous scientific reputation of Germans was due to exploiting the work of others. Also contributing to the anti-German iconoclasm was Sir James Crichton-Browne, known both for his expertise in neurology (he once held the prestigious position of Lord Chancellor's Medical Visitor in Lunacy) and his shoulder-wide sideburns. He expressed gratitude that the war would "pull down from its pedestal and shatter for ever the notion of the German super-man in science, literature, art." Berlin had become a home only to sterile thought; there was no longer any need to hear what it had to say about the world.

WITHIN BRITISH SCIENTIFIC journals the war became a regular feature. Reports of scientists killed or wounded at the front; relatives missing in action. Meetings of the BA evolved from celebrations of scientific internationalism into "a symbol that the Empire was united and determined in the face of the common enemy, and not to be dismayed by his aggression." The 1915 meeting was repeatedly disrupted by warnings of Zeppelin raids.

Eddington found little support for his vision of international science in either London or Cambridge. Isolated politically, he hoped for sympathy from the still-neutral Americans. He wrote to Annie Jump Cannon at Harvard looking for assistance. He had met Cannon at a meeting of the International Solar Union in Bonn, Germany, years before. She was one of the greatest astronomers of the day—she catalogued by hand more than a quarter-million stars—and was the first woman to ever receive an honorary doctorate from Oxford. And as a deaf woman in an age of strictly gendered science, she knew something about being excluded. Eddington lamented:

> It is very sad after the jolly days in Bonn, that this division should come between us and our German colleagues. If only there was mutual respect between the combatants, it would be a less depressing outlook; but I am afraid the contempt and hatred of Germans has increased over here very much in the last three months.

He hoped that American astronomers would be allies in rejecting the barriers being thrown up around German science. They were not steeped in British war propaganda and could still talk to their German colleagues. Surely a noncombatant country would have a more balanced view?

Not many Americans were on Eddington's side. W. W. Campbell at the Lick Observatory in California declared that Germany had blood on its hands. And, he asserted, it was not the case that "science is thicker than blood." Germany, "the most scientific of all nations, has prostituted science to the base ambition" of military power. Eddington had found no sympathetic ears on the far side of the Atlantic either.

Eddington was split from patriotic science by his Quaker beliefs. For the Quakers, the most important factor in relationships was understanding an individual's responsibility to their own divine nature. A group might take an immoral action, but a person could be judged only by their own deeds. From Eddington's point of view, the German scientists were themselves victims of the war. Specifically, they were victims of the perverse systems of militarism and imperialism. They needed to be rescued from those systems, not blamed for them.

The Society of Friends' war-relief efforts were organized around these same assumptions. It did not matter to them on which side of the trenches someone was suffering. Everyone deserved compassion. So they ignored national boundaries, much to the dismay of patriots. They ran refugee camps in France but also supported displaced Germans in Russia. In Britain, the Quakers set up the Emergency Committee for the Assistance of Germans, Austrians and Hungarians in Distress. That group, organized to aid enemy citizens detained when the war broke out, was led by close friends of Eddington's.

This was seen as sympathy for the enemy by many people in Britain, and the Quakers were roundly criticized. It was even suggested that this violated the Treason Act of 1534. The Emergency Committee received death threats and three of its members ended up in prison, including Eddington's friend and fellow Cambridge scientist Ernest B. Ludlam.

The anger toward pacifists like Eddington was rooted in the idea that if someone was opposed to the war, they were necessarily supporting the aggression of Germany. The Friends moved to correct this—they were not "pro-German" by any means. They did not approve of German military aims, policies, or methods. Rather, they thought that inclusion and tolerance were more likely to end the war than violence. Building bridges with "the enemy" would, hopefully, give rise to a new internationalism that could prevent future wars.

When an Allied astronomer imagined the militarized, dastardly Prussian scientist, he could hardly have done better than Fritz Haber. Haber wore his military uniform, complete with rank, to the laboratory every day. His face, sword-scarred from youthful dueling, always sported a cigar. He enjoyed being a commanding authority and both scientific and military staff constantly followed him around. He was known for oscillating between manic activity and deep depression. Key to his success in chemistry was working closely with his wife, Clara Immerwahr. As was typical for scientific marriages of the time, she received little recognition. Their relationship was known to be a tense one.

Much like Einstein, Haber grew up in his family's factories—in his case, dye and paint manufacture. His chemical expertise came early. He became fixated on nitrogen as a young man when he realized Europe's dependence on fertilizer. His breakthrough of ammonia synthesis was originally small-scale; nonetheless he was able to sell it to BASF in 1908 for 10 percent of the net profits and long-term research funding. Industrial production only began in 1913 with the help of the brilliant chemist Carl Bosch (who would later found and run the conglomerate IG Farben). That 10 percent share quickly made Haber famous and rich.

Also like Einstein, Haber was born to a Jewish family. He converted to Christianity as an adult purely for professional and social reasons—it was very difficult for a Jew to secure academic positions. His son later speculated that Fritz's intense patriotism was a way of compensating for his Jewish origin. Even his colleagues were sometimes surprised by his "uncritical acceptance of the state's wisdom."

While solving the nitrate crisis for German munitions manufacturing (and thus keeping his country in the war) Haber was already pondering what else his chemical expertise could offer the war effort. Most of the combatants had made minor efforts at using shells filled with tear gas. Haber's experiments found those to be nearly useless. He thought there was a better possibility: chlorine.

Chlorine is a green-yellow gas at normal temperatures. When inhaled it combines with moisture to form hydrochloric acid and begins burning the lungs. In sufficient quantity it could easily incapacitate or kill. There was another feature of chlorine gas that made it appealing to Haber: they had an enormous amount of it. It was made as a byproduct of the synthetic dye industry that Germany dominated. Essentially, BASF and Bayer already had vast chlorine factories up and running. They had a great deal of experience producing, storing, transporting, and handling the gas. It was perfect.

Haber had personal connections directly with the German General Staff and quickly received support. He suggested releasing the gas from massed cylinders to create a poisonous cloud, which would then (hopefully) drift toward enemy positions. Chlorine is heavier than air, so it would fill the trenches and force infantry out of their fortifications. The process of turning industrial waste into a weapon went fairly smoothly. Haber was assisted by Otto Hahn, who had returned from the front when his unit was so badly mauled that it was no longer an effective fighting force. They set up a test on April 2. It was so successful that Haber was slightly gassed himself. Over the next two weeks more than five thousand cylinders, each about the size of a person and weighing two hundred pounds, were filled and hauled to the front. Then, they waited for the wind to be right.

Haber's invention was first used on the battlefield at the Ypres salient, where British troops had been dug in since the end of the war of movement. The 150 tons of chlorine released that day formed a greenish cloud somewhere between 30 and 100 feet high, creeping forward at about one mile per hour. The defenders were French punishment battalions and Algerian troops. Their first experience of chlorine gas would have been a peculiar smell of pepper and pineapple, then they suddenly would have begun coughing and choking as their lungs filled with fluid. Drowning on dry land, they abandoned their positions and fled to the rear. The Germans were unprepared for their success and made only modest gains. We do not know the casualties from that initial use of chemical weapons, possibly around 5,000 killed and 10,000 injured.

The actual physical damage of gas attacks was secondary to the psychological effects. Haber himself praised the novel weapon for the way it "troubles the mind with fresh anxieties of unknown effects and further strains the soldier's power of endurance at the very moment when his entire mental energy is required for battle." The fear of suffocation was, for whatever reason, greater than the fear of bullets and shrapnel.

The general staff was quite impressed. Haber was celebrated as the key innovator for this and became essentially the sole adviser to the government on chemical warfare. He was promoted to captain, though he never commanded troops. His subordinates grew rapidly nonetheless—he was eventually in charge of more than 2,000 personnel.

The Allied reaction to this new weapon was panic. No counters or defenses existed for this totally unexpected development. As a stopgap, soldiers were instructed to breathe through urine-soaked socks during a gas attack. Even the Germans had no defenses in place against chlorine. In his rush to get gas to the front, Haber spent little time thinking about that. Ironically, the first gas casualties were the German soldiers setting up the initial attack on April 22: they were bombarded by the French and some of the cylinders cracked.

Gas quickly became seen as a distinctly *scientific* weapon. On both sides of the trenches it was portrayed as a singular product of Haber's genius. It wasn't—it was much more a product of the chemical dye industry—but that was the way it came to be seen. Chemists were responsible. On the German side, that meant rewarding Haber; on the Allied side, that meant anger toward him. In response, French and Belgian scientific academies quickly expelled all of their members from enemy countries. The Chemical Society of London was slightly more targeted—they only threw out scientists involved with weapons work, such as Nernst and Ostwald. Their statement: "The Chemical Society considers that it is neither compatible nor consistent with its loyalty to the Crown, whence the Royal Charter under which it works was derived, to retain any alien enemies upon its List of Honorary and Foreign Members." The Prussian Academy nearly ejected all its foreign members in retaliation (Planck intervened at the last moment). Nonetheless, Allied scientists essentially saw no more possibility of working with their Central Powers colleagues. The explicitly scientific nature of chemical weapons, on top of the Manifesto of 93, made cooperation across the lines of battle unthinkable. International science was dead.

Even in Berlin, not everyone was delighted with Haber's success. His wife, Clara, was outraged. She had not become a chemist to kill and maim. Now she was complicit in mass destruction. The night of May 1 the couple had a terrible argument. Sometime after midnight Clara took Fritz's army pistol and shot herself in the chest. Stunned, Haber threw himself into his work. He left for the eastern front the next morning to plan the first gas attack against the Russians. Einstein heard about the suicide some time later. He tersely informed his estranged wife, Mileva, about the death without introduction or explanation: "Mrs. Haber shot herself two weeks ago."

Haber's success with chemical weapons became a model for many German scientists wanting to contribute to the war. Scientists often took whatever they were working on before the conflict and adapted it to new battlefield needs—if you studied electromagnetism, now you designed radio antennas. If you were an expert on kinematics, now you performed ballistic calculations. Astronomers often had training in meteorology to plan for clear skies; now they predicted breaks in the rain to allow for infantry offensives. As with Haber, this was usually done on the scientists' own initiative—early on, the government thought the war would be too short to need their technical expertise for large projects.

British scientists had the same problem. In the fall of 1914 the Royal Society of London, puzzled at the lack of government inquiries, formed a War Committee to *offer* scientific help for the war. The War Office did not reply for two months—an apocryphal tale said that their response was that they already had a chemist. This is not to say there were not substantial problems about which scientists might have some advice. As in Germany, the British government was completely unprepared for the industrial and technical problems caused by the disruption of trade. Germany's dominance of the chemical industry meant that many important chemicals had never before been made in the UK. It was found that German patent filings did not give nearly enough information to duplicate a process. Britain was almost completely reliant on Germany for anesthetics and analgesics, not to mention optical glass and precision instruments. Factories were scrambling to find new suppliers, and the Royal Society chemists did their best to help. The government was barely involved in solving the initial crises. There was little direct state involvement in science before the war; it wasn't clear to Whitehall that the fighting called for any change in that.

Around the time of the *Lusitania* the Royal Society decided to go public with their frustration over the lack of government support for what science could do for the war. Their formal request produced £65,000 of funding for two years and the formation of an Advisory Council for Scientific and Industrial Research, eventually to become a Department. It did very little. Partly this was a cultural issue—war seemed to be no place for fuzzy-headed professors—and partly organizational. Research efforts were fragmented throughout Britain. The Admiralty and the Ministry of Munitions each had their own boards pursuing new ideas; medical research was completely separate. The military had a Chemical Products Supply Committee and one for Explosive Supply; neither had a chemist.

British scientific societies continued creating their own committees and sub-organizations devoted to the war. They desperately tried to get the attention and support of people in power. They felt ignored compared to their enemy counterparts—just look at Haber's Kaiser Wilhelm Institute. There was nothing like it in an Allied country. It seemed to them that in Germany and Austria scientists were respected and valued. In Britain, apparently not. A committee of major scientists was formed to address this "Neglect of Science." As part of this effort H. G. Wells launched a newspaper campaign demanding that the government take advantage of their scientists just as the Germans did. "On our side we have not produced any novelty at all except in the field of recruiting posters."

While the military did not think much of scientists, the reverse was not true. The Royal Society Council decreed that any scientist who was killed at the front would receive the singular honor of being made a Fellow. Elevation to Fellow of the Royal Society (or "made FRS") was a mark of enormous distinction normally reserved for scientists making major contributions to their field. They had equated scientific excellence with military service. British scientists wanted them to be interchangeable.

Those who wanted to use their technical expertise to support the war often had to find their own projects. The father-son Nobel Prize–winning physicist team Lawrence and William Henry Bragg shifted their work from interpreting X-ray waves passing through crystals (similar to Moseley's research) to interpreting sound waves coming from German artillery. As far as the mathematics goes, waves are waves. They built and operated devices that would use the sounds of cannon fire to locate enemy positions well behind the lines, which could then be targeted by friendly guns. The Braggs were so effective at this that eventually German troops were forbidden from shooting during certain periods, lest they bring down counterartillery fire. Despite their success, the Braggs never stopped complaining about military resistance to scientific ideas. Because they needed to operate the equipment themselves, they actually lived near the front. In an effort to maintain a semblance of scholarly life, they held monthly scientific seminars. By the end of the war about half of British scientists were doing war-related work, even if it wasn't coordinated with an official military program.

The sudden appearance of chemical weapons on the battlefield refocused the British government on the importance of science (as did the "shell crisis" of the time, when an ammunition shortage led to political upheaval). Laboratories were set up at the front to take samples of German weapons and develop British equivalents. Their system for weapons development was badly organized and chaotic, which slowed progress. Much of the gas was produced in small academic labs. By July, Britain had her own gas units, though they weren't used until the Battle of Loos in September. The cylinders were cracked and corroded and the attack wasn't very effective.

The biggest problem was simply producing new chemists. The British system of science education was not set up to produce large numbers quickly. Those who were trained often did not have the large-scale industrial expertise necessary. Australia sent about half of their total chemists (about a hundred) to London to help expand the lists. Margaret Turner, a scientist at the Chemical Laboratories at Aberystwyth, asked to help. She was denied; women could work in munitions factories but not as scientists.

This attempt to make lists of scientists immediately clashed with an attempt to make lists of soldiers. At the same time that scientists were beginning to be recognized for their technical skills, the government created the National Register of all persons between age fifteen and sixty-five. This was an obvious step toward mass conscription. As men were asked to attest that they would be willing to serve, an unusual situation occurred: what if a man attested that he would serve, but his employer wanted to prevent him, say for a crucial munitions job? A series of tribunals were set up to screen these sorts of requests—someone removed from the conscription rolls was said to have received an "exemption." The usual standard was whether this person's civilian work was of "national importance." Soon they found themselves handling a tricky question: was the work of a scientist, by itself, of national importance? There was no clear guidance given for this, and decisions were chaotic and arbitrary. One industrial chemist was not granted an exemption, but a pharmaceutical chemist was. After the system was well under way, of 330 chemists who had sought exemption, 94 were refused.

Eddington was a good target for any conscription program, so Cambridge University preemptively applied for an exemption for him. They argued that if he were taken as a soldier, the observatory would have to be shut down, which would damage the scientific work of the country. Beyond the scientific value of Eddington's work, the university was surely engaging in some advance public relations. Eddington's pacifism was well known and they surely would not have wanted to deal with the embarrassment of having one of their professors publicly refuse to fight.

The Royal Society received a steady stream of requests for help from scientists looking for exemptions based on their work. They decided not to take any action to support these. The Society was still trying to convince the government that scientists were properly patriotic, and they wanted no part of anything that suggested scientists were not eager to serve in the military. Britain remained deeply ambivalent about whether science was important in wartime.

LIFE FOR SCIENTISTS in Germany was perhaps not quite so rosy as H. G. Wells had imagined. There had been no deliberate effort to mobilize scientists for war work there either. They, too, had no clear guidelines exempting scientists from military service at the beginning of the conflict. As in Britain, many scientists from the Central Powers decided their best contribution would be holding a rifle. We have already heard about Nernst's and Hahn's adventures. The Austrian physicist Friedrich Hasenöhrl was killed in October 1915 at age forty. The not-yet-famous philosopher Ludwig Wittgenstein enlisted in the Austrian Army, hoping that "standing eye to eye with death" would bring him some enlightenment. Erwin Schrödinger, bereft of cat, enjoyed watching the electrical discharge of St. Elmo's fire dance on the barbed wire.

There was plenty of science-based work to be done near the front, too, like the Braggs'. Lise Meitner, the Austrian physicist who would later go on to introduce the world to nuclear fission, left her positions in Berlin to become a radiological technician at a military hospital. Using X-rays for medicine was still a cutting-edge technology and skilled operators were needed. A year into the war she had taken two hundred X-rays, assisted in the operating room, and fixed the hospital's electrical system.

Meitner returned to Berlin in the middle of the war. She had overcome tremendous obstacles as a woman in science (she was once denied entry into a chemistry lab on the grounds that her hair was too flammable). She complained that she had been in the field for so long that she "no longer knew what physics is." Meitner felt strongly that it was the duty of German science to support the war, and was pleased to attend the patriotic evening gatherings of professors common at the time.

One of these gatherings was at Planck's house, where Einstein happened to be in attendance. He shared his views on the war and Meitner was not impressed: "Einstein played the violin and, in passing, offered such deliciously naïve and peculiar political and military opinions. The very fact that an educated human being exists who, during this time, does not consult a single newspaper, is surely a curiosity." No one was impressed with his pacifism. It read not as idealistic but foolish.

He was still active in the BNV. He and Elsa regularly attended meetings that spring. It is an interesting note of Einstein's pre-fame life that no one recognized him. Walther Schücking, a prominent pacifist at one of the meetings, wrote himself a note to remember who the eccentric physicist was—apparently he had discovered some law regarding the unity of time.

The group had smuggled a censored book, Richard Grelling's *J'accuse*, into Germany. It explained Germany's responsibility for the war in seditious terms. They ran a kind of lending library for it—a member could borrow it and then return it within forty-eight hours. Einstein took it out, read it, and returned it promptly. He might have hesitated had he known that he was being investigated by the police for his political loyalty after he sent a poorly phrased postcard through the mail. The police decided that he had been "until now not politically active" but was now "a supporter of the peace movement." He was left alone—politically unreliable but not dangerous.

Einstein had been growing closer to Elsa, though her bourgeois lifestyle still did not seem very inviting. After Mileva left, he moved to an apartment on the Wittelsbacherstrasse, near where his cousin lived. He knew she wanted to marry—the role of the professor's wife was appealing to her in a way it never was to Mileva—but he was wary of entering into a new marriage so soon after his last. She fulfilled wifely duties nonetheless, trying to get him healthy food despite the privations of the blockade. When she was out shopping for him she would have been one of many women on the streets. As men were drawn away to the battlefields women increasingly took on the everyday tasks of running the city. Many saw this sudden visibility as a kind of empowerment—Marlene Dietrich called wartime Berlin "a woman's world."

Liberated women were not the only unusual sights on the capital's streets. Some 70,000 Eastern European Jews had fled the fighting and taken shelter in Berlin. These orthodox Jews were very different from the largely assimilated communities already there, who rarely ate kosher food or visited the mikvah. Many of these secular Jews, including Einstein, were somewhat taken aback by these kinsmen who seemed so unfamiliar and alien. They were also shocked by the rising anti-Semitism that greeted the refugees (anti-Semitism increased in every country involved in the war). Right-wing groups accused the refugees of bringing crime and shirking their duty to fight; in fact, Jews were overrepresented in the German military, using the war as an opportunity to demonstrate their wholehearted citizenship. Even within a country's borders, citizenship could be complicated.

Planck, Einstein's mentor, had begun having more complicated thoughts about the war as well. We can see the hand of Lorentz, Einstein's other father figure, in his evolution. Everyone trusted Lorentz, so when he contacted Planck about conditions inside German-occupied Belgium, notice was taken. They met in person in 1915 and he persuaded Planck to repudiate the Manifesto of 93. Planck wrote an open letter doing so, which Lorentz had printed. In the letter Planck demurred that signing the manifesto had been only a defense of Germany and not any specific actions taken by any individual German. He pleaded that international values were compatible with love for one's own country, and that perhaps the scientific community could come together once again. This should not be mistaken for a turn to pacifism, though—he wrote that scientists should withdraw from the field and let cannons speak instead of manifestoes.

Lorentz tried to similarly soothe the fiery Wilhelm Wien, to less success. Trying to build on his moderate victory with Planck, Lorentz sent copies of the repudiation to physicists in enemy countries—the elder Bragg, Joseph Larmor, Oliver Lodge—to help calm their anger. It was not well received. With the bodies of prodigies like Moseley constantly coming home, it was not a time for healing.

CHAPTER 6

A Vital Victory

"An unscrupulous opportunist."

EDDINGTON LOVED A good puzzle. Not just crossword puzzles, though his obsession with them was legendary. He felt puzzles—mysteries—were the real heart of science. Science wasn't about celebrating what we knew, it was about exploring what we didn't know. It wasn't about being sure you were right; it was about finding a new puzzle to solve.

So when he became co-editor of the Royal Astronomical Society's journal *The Observatory* in 1913, he instituted a new feature titled "Some Problems in Astronomy." Each issue had an astronomer present some "perplexing" part of the field—something that was not well understood, or for which there were no satisfactory explanations. Eddington wanted everyone talking about the exciting puzzles, not the stuff we already understood well. These included the nature of the spiral nebulae, why some moons orbit counterclockwise, whether the dark spots in the Milky Way are actually gaps, and the precise shape of the Earth.

Eddington wrote several of these columns himself (more than any other contributor). He speculated about cometary orbits, why stars were certain colors, and, in February 1915, the nature of gravity. He remarked that gravity remained as much of a mystery as it had been in the days of Newton. They had the formula, but still no one had any idea what gravity really *was*. Many possible explanations had come and gone—electrical inequalities, particle bombardments, ether squirts. Eddington was disappointed that none had led anywhere. Gravity still seemed to have no connection to any other force of nature. Most frustratingly, none of the theories had a way to test them and decide whether they were worthy of further investigation.

Except, he noted near the end of his essay, that there was a prediction that light would change its speed in a gravitation field. This suggestion had been offered by a German professor named Einstein. Eddington knew about it from the 1912 eclipse expedition, where C. D. Perrine had failed to test it. Eddington's brief discussion here referred only to Einstein's 1911 paper, suggesting that he was completely unaware of any of the work done since then. He had learned almost nothing about Einstein since he first heard the name in 1912.

Eddington's ignorance of what had been happening in Berlin was not at all unusual. It is hard today for anyone to remember a time when Einstein's name would not have been immediately recognized. Einstein's name appeared only once more in the whole "Problems in Astronomy" series, when he was mentioned in the same breath as Lorentz with the comment that neither of their theories were well supported by observational evidence. One of the "Problems" columns even focused on whether there was a redshift in the solar spectrum—one of the primary tests of general relativity—without a single mention of Einstein or his theory.

The handful of British scientists who knew about Einstein before the war generally thought of his ideas as contributions to well-established electromagnetic theory, not a revolutionary reworking of the fundamentals of knowledge. One of these was Ebenezer Cunningham, a physicist at St. John's College in Cambridge. Cunningham was for the most part trying to elaborate on Einstein's 1905 theory, and was actually working on a textbook on relativity when the war broke out. We don't know whether Cunningham and Eddington discussed relativity in this early period—they were interested in it for very different reasons, in different contexts, and wouldn't have had much to say to each other. Interestingly, Cunningham was also a pacifist (though not a Quaker). As with Eddington, the arrival of the war changed his work completely.

So, by mid-1915, a handful of people in Great Britain were vaguely aware of Einstein. They didn't really know what his ideas were about. What they did know was between four and ten years out of date. Einstein was a footnote in other people's projects. Eddington's attention had been caught by how empirical evidence for relativity "would mean that gravitation has been pulled down from its pedestal, and ceases to stand aloof from the other interrelated forces of nature." But as far as he knew, nothing interesting had been happening with the theory. In any case, there was a war on.

BY THAT SUMMER Einstein's personal war was not unlike the western front. It had been in a stalemate for some time. His early advances, like the Germans' into France, had stalled. While he held ground, there were no signs of a breakthrough. He wasn't sure whether total victory was possible—was there a brilliant maneuver that would lead to triumph? Was there more beyond the Entwurf?

It is not always obvious when a piece of science is finished. The crisp form of a theory you see in a textbook usually only appears decades after its discovery. The classic laboratory experiment you do in school, with its clear right and wrong results, was profoundly confusing to the first person who performed it. Everything makes sense when you look back—but if you're not sure if your theory is complete right now, how would you know?

One of the things you do with a possibly complete theory is try to use it to solve problems. Does it give a better explanation for old problems than the theories we already have? Does it help solve a problem that we haven't made any headway on? Does it tell us about a problem that we didn't even know was a problem? Einstein was trying all of these. It's much like when you have a new wrench in the workshop—you try it on all the stuck bolts that you have sitting around. Hopefully it works better on them than the old wrenches and still works just as well as them on the old projects. Most of all, it should let you build something new that you couldn't before. Did the Entwurf do all these?

The stuck bolt of physics was the orbit of Mercury—a known problem that couldn't be solved by the old tools. The Entwurf didn't do much better. The old projects were showing that relativity was similar enough to the Newtonian theory and the laws of conservation of energy and momentum. The new projects were covariance and the equivalence principle (which could hopefully be demonstrated through Einstein's equations), and the deflection of light and the gravitational redshift (which had to be tested empirically, impossible from wartime Germany). So one of the things he spent the summer of 1915 doing was revisiting the old and new projects. Did the Entwurf in fact do everything he wanted it to?

He was expecting reassurance that the work he had already done was sound. A great breakthrough wasn't really on his mind. But that summer he made a new friend, a moment that would set in motion a chain of events that would completely reframe all the work he had been struggling with for a decade. That friend was David Hilbert, a distinguished professor at the University of Göttingen and one of the most influential mathematicians of the twentieth century. His close-set eyes gazed out from behind a pince-nez and above a neatly trimmed beard. His somewhat informal dress (by the standards of German professors) meant that in his early years of teaching, students often did not realize he was the instructor.

Hilbert came to general relativity through a roundabout route. His early career focused on geometry, which he originally thought of as being just an elaboration of sensory experience—Pythagoras's theorem should come from actually inspecting triangles, for instance. He gradually became interested in an "axiomatic" approach, in which the actual physical referents of geometric statements were irrelevant, and only their logical relationship mattered. So when Euclid said "two points define a line," points and lines didn't actually matter—you could replace them to get "two jabberwocks define a cromulence," and the logical structure of the laws of geometry would still be true.

Hilbert was interested in whether it would be possible to derive entire sciences just from these logical structures, without any physical references. His project was enormously ambitious and he had tremendous optimism: "There is the problem. Seek its solution. You can find it by pure reason, for in mathematics there is no *ignorabimus*." That last bit of Latin—*ignorabimus*—meant something like "we can never know," a famous declaration of the limits of science. Hilbert rejected this flatly. He had great confidence in the power of pure reason and also thought that scientists should challenge themselves by tackling vast problems. He was fine with science having "loose foundations" as long as everyone was exploring boldly.

He thought that the electromagnetic theory of matter (ETM for short) was an excellent candidate for "axiomization"—that is, that he could figure out the basic properties of matter from purely logical and mathematical principles. This was how he encountered Einstein's work in 1912 or so. Einstein's efforts to ground relativity in a handful of elegant principles looked very promising to Hilbert. Predicted phenomena such as light deflection suggested there was some connection between matter (via gravity) and electromagnetic energy (the essence of light), which might help crack open his ETM ideas. He decided to include some of Einstein's work in his physics seminar, which only deepened his interest. He then teamed up with his mathematics colleague Felix Klein to invite Einstein to Göttingen to deliver the Wolfskehl Lectures from June 28 to July 5.

Göttingen, both university and town, were ancient and elegant. The university was the most visited school in Europe, drawing international students from around the world. It produced not just top-class mathematicians but also the Brothers Grimm and Otto von Bismarck. Cutting-edge research blended with antique traditions—doctoral graduates were drawn across town in handcarts to kiss the statue of the goose girl.

Einstein gave six two-hour lectures on relativity (specifically, the state of the Entwurf) during his week there. Presumably his unimpressive delivery did not compare well to Hilbert's famously clear teaching. Nonetheless, Einstein and Hilbert developed an immediate rapport. Einstein said he found himself "enchanted." He called Hilbert a "great man!" and "a man of astonishing energy and independence in all things." He was impressed not just with Hilbert's mathematical virtuosity—he was one of the few people of the time who actually understood the project of general relativity—but also his political leanings. Hilbert had refused to sign the Manifesto of 93 and was public about his opposition to the war. This combination of scientific and political comradeship deeply affected Einstein. He found himself appreciating Hilbert as someone who, as he put it, could rise above the fray. The respite from being surrounded by imperialism was remarkable: "Berlin is no match for Göttingen, as far as the liveliness of academic interest is concerned."

He found that his conversations with Hilbert had "clarified very much" the state of relativity. Klein, who was present at most of these, had a slightly different take: he reported that Einstein and Hilbert were often talking past each other and not paying that much attention to what the other was actually saying. With a verbal shrug, Klein said this was fairly common for mathematicians. And there is an important sense in which he was right—Hilbert was interested in the ETM, Einstein was interested in the nature of time and space. They were looking at the same sets of equations but for different reasons.

Nonetheless, both Hilbert and Einstein came away from their debates stimulated and, though they probably did not know it at the time, armed with everything they would need for their battle plans against the problem of relativity. They both began working on a better, more complete version of the Entwurf. As the two parted, on good terms and as good friends, the stage was set for one of the great scientific races of all time. They both had the keys to general relativity; the question was, who would realize it first?

THE TRIP TO Göttingen got Einstein thinking about how much the war had isolated him. He had used to travel quite a bit to see scientific collaborators, but it was much harder now. He commented that his "truly profound colleagues"—meaning Planck and Hilbert—still wanted to maintain connections with scientists in other countries. "Hilbert now regrets doubly, as he told me, having neglected to maintain international relations better. Planck is doing everything to bridle the chauvinistic majority at the [Prussian Academy]."

A planning meeting at the BNV seems to have prompted Einstein to contact Lorentz with a new plan. In a moment of big-tent ecumenism, he contended that academics were innocent of the war (quite a change from his earlier reactions to the Manifesto) and that solidarity among scientists was more important than assigning blame. If he didn't live in Berlin, he wrote, he would be contacting like-minded scientists in Britain and France to gather in a neutral country and nurture personal connections. But he did live in Berlin under the blockade. Worse, he had few contacts of his own. Worst of all, Einstein admitted, he just wasn't very good at talking to people. He hoped Lorentz could take on this project of gathering international supporters. Surely there must be other scientists battling "nationalistic blindness" in other countries?

Lorentz declined to take up the project, apparently because he thought the mood of British and French scientists was already too far gone to make a real difference. Einstein was disappointed, though not particularly surprised. He tried to explain the political tenor of Berlin, saying that the scientists were generally level-headed—it was the historians and philologists that were "chauvinistic hotheads." He defended Planck by saying that the elder man had signed the Manifesto of 93 without reading the text. In the end, though, Einstein said he had realized that apparently advanced societies were really just disguised oligarchies. The powerful would forever be able to make the people hate one another.

Einstein's pessimistic mood may have been shaped by his reading material at the time—Tolstoy's *Christianity and Patriotism*. He saw little opportunity to convince the masses not to fight for their masters. Sourly, he wrote, "no education and intellectual cultivation seems to be able to protect against this wretched madness." He confessed that he had recently shared a train car with soldiers and "relished their stupidity and crudeness."

Einstein probably encountered those soldiers on his journey to or from Göttingen. He had also hoped to take another trip that summer. Relations with his oldest son, Hans Albert, had become strained since Mileva took the boys back to Zurich. They had recently exchanged emotionally charged letters in which Hans Albert declared that he had no interest in seeing his father. Einstein took that at face value, canceled his trip, and took Elsa and her daughters on a vacation to the Baltic.

They were all delighted to be out of Berlin, even for a short time. The effects of the British blockade were growing rapidly, and were generally magnified in the imperial capital. Rationing had begun in February 1915. Each Berliner received one half pound of bread per day. Meat, eggs, and fat were distributed by ration card. Potatoes were soon rationed as well, leading to the violent protests known as the *Kartoffel-Krawalle* (the potato riots). By the fall, the government was enforcing meatless and fatless Tuesdays and Fridays (extended to Sundays for bars and restaurants). Elsa had been cooking lunch regularly for a group of poor women, trying to spread her family's better-than-most resources.

At the universities, paper, ink, and coal were all heavily restricted. Library orders were canceled as finances entered a crisis. Government support had been severely cut, and the precipitous drop in enrollment (75 percent fewer students by summer 1915) came with a proportional decline in student fees. The German High Command at one point considered simply closing all the universities.

Patriotism was widely encouraged and often enforced. Foreign movies were banned and replaced with nationalistic films. A replica of a front-line trench was set up in a park, a common sight in the combatant capitals. The rising toll of the war—in lives and money—was becoming increasingly clear, and there were quiet conversations about the goals and eventual outcome of the war. Was it still possible to emerge from the conflict as an imperial power the equivalent of Great Britain? The great political split was between those who called for a German victory that resulted in territorial gains and those who called for a German victory without any annexation. Einstein hoped that Germany would lose the war but signed a petition for the latter anyway (as did Planck and Hilbert). A return to prewar conditions would at least be better than a revitalized Germany dominating Europe. The petition garnered 141 signatures, one-tenth of the signatures attached to a pro-annexation petition.

But there were widespread doubts about the conduct of the war that had emerged after the first year of fighting. The staggering death toll and virtually nonexistent gains in the west made optimism difficult. Even the nationalists in Einstein's circle had begun to moderate their enthusiasm. Both of Nernst's sons had been killed. One of Planck's sons had been killed; one was a prisoner of war.

EINSTEIN MANAGED TO reconnect with his own sons in September. He decided to make a trip to Zurich regardless of the earlier tensions. International travel was far from easy, even to a neutral country. This was aggravated by Einstein's habit of forgetting (perhaps intentionally) to register his absence from Berlin with the police, as was required.

The train to the Swiss border took a couple of days. During a stopover in Heilbronn he wrote to Elsa to excitedly report on how much better the food was there than in Berlin: "Milk and honey may not be flowing anymore, but there is still a decent trickle. You cannot get any fruit here, but there are vegetables." To pass the time on the journey he read Spinoza's *Ethics*.

Once in Zurich he hiked with Hans Albert, much to both of their delight. Albert's unflagging friend Besso helped mediate between Einstein and Mileva, setting up schedules where the estranged parents did not have to speak to each other. In addition to his family visits, though, Einstein had another job. The BNV had been very interested in allying with Romain Rolland, the Nobel Prize–winning French pacifist writer who had taken up residence in Switzerland. Rolland was influential among antiwar thinkers across Europe but had refused to join any organizations. Einstein had written to Rolland earlier and met him in person after discharging his family responsibilities.

They had a vigorous conversation about the war. Rolland described Einstein as both "very vivacious and serene; he cannot help giving the most serious thoughts a jocular form." He asked whether Einstein voiced his criticisms of Germany to his friends in Berlin. Einstein replied that he did not, and instead asked many Socratic questions in order to upset their peace of mind. "People don't like that very much."

After the meeting, while he was still in Switzerland (and thus free from the German censors), Einstein wrote Rolland a passionate letter. He described the dire straits in which the BNV found itself: "It is being harassed by the inspection authorities and being condemned (on the whole) by the press. . . . In many cases intellectuals have completely lost their composure." The "war of words" had started among Berlin writers. Einstein luxuriated in these few days of unfettered conversation.

On his way back to Germany, Einstein crossed at Konstanz. The German border authorities turned him back. He had failed to file the proper paperwork for his passport while he was in Zurich. Humiliatingly, a soldier escorted him back to Kreuzlingen in Switzerland, where he had to wait for a few days until the bureaucracy was sorted out. Strangely, he found himself trapped in a place he loved and where he was free to talk, seeking permission to return to the patriotic bustle he hated.

He finally made it home on September 22 after twenty hours on the train. His travel weariness was wiped away when he saw on his desk a new scientific paper from Lorentz. He read it eagerly. The paper involved applying what is known as Hamilton's principle in gravitational and electromagnetic fields. The principle, named after the formidable Irish mathematician Sir William Rowan Hamilton (1805–1865), was a powerful alternative tool used for analyzing the motions of bodies. In classical Newtonian mechanics one can take a given object, figure out the forces acting on it and how those forces change, and then calculate the path it will take. This can quickly get tedious. Instead, Hamilton's principle ignores the specific forces at work and analyzes the energy the object has, including how that energy changes back and forth between kinetic (energy of movement, easily seen) and potential (hidden energy, ready to be unleashed when conditions allowed). The principle basically says that objects will move in the way that requires the least expenditure of energy over the least amount of time—as one famous physicist put it, "Nature always uses the simplest means to accomplish its effects." "Simple" here means something like "thrifty"—minimal effort, done quickly.

So to use Hamilton's principle, you take an object in a particular environment and set up your equations to express this thriftiness. Out will pop the trajectory—the path—that your object takes. This makes many problems much, much easier, at the cost of a high level of abstraction. You can no longer talk about a specific force pushing your object, only a general sense of energy shifting around. Lorentz's paper applied this method in a new way to objects subject to gravitational and electromagnetic forces. This was similar to some of the methods, called Lagrangians, that Einstein had been using to understand the conservation laws in his Entwurf theory. He immediately sat down to see if Lorentz's methods could help, firing off a letter to Lorentz describing how "delighted" he was.

This was a new wrench that Einstein was checking against old bolts. He was already happy with how the Entwurf handled conservation laws, but a new wrench was a new wrench—he couldn't *not* try it out. It seems that while he was checking out the conservation laws he decided to take a look at some other old bolts too. Specifically, rotation. The problems of how rotation worked with the equivalence principle and covariance were some of the initial steps in his attempts to generalize relativity, so he expected no problems. But in retracing those original steps, he found something extraordinary. He found a mistake.

It actually wasn't a new mistake. Back in 1913 when he and Grossmann had been putting the Entwurf together, Grossmann had pointed out that the equations were not covariant for certain kinds of rotation. That was when Einstein formed his "hole argument" explaining why that was not really a problem. But now that he studied the initial problem with two years of perspective, something looked different. He wasn't sure exactly what, though. Was the Entwurf wrong? On one hand, despair was creeping up on him—had he been on the wrong path this whole time?—on the other he said that the possibility of something new "electrifies me enormously."

After a few days of intense thought, he wrote to Freundlich, someone physically close by who was familiar with relativity's foundations. The rotation problem wasn't really one of mathematics. The calculation involved was extremely simple, just a few lines. And he had seen it all before, there was nothing new. Einstein realized he just needed to think about it in a different way. He also realized that he couldn't do it. He told Freundlich that he was stuck "in the same old rut." He had been staring at it too long to shift his perception. "I must depend on a fellow human being with unspoiled brain matter to find the error. If you have time, do not fail to study the topic."

We do not know if Freundlich offered anything useful. Einstein threw himself into the problem, though. In early October he decided to try to re-derive the entire Entwurf theory to try to find the source of the difficulty. He and Grossmann originally crafted the Entwurf from a mix of mathematical and physical reasoning. This time he decided to try a purely mathematical approach, perhaps using the new Hamiltonian tools.

This was the beginning of what Einstein would later call the most intense and stressful period of scientific work of his entire life. Nonetheless, on two occasions in October he put relativity to the side. The first was when the Central Organization for a Durable Peace, a pacifist group based in The Hague, contacted him. They were setting up a Large International Council and invited Einstein to serve. He agreed to do so, and sent a postcard to that effect.

The second was a letter from the Berlin Goethe League, a well-known cultural organization. The League wanted to publish a patriotic album consisting of essays by intellectuals that would help reassure the public of the war's righteousness. Inexplicably, they asked Einstein for a contribution. Presumably they expected a conventionally professorial statement of nationalism. Instead they got a three-page screed mocking the very notion of patriotism and rejecting any kind of war. He decried nationalism as a tool for encouraging "animal hatred and mass murder." He called for a new political order in Europe that would prevent all future conflict. The League was, to say the least, surprised.

Einstein somehow wrote this essay, "My Opinion of the War," while furiously reworking the foundations of general relativity. His problem was essentially this: he had rediscovered a problem in his theory (the lack of general covariance). But he had already come up with a neat explanation for why that wasn't really a problem (the hole argument, which said you shouldn't expect general covariance). So he had to figure out why the hole argument was wrong, and come up with a better way to solve the initial problem.

Remember that the core of the hole argument was Einstein's loyalty to Mach. He had been convinced that coordinate transformations—which could represent the different measurements two observers make in different locations—had to have a physical, tangible meaning. That was the whole point of Machian positivism, that measurements were attached to specific activities done with physical measuring devices.

As he looked back on his earlier arguments, though, Einstein realized he was now seeing things a little differently. He was no longer convinced that the meaning of the numbers had to be quite so physical. Perhaps the changes induced by the coordinate transformations were only mathematical—only an illusion. If he was willing to give up the Machian notion that all measurements had to be associated with actual physical devices, then they could just be mathematical representations. The differences in measurement could be easily accounted for with the right equations, and every observer could agree on the nature of gravity. Einstein realized that if he released himself from what he called his "fateful prejudice"—Mach's insistence that every number be a physically possible measurement done with rods and clocks—he could have a generally covariant theory after all. He just had to loosen the Machian goggles through which he had been looking. Relativity could again become the theory he had first imagined—laws of nature truly independent of people, place, or movement. It could transcend the merely personal.

This was a seismic shift in the way Einstein saw science. Relativity had begun in 1905 by thinking deeply about the actual physical processes of measurement and their significance—positivism. He used those results to dismiss unseen entities such as the ether as metaphysical and superfluous. But now Einstein was willing to give the equations a life of their own, even if they referred to things like space-time that could not be seen and directly measured. The deepest truths of the universe needed to be accessed by abstract mathematics, not by empirical experience. We might call him a rational realist by November 1915—he thought the constructs of mathematics and logic could reveal the true nature of things, and transcended what humans could actually see. He was now more like Plato than Mach.

We are not used to hearing about geniuses changing their minds. But this is precisely what happened here. In the cartoon versions of the way science works, Einstein should have changed his mind because of a new experiment or a more precise calculation—some new piece of evidence. That's not what happened, though. Instead he just looked differently at what he already had. The temptation is to try to give a rational explanation for Einstein's shift, some concrete logical justification. There wasn't any. Scientists are just like everyone else—they don't always have a good reason for the decisions they make. Einstein made an intuitive leap without a particularly good reason to do so. He saw a way to make general relativity into the kind of theory he wanted, and he took it. In his own words, Einstein was an "unscrupulous opportunist."

ONCE EINSTEIN RELEASED himself from his "fateful prejudice," everything began falling into place. Now, convinced that general covariance was possible, he went back and reexamined all the candidate equations he had discarded in 1912–1913. Since then he had accumulated a couple of new tools to use to evaluate them. He had his new non-Machian way of thinking about coordinates, and he had the Lagrangian methods that he had been working with for the previous year. He unburied one tensor that he had given up on early in the Entwurf process because he couldn't make it equate to the Newtonian limit. It had covariance and equivalence, but the calculation to reduce it to Newton's law was so complicated he had thrown in the towel. But now his Lagrangians made that calculation much easier, and he found that it did in fact correspond to Newton's law. It worked.

He realized that Hilbert would never have made the same mistake he did, and the Göttingen mathematician might well have already discovered these equations. Desperate to keep priority on his own pet theory, Einstein assembled hastily what he had so far and submitted it to the Prussian Academy on November 4. In a moment of humility unusual for any scientist, he took the podium and began describing where he had gone wrong. He rejected the Entwurf publicly, saying that it had been an "absolutely impossible" approach. He admitted that he had "lost faith" in the Entwurf equations and tried to restart the project from zero. He described how he had missed the correct solution in Zurich and only now realized his mistakes. As he put it to a friend: "I have immortalized the final errors in this struggle in the Academy contributions."

The story he told the Academy of his search for general relativity was not quite accurate. He described his search as solely a mathematical one, largely ignoring the physical considerations that had guided him. This certainly made for a cleaner story and was easier for the audience to follow. Einstein scholars Michel Janssen and Jürgen Renn call this the "arch and scaffold" strategy. The scaffold of the Entwurf had all kinds of physical reasoning and messy techniques, but it allowed Einstein to build the arch of his field equations. Einstein then dismantled and tried to hide the scaffold, so only the beautiful arch remained. Generally he stuck to this version of the story for the rest of his life, making the quest for relativity sound much neater and more straightforward than it actually was. As often happens, a confusing struggle was edited to make a better story.

The meeting ended with his paper incomplete, though he felt he had made a huge step forward. Thrilled, he wrote to Hans Albert: "In the last few days I completed one of the finest papers of my life; when you are older I'll tell you about it." Three days later Einstein sent a brief note to Hilbert attached to his new equations. He was careful to note that these changes were based on work he had done four weeks before, just in case Hilbert was close. Sommerfeld had let Einstein know that Hilbert had found "a hair" in the soup of relativity, and Einstein was deeply worried that his fix had been anticipated. "I am curious whether you will take kindly to this new solution," he wrote.

Amazingly, in the midst of this race with Hilbert, finally having broken the logjam preventing his progress on relativity, Einstein again put his equations down. He had received a letter from the Berlin Goethe League regarding his essay on the war. They said they would indeed like to publish it, if they could only remove a few paragraphs. Calling for new forms of political organization was fine; attacking patriotism itself was perhaps a bridge too far. In his response he expressed his opinion that "the glorification of war" must be opposed by "all genuine friends of human progress." In the end he agreed to the publication of the edited version.

He wrote his response on November 11, the day he continued his presentation to the Academy of the new general relativity equations. Dropping the politics and picking up the equations, he began by correcting a few errors he had made the week before. He remained very concerned about establishing his priority, stressing that these "new" results had really been discovered in 1912, he just hadn't realized it. Letters began flying back and forth between Einstein and Hilbert as each strove to inform the other of any progress that had been made. Einstein particularly noted that he had found something that bore on the ETM theories that Hilbert was so interested in, a tweak that surely did not go unnoticed.

Hilbert casually let Einstein know that he was planning on presenting all of his work on general relativity at a seminar in Göttingen on November 16, a Tuesday just three days away. He invited Einstein to be present for it, even suggesting two trains that would get him there on time. He would be happy to host Einstein at his home so they didn't even have to worry about a hotel reservation. Einstein did not respond until Monday, demurring that he was tired out and suffering from stomach pains from his increasingly blockade-damaged diet. He asked Hilbert to keep sending his new work regardless.

Einstein was now extremely pleased with his equations, and it was time to start testing them on the old stuck bolts that had been bothering him. He sat down to calculate the perihelion shift of Mercury with his new tools—and the bolt moved. His calculation said the shift should be 43 arc-seconds (about 1/100 of a degree) per century; astronomers had measured the shift at 45 ±5 arc-seconds. He had it—clear, empirical evidence for his theory. The same day he declined Hilbert's invitation he wrote to his friend that he had solved the problem of Mercury: "Imagine my joy!" And then to Besso describing his victory:

> In these last months I had great success in my work. *Generally covariant* gravitation equations. *Perihelion motions explained quantitatively.* The role of gravitation in the structure of matter. You will be astonished. I worked horrendously intensely; it is strange that it is sustainable.

It was a huge step. He wrote an anxious letter to Hilbert to let him know (and to remind him that the equations were actually three years old): "Today I am presenting to the Academy a paper in which I derive quantitatively out of general relativity, without any guiding hypothesis, the perihelion motion of Mercury discovered by Le Verrier. No gravitation theory has achieved this until now." Einstein's marking of his territory was less than subtle. Hilbert sent his "cordial congratulations on conquering perihelion motion." He was amazed that Einstein had completed that calculation so fast—it was extremely difficult. Einstein had the advantage over Hilbert because he had already done a very similar calculation back in 1913, so was able to do this one quite rapidly.

The presentation Einstein mentioned was the first one he delivered orally that fall. This was probably because he hoped to get the attention of one crucial astronomer in the audience—Karl Schwarzschild. Schwarzschild had been the director of the Astrophysical Observatory in Potsdam, outside Berlin, and was well known for his powers in mathematical physics. He was admired worldwide for his research in stellar atmospheres and motions as well as for his outgoing personality. As a loyal Frankfurter he'd enlisted in the German Army immediately after the outbreak of war, despite being over forty years old. He happened to be on leave on November 18 and was in the audience at the Academy—he was good friends with Hilbert and had heard a lot about relativity. Schwarzschild was exactly the person Einstein needed to impress if he was ever going to get astronomers on his side, so he put on his best show.

He needed more empirical evidence, and that meant persuading astronomers to conduct the completely new predictions of the gravitational redshift and deflection of light. Einstein reworked both of those predictions with his new equations and had a bit of a shock. The redshift was unchanged. The deflection, however, had been totally wrong. With his new methods he found that the correct prediction was about 1.7 arc-seconds, twice what he had predicted in 1911. So if the Brazil eclipse had not been rained out, and if Freundlich's Crimea eclipse had not been stopped by the war, they would have been looking for *the wrong deflection*. If their measurements had been accurate, his theory would have been disproven before it even started. We can only imagine Einstein's sigh of relief as he realized that tragedy had turned into the greatest of luck.

These few weeks were Einstein's most productive and exciting time of his entire life. It was the culmination of, essentially, all of his career since the patent office. He wrote to Paul Ehrenfest, "I was beside myself with joy and excitement for days." The results were all that he had hoped for: "The theory is beautiful beyond comparison."

Just two days after he presented his Mercury results, though, Hilbert presented the full set of correct equations for general relativity—which Einstein did not yet have—at the Royal Academy of Sciences in Göttingen. Hilbert credited Einstein with providing a starting point for his work, though his paper suggested that Einstein had merely formulated the right questions and Hilbert had found the answers. He had, somehow, gotten to Einstein's end goal in a matter of weeks instead of years. Certainly a large part of this was that Einstein had already done much of the work by the time they met in the summer of 1915. Another large part was simply that Hilbert was a better mathematician and was never hamstrung by Einstein's various dead ends.

Five days later, Einstein presented his version of the final equations. He was not happy about being scooped by Hilbert. His friend, whom he had named as being the one person in the world who truly understood relativity, had claimed Einstein's results for his own. Bitterly, Einstein wrote, "In my personal experience I have hardly come to know the wretchedness of mankind better than as a result of this theory and everything connected to it. But it does not bother me." Hilbert quickly realized he had gone too far, and revised his paper on December 6 to give priority to Einstein. He may have written an apology directly to Einstein as well. On December 20, Einstein wrote a short note hoping to bring their rivalry to a close:

> There has been a certain ill-feeling between us, the cause of which I do not want to analyze. I have struggled against the feeling of bitterness attached to it, and this with complete success. I think of you again with unmixed geniality and ask you to try to do the same with me. Objectively it is a shame when two real fellows who have extricated themselves somewhat from this shabby world do not offer each other mutual pleasure.

Finding a like-minded person in politics and physics was too precious a thing to waste on a priority dispute. Einstein knew he had trouble making friends. Once he took someone into his inner circle he would go to great lengths to keep them there. If he could still be friends with Haber after Ypres, he could still be friends with Hilbert after Göttingen.

In priority disputes such as this, there are two major issues. One is the question of independence: was each person really working independently of the other? Or did someone take another's work without acknowledging it? In this case it can be difficult to determine. Einstein and Hilbert spent a month communicating directly with each other, informing their rival of every small bit of progress. So we probably cannot consider their work to be independent. As irritated as Einstein would feel about it, there was a sense in which they were collaborators rather than competitors. The second issue is equivalency. Did they really find the same thing? It seems in this case that the answer is "not quite."

Hilbert realized that there was a problem with the way his field equations dealt with energy conservation. He held off publishing his results until 1916, when he felt he had fixed the problem. So Hilbert was first over the finish line but with an incorrect result. If priority is important to us, we can still feel comfortable saying relativity belongs to Einstein.

THE REACTION TO Einstein presenting his masterpiece was mixed. Schwarzschild had been skeptical earlier—he had called evidence for the theory "rather fishy." But in December he wrote to Sommerfeld quite impressed by the calculation of Mercury's orbit. That felt like real science to him: "That is something much closer to astronomers' hearts than those minimal line shifts and ray bendings." It was scarcely a proof, though. Max von Laue was not particularly moved. He said the Mercury prediction was merely the "agreement of two single numbers." Surely that was not enough to change the "whole physical world picture in its foundations."

Einstein was not dissuaded. Most of his letters that month consist of him taking a victory lap, informing old friends and colleagues of his success both with covariant equations and the empirical support from Mercury. He sent his completed papers to Sommerfeld, writing that "it is the most valuable finding I have made in my life. . . . The result of the perihelion motion of Mercury gives me great satisfaction. How helpful to us here is astronomy's pedantic accuracy, which I often used to ridicule secretly!" He delighted in how the four papers he submitted to the Academy tracked the final emergence of the theory like a film: "As you read [the papers] the final stage in the battle over the field equations is being fought out before your eyes!" Einstein described his Berlin colleagues as all but one "trying to poke holes in my discovery or to refute the matter . . . Astronomers, however, are behaving like an ants' nest that has been disturbed."

He was also pleased that people were taking the larger implications of the theory more seriously. Moritz Schlick, one of the founding fathers of the Vienna Circle, had just completed the first major analysis of the philosophical significance of relativity. Einstein was pleased that Schlick had noticed the importance of Mach and Hume in the theory: "It is very possible that without these philosophical studies I would not have arrived at the solution." He stressed that the most important features of relativity were that it agreed with all previous experiments and theories (that is, Newton's laws) and was generally covariant. Now that we had covariance, the true nature of the universe was that of space-time, not our ordinary experience. He declared, rather opaquely: "Thus time & space lose the last vestiges of physical reality. There is no alternative to conceiving of the world as a four-dimensional (hyperbolic) continuum of 4 dimensions." Time and space, the most elemental tools we use to organize our experiences of the universe, were illusions. In their place was a new kind of reality, just at the edge of human comprehension.

In the four-dimensional universe of relativity, space and time were a single thing. We limited humans, though, think we see a changing mix of the two—the apparently separate phenomena of time dilation and length contraction were actually just different perspectives on the single entity of space-time. The laws of physics were the same for anyone, anywhere. Gravity was no longer a force between objects, as it was for Newton, but something far stranger. For Einstein, objects naturally moved through space-time on the shortest line between two points (a line called a geodesic). What counted as "shortest" could sometimes be altered by the presence of large masses like planets or stars, similar to how a picture drawn on a piece of cloth becomes distorted as you stretch the fabric. This "shortest" line was essentially still straight in four dimensions, though it could look curved to three-dimensional beings like us (much as how the path of an international flight, straight on the Earth's round surface, looks like an arc on a flat map). Our brains see that "curve" as an object being pushed or deflected by an invisible force—what we call gravity. Gravity, then, was simply a side effect of our limited perception of objects moving across a four-dimensional continuum. Newton said an apple fell because it was pulled by an invisible force generated by a massive body; Einstein said an apple fell because it was trying to find the shortest path through space-time as distorted by a massive body. Did this make gravity less mysterious? Should it be removed from Eddington's list of perplexing phenomena?

All of this was summed up in one simple, elegant equation (note that there are several different ways this can be written):

$$G_{uv} = 8\pi T_{uv}$$

On the left we have the curvature of space-time, the way it is stretched or distorted. On the right we have the arrangement of matter and energy. The presence of matter and energy creates curvature; curvature controls how matter and energy move. The two sides feed and depend on each other. As equations go, this is fantastically concise and clear—assuming you are familiar with non-Euclidean geometry, of course. This short expression actually represents ten separate equations all tangled together. What counts as "simple" in Einstein's world can be somewhat complicated.

There was still quite a bit of work necessary for Einstein to make his theory clear and persuasive. The cost of uniform laws of nature was this new bizarre universe that was not easily understood or accepted. This required some more empirical evidence—something that could be seen, something more tangible than equations on a page. Mercury's orbit was good, but that was known before he had created the theory. He wanted something new and unexpected to convince waverers like Planck—either the redshift or light deflection. Those were beyond what he could do himself.

He knew he could always lean on his oldest ally, Michele Besso. He hoped they could discuss the new theory in person soon in Switzerland. He was planning a visit, but the border was almost constantly closed. He did not mention that the reason he would be in Bern was for a meeting of the Anti-War Council, to which he had been elected. "In these times everyone must do whatever he can for the community as a whole, even if it is only slight and ineffectual." He finally had a victory with relativity; a victory against the war would have to wait.

CHAPTER 7

To Cross the Trenches

"Your theory still seems to be almost entirely unknown in England."

IT WAS HARD to move around Europe in those years. Hard for people, certainly—trenches ran continuously from the North Sea to the Swiss border. Hard for ideas too. The papers on which scientific theories were written were trapped by the barbed wire and the naval blockade, unable to escape their home countries. We think of equations and hypotheses as immaterial, transcendental, unrestricted by mundane concerns like borders and checkpoints. Einstein learned the hard way how false this was. His triumph with general relativity in 1915 was unknown in most of the world. The articles announcing his theory were halted by machine guns and artillery just as decisively as the advancing armies were. For his theory to cross borders, Einstein needed allies and hard work. Relativity was tough to learn not just because of the formidable mathematics but because science was deeply entangled in the war.

Besides cutting lines of communication, the endless fighting questioned what it meant to be a scientist at all. This new kind of total war put pressure on the most essential points of what scientists should do, write, and think. Could Einstein pursue the truths of the universe while mired in questions of politics and patriotic strife?

By 1916, Einstein needed help, and he knew it. It is usually said that Einstein finished the theory of relativity in November 1915. This is true, in a sense. By then he had found his field equations—the mathematical statements that summed up the principles and concepts on which the theory was based. But theories are more than just equations, and there was work to be done in understanding, refining, and applying. By January he wanted nothing more than to sit down with his friends and talk about relativity. Unfortunately he had few colleagues who both understood his work and—perhaps more important—with whom he felt he could talk freely against the background of the Great War.

Two of the members of this rare group worked in the Netherlands, an island of neutrality in a continent at war. They were professors in Leiden, site of the oldest university in the Netherlands and a scientific and cultural center for centuries. A bustling city crossed by small canals about fifty miles north of Belgium, it was the professional home to Einstein's friends Paul Ehrenfest and Hendrik Lorentz. Einstein's intimate correspondence with them had tracked his scientific success and failures for years. It was to Ehrenfest that he declared, "[I was] beside myself with joy and excitement for days" after finally cracking the covariance problem (the key to making relativity truly universal). To Lorentz he confessed that all his papers on gravity had so far been a "chain of wrong tracks." Joining Einstein's relativistic clique in Leiden was Willem de Sitter, the director of the Leiden Observatory, striking with his pointed beard and prominent ears. De Sitter was a skilled mathematical astronomer, and his interest in problems of gravitation pulled him into Einstein's orbit. However, it was not just his friends' skill in physics that Einstein relied on. He also reveled in their like-minded politics. In imperial Berlin he never felt he could talk openly. One of his favorite parts of visiting the noncombatant Netherlands was that he could "walk around without a muzzle."

In January 1916, Einstein was struggling with certain formulations in relativity. He more than once regretted skipping mathematics classes back in college. Lorentz, he hoped, could help. Einstein had confidence in the basic formulas, but his own presentation and derivation of them, he felt, was "abominable." He had been impressed with

Lorentz's ability to explain covariance and hoped to absorb some of the elder scientist's wisdom. He worked hard to support his friends' interest in his theory, even providing them with step-by-step calculations to make sure everything was understood. At the end of January he sent a walkthrough of relativity to Ehrenfest, several pages of dense handwritten formulas. He assured Ehrenfest that he would have no more problems with relativity now that it was all laid out nicely. Einstein asked him to show the material to Lorentz and then return it—it was, he felt, the best summary of his theory yet. He needed the letter back to use it as the basis for a publishable paper.

Einstein hoped to go to Leiden to visit them in person early that year despite wartime restrictions. One of Einstein's friends jokingly asked if relativity might allow Leiden to come to him, instead of vice versa. Although the Netherlands was neutral in the war, the border was often closed. The Dutch had watched Belgium crushed by the kaiser, and they were not at all confident that their neutrality would be respected. The fighting felt close—one could hear artillery fire from many parts of the country. Their army was mobilized throughout the war just in case. Ehrenfest belonged to a home-guard unit that was never called up. Traffic across the border was carefully controlled by both sides.

Even to Einstein with his Swiss passport, a visit was challenging to plan. Worse, the police had been investigating Einstein as a politically suspect person. He had foolishly written some of his pacifist correspondence on a postcard, and an alert mail carrier notified the police. Their official investigation concluded in January 1916 that he had only recently become politically active but was definitely "a supporter of the peace movement." They decided to let him travel for the moment (his movements were eventually restricted in 1918). But even when cooperative, the German bureaucracy was a formidable opponent. Einstein wondered if maybe he should just wait until peace came. Surely the war could not go on much longer.

For most Germans there were few signs that the war would be ending soon. Karl Schwarzschild, finished with the leave where he had heard Einstein speak, had already arrived at his second military post of the conflict. His abilities were first put to use constructing weather forecasts in Belgium and then calculating artillery trajectories on the Russian front. The earthworks and explosions of the battlefield could hardly have been more different from the calm parks and palaces that surrounded his former institute in Potsdam.

That said, for a theorist like Schwarzschild, a fortification was a decent place to do science. The German trenches were famously well appointed. Advancing British troops incredulously reported concrete floors, heating stoves, running water, and electric lights. Those deep, comfortable bunkers provided German soldiers almost complete protection from bombardment and shelter from the east European winter. Schwarzschild would have had plenty of opportunities to scribble and calculate. He did not need a laboratory, just time and focus.

Schwarzschild's trenches represented the farthest reach of the theory of general relativity. He was able to keep up on scientific developments back in Berlin without too much difficulty. Soldiers were able to receive regular mail shipments, and he was able to travel there in person when his unit was on leave. Upon receiving copies of Einstein's papers on general relativity, he sat down to study this strange new theory. He quickly discovered Einstein's "abominable" derivations and wondered if he could do better.

The core issue was the calculation of the orbit of Mercury, which Einstein's theory matched with observation so nicely. Schwarzschild immediately noticed something important. Einstein's calculation of the orbit of Mercury was only an approximation. This is a standard practice for physics problems that require large amounts of calculation. When it becomes clear that an exact answer will be labor-intensive, there is a strong incentive to find a shortcut. Generally the shortcut is an answer to a problem that is just about the same as the one you are really interested in, but much easier to solve. If you can present an argument that *just about the same* is *good enough*, then your colleagues will often give you the benefit of the doubt. And this was indeed where Einstein found himself. He had a *good enough* explanation for the wobble in Mercury's motion. Schwarzschild set himself the task of finding the precise answer, starting from Einstein's equations, sheer mathematical skill, and a bit of chutzpah.

A German trench similar to where Karl Schwarzschild performed his calculations BRETT BUTTERWORTH COLLECTION

In short order he was able to demonstrate the needed precession of 43 arc-seconds per century. He was pleased that the derivation was just a few pages long and was impressed that such a concrete, observational prediction came out of the abstract heights of Einstein's principles. Schwarzschild wrote to the theory's author to alert him to the achievement. The calculations were literally done when he had a break between firing cannons at the tsar's armies. Fighting on the eastern front was fairly quiet at the time, with the Russians having been pushed out of German territory over the course of the year. One of his letters to Einstein suggests that he found mathematical physics a welcome distraction: "As you see, the war is kindly disposed toward me, allowing me, despite fierce gunfire at a decidedly terrestrial distance, to take this walk into this your land of ideas." Einstein was surprised. The calculation was so simple. How had he missed it?

Einstein was even more startled by Schwarzschild's next letter a week later. This short note emerged from the trenches to do something that Einstein was unsure was even possible—it presented an exact solution to the equations of general relativity. The significance of this requires a bit of explanation. General relativity is described by mathematical entities called *differential equations*. These are a little stranger than the equations you learn to solve in high school algebra. In an equation like $x + 3 = 7$, we say the solution for x is 4. But for a differential equation the solution is a whole other equation—in fact, usually a whole group of equations. If someone hands you a random differential equation, it is pretty straightforward to find an approximate answer—a *good enough* answer. Unfortunately, if you want an *exact* solution there is often no easy way to find the answer. It is even possible that there is *no* exact solution. Finding exact solutions to complicated differential equations is a trying task.

So when Einstein "finished" general relativity in November 1915, what he actually had was a group of ten differential equations that might or might not have solutions. If they did, he didn't know what they were. It would be reasonable at this point to voice an objection like this: what good are the field equations, then, if we still need further solutions? This happens a lot in physics. When you articulate a general principle (like $F=ma$, Newton's second law) as a differential equation, you are making a statement about what kinds of things are allowed to happen in the world—what kind of processes nature prefers to do and what patterns you might expect to see. It is a claim about how the universe likes to behave in general. But for a specific example of that principle in action, you need a solution. Newton's second law tells us the general rules for accelerating bodies; a solution to the second law tells us the actual trajectory a specific body will take in a certain situation.

Imagine that you are a scientist studying highways. Your differential equation might be a statement like "rest stops are always found along highways" or "cloverleaf interchanges appear where highways meet." Those are important things to know about highways. They are not, however, the same as the map of an actual highway. You know the actual highway will obey your principles, but someone still has to go out there and make the map.

Einstein wasn't sure there was a map for general relativity. He was confident the differential equations—the general principles of how space-time should behave—were correct without knowing what an actual piece of space-time might look like. He was just working with approximations, with the *good enough* answers. Schwarzschild was the first to find an exact solution to Einstein's equations. He found a possible real world within the kind of universe that Einstein had imagined. Nowadays in the twenty-first century we have many solutions to the Einstein equations. This was the first.

WHAT WE NOW call the Schwarzschild solution applies to a very specific physical situation: a perfectly round object sitting by itself. His solution describes the precise shape that space-time takes around such an object, which then lets us find exact trajectories for things moving in that area. Physicists love perfectly round objects because they make many calculations much, much simpler. Spheres have a property called rotational symmetry, which means they look the same from any angle. It is impossible to overemphasize how much easier this property makes physics, and how much physicists like to use it. There is an old joke in which a dairy farmer asks a physicist for help increasing milk production. The physicist pulls out a blackboard and begins, "Assume the cow is a sphere. . . ."

Fortunately for Einstein and Schwarzschild, most astronomical bodies look a lot like spheres, so this simple solution is extremely helpful. Einstein was enormously impressed (and grateful). He read Schwarzschild's paper verbatim at the next meeting of the Prussian Academy. Few in attendance would have appreciated the significance. He knew he still needed an empirical test to win over the skeptics. He hoped that Schwarzschild, as an astrophysicist, might be able to help with a test as well. The contribution to the Mercury problem was good, he wrote, though "the question of light deflection is of most importance." It was on this observation—whether a beam of light was deflected by gravity—that he thought his theory would rise or fall.

Einstein tried to enlist Schwarzschild in his endless task of winning time for Erwin Freundlich to undertake new deflection measurements. Outside of Schwarzschild and Max Planck (both of whom were somewhat lukewarm), relativity had few enthusiasts. Even Max Born, with whom Einstein got along well, called the theory "frightening" and best "enjoyed and admired from a distance." Astronomers in particular were generally uninterested in this strange theory that seemed to have little significance for their work. But Einstein knew he would eventually need those same astronomers to search for the empirical evidence for relativity. Freundlich was thus an important ally. Einstein had been working diligently to get him liberated from his tedious ordinary duties at the observatory in order to get the young man focused on Einstein's own project.

Freundlich was still despondent about his failure to test the deflection at the solar eclipse in 1914. As an alternative, he thought he might be able to measure the light deflection due to Jupiter's gravity. It would be spectacularly difficult (about a hundred times harder than measuring the deflection from the sun). Nonetheless, an increasingly anxious Einstein declared to Schwarzschild, "It *has* to work!" Freundlich's boss, Otto Struve, had no interest in losing his assistant to Einstein's odd ideas and blocked all efforts. Einstein hoped Schwarzschild, as a fellow astronomer, might be able to persuade Freundlich's famously truculent supervisor to be more reasonable. With typically Einsteinian bluntness, he described Freundlich as not "a very great talent" but merely the first astronomer to understand the importance of general relativity. And that "is why I would regret it deeply if he were deprived the possibility of working in this field." Schwarzschild commiserated, but replied that Freundlich had so alienated Struve that the task was hopeless. Also, he informed Einstein, Jupiter would be too far south for precise observations for the next several years. This was only one of Einstein's many attempts to gain support for Freundlich. He complained to Arnold Sommerfeld that he simply lacked contacts in astronomy. To David Hilbert, he described his fruitless efforts as a "foiled assault on the astronomical fortress defended by an invincible phlegm."

In the same letter that Schwarzschild dashed Einstein's hopes for Jupiter, he noted that he had found a strange feature of his solution to the field equations. If enough mass were packed into a small enough place, a sort of "closed-off" pocket of space-time would form. He thought of this as a mathematical oddity, a quirk of the equations that probably had no physical significance. Later generations of physicists would decide that this closed-off pocket was quite real and give it a special name—a black hole. Neither correspondent realized the implication of this at the time.

Einstein, in particular, was distracted by other matters. On the same day he replied to Schwarzschild he finally penned his dreaded letter to Mileva: a proposal of formal divorce. He had been hoping to put this moment off as long as possible, to avoid any further painful personal communications with her. But he knew that marriage to Elsa was on the horizon, and that could not happen without a divorce from Mileva. Their war, at least, would soon come to a resolution.

IN BRITAIN THE nation began to resign itself to a long war. A peculiar symbol of this resignation hung in the Cambridge Observatory. It was a poster about fifteen inches high. It read: "Defence of the Realm (Consolidation) Regulations: List of Male Employees between the Ages of 18 and 41. NB—this List must be posted in some conspicuous place on the premises in or about which the persons are employed." Eddington's name and address were listed, along with a short statement explaining why he was not fighting at the front. He would have walked by it every day as he crossed under the building's Greek columns. Indeed, he would have been the only one who saw it—every other member of the observatory staff was already serving with the army. Eddington's solitary presence in the observatory was a reminder of both the difficulty of reconciling mass conscription with a free state and the strained position of science in a country at war.

The failure of the semivoluntary Derby Scheme had made mass conscription inevitable. By the end of January 1916, the Military Service Act became law. All unmarried men between eighteen and forty-one were drafted into the armed forces. In an attempt to keep unrest from flaring up, Ireland—then still part of the United Kingdom—was exempt. The political left was furious over the entire act. Two Quaker Members of Parliament, Arnold S. Rowntree and T. Edmund Harvey, ensured that the final bill contained a "conscience clause." This was a provision that allowed men with principled opposition to military service to claim "conscientious objector" status and thus be exempted. This was built on the long tradition of Quakers and similar religious groups being excused from military service (William Pitt released them from the militia during the Napoleonic Wars). The conscience clause was widely condemned and was commonly called the "Slackers' Charter."

Quakers like Eddington had been preparing for this moment since the beginning of the war. There was no question that they would refuse to fight on the front lines. There was still a serious debate, though, on the question of alternative service. Would it be acceptable for Quakers to take noncombatant jobs within the military? What about agricultural work, since that might free up another man to go fight? John William Graham, a respected Quaker leader and Eddington's mentor from Manchester, pointed out that in a sense it was impossible not to help the war in some way simply by being a British citizen. The national Quaker organization stated their position clearly: "We regard the central conception of the Act as imperiling the liberty of the individual conscience—which is the main hope of human progress—and entrenching more deeply that Militarism from which we all desire the world to be freed."

Now that conscription was law, the government had to formalize some of the poorly planned institutions that had developed in the recruiting process. Most important of these was an expansion of the tribunal system that had been part of the Derby Scheme. By the end of the war there were about 2,000 tribunals, most often with five members. They were formed on a county or city basis, and usually were composed of upstanding local citizens or low-level government officers asked to volunteer (there was a Central Tribunal in Westminster to deal with difficult cases).

The tribunals formerly had to deal with a small trickle of exemption requests—the cobbler who wanted to stay with his business—now it was a flood. Some 1.2 million men were immediately conscripted. Of those, 750,000 applied for exemption. The Military Service Act allowed exemptions for work of national importance, serious hardship (say, taking care of an infirm relative), ill health, and conscientious objection. The government gave the local tribunals almost no guidance on how to apply these rules. The general assumption was that every man was most valuable as a soldier; beyond that, no one really knew. Chaos reigned and decisions were often made based on local concerns rather than national ones. It is interesting to note that most of the tribunal records were destroyed after the war—it was generally felt that the process had been poorly planned and embarrassing for the government.

Tribunals could dismiss an application and send the man directly into the military. An applicant could be granted complete or absolute exemption and never be bothered by anyone again (this rarely happened). They could be granted conditional or partial exemption, say, for a certain period of time to close a business. A common outcome for Quakers and other religious objectors was to be assigned alternative service that would support the war, such as working in agriculture or the ambulance corps. Refusal to accept alternative service meant being treated as a soldier disobeying orders, which could mean prison or corporal punishment. Without clear instructions, the tribunals had almost complete power to decide someone's fate. If a tribunal did grant some kind of exemption, that decision could be challenged by the military representative in attendance, starting the whole process again.

Most men simply filed for exemption as a default response, so the tribunals were clogged with work. The Banbury Local Tribunal had 40 percent of applicants asking for exemption on domestic grounds (who will take care of my children?); 40 percent on employment (my haberdashery will close without me); 10 percent on both; and 10 percent on grounds of conscience. Most were dismissed quickly—no, making black pudding is not work of national importance. Those who claimed exemption for more than one category were quite suspicious. They were surely just looking for an excuse.

That particular issue loomed over Eddington. His university had filed Form R.41 for him, arguing that his scientific work was of national importance. It was no surprise that he was granted an exemption on those grounds—the university had enormous influence on the local tribunal. Eddington would not have to serve in the army, at least for the moment. But this was not enough for him. His refusal to fight was *not* based on the idea that his work was of national importance. His refusal to fight was based on his deeply held religious objection to war and violence. He filed his own Form R.41, checking the box for conscientious objection. Because the records are fragmentary, we do not know exactly what happened next. We do know that the local tribunal never processed that second form, and that there was no official record of Eddington's conscientious objection to the war. As far as the government was concerned, his astronomy was part of the war effort. He had to carry his exemption papers at all times. The poster was hung outside his office.

It seems likely that the university intervened to prevent Eddington's conscientious objection from being processed. They were very concerned about any appearance that their faculty and students were not fully in support of the war. And the university had good reason to worry about pacifist professors. In April 1916 (soon after the conscription system was expanded to include married men), an anonymous anticonscription pamphlet titled *Two years' hard labour for not disobeying the dictates of conscience* began circulating. The pamphlet presented the case of a Quaker who had been sent to prison for refusing conscription. Under Section 27 of the Defence of the Realm Act, which prohibited interference with recruiting or discipline, this pamphlet was illegal. It soon came to light that the author was Bertrand Russell, the famous Cambridge mathematician and socialist. The chisel-faced Russell, whose area of expertise would drift into philosophy as the decades wore on, had soured on the war quickly. The son of his longtime collaborator Alfred North Whitehead was killed in action with the Royal Flying Corps, and his apprentice Ludwig Wittgenstein was sent to the front (on the enemy side). Russell decided to freely admit that he wrote the document once it became clear that the state would arrest *someone* for it. A precise and influential speaker, at his trial he elegantly argued for the unjustness of both the act and conscription itself. He was found guilty and given a £100 fine, which he refused to pay, hoping that he would then be imprisoned. Instead, his library was confiscated and the books sold to cover the fine. He was stripped of his fellowship and even denied access to his rooms. Later in the war he was sentenced to six months in prison for further pacifist activity.

The affair was enormously frustrating for the University of Cambridge in particular and British intellectuals in general, who were generally still working hard to justify their usefulness to the war. Cambridge was seen as a hotbed of pacifism (it had more peace societies than any other university). To combat this perception the university banned some student peace groups and actively helped in recruiting. Even beyond official support for recruiting, many students and scholars felt called to fight as guardians of civilization after Louvain. One Cambridge don praised any student who enlisted: "He will carry with him into the field the memory of that martyred city, whose ashes cry aloud for the vindication of true culture against the barbarity made possible and said to be sanctioned by a false *Kultur*." There were fewer and fewer students attending classes every day. Enrollment plummeted from 3,263 in 1913 to 398 in 1917.

The legendary Cavendish Laboratory—the home of James Clerk Maxwell and J. J. Thomson—was partially converted into a barracks. The rest of the lab worked on signaling, acoustics, wireless transmission, high explosives, and other projects of (hopeful) military importance. Soldiers marched on the not-to-be-trod lawns. Russell described it sadly: "The melancholy of this place now-a-days is beyond endurance—the Colleges are dead, except for a few Indians and a few pale pacifists and bloodthirsty old men hobbling along victorious in the absence of youth. Soldiers are billeted in the courts and drill on the grass; bellicose parsons preach to them in stentorian tones from the steps of the Hall."

"Pacifist" became a term of abuse, and Eddington's life was tense. Other Fellows of Trinity began to avoid him. Other physicists pressured him to do war work. He found refuge in the silence at the Friends Meeting House on Jesus Lane even as other members there were hauled before tribunals and sent to work camps or prison. The tribunals saw their job as determining whether a conscientious objector was *sincere*. This often meant presenting increasingly outrageous scenarios—what if a German attacked your mother? What if a German "refused to sheathe his sword until he should have imbrued it with the blood of your deceased wife's sister"? The Quakers were increasingly seen less as principled moralists and more as simple shirkers.

Eddington was run ragged trying to do all the work of the observatory by himself while also conducting his own research. As secretary of the Royal Astronomical Society he also had significant administrative duties. It was in this capacity that he received a curious letter from the director of the Leiden Observatory, Willem de Sitter. In late May or early June 1916 (we do not have the original letter) de Sitter decided to try to spread word of Einstein's work across the English Channel. In mid-May, Einstein had produced a new summary of general relativity that emphasized the theory's astronomical consequences, which probably spurred the Dutch astronomer's attempt to find new converts. Interestingly, he did not include any of Einstein's actual papers, just his own summary of the theory and its implications. Since the letter is lost we do not know exactly what he said. But we do know that this was the first appearance of general relativity in an enemy country. The theory had leaped the trenches.

EINSTEIN WAS VERY, very lucky that it was Eddington who opened that envelope. At this point in the war few British scientists would have been willing to even think about a German theory. Eddington, the pacifist and internationalist, was. And even further, he was one of the handful of people equipped to understand even the rudiments of general relativity. Its daunting mathematical framework was quite familiar to Eddington. While an undergraduate, he had been coached for his exams (the famous Mathematics Tripos) by Robert Alfred Herman, a mathematician fascinated by differential geometry. He made sure all his students were skilled in this exotic subfield, which seemed to have few practical applications at the time. De Sitter had chosen as a correspondent perhaps the one Briton both willing and able to think about Einstein.

Eddington's June 11 reply to de Sitter noted that "Hitherto I had only heard vague rumours of Einstein's new work. I do not think anyone in England knows the details of his paper." The embargo on scientific communication between the warring countries meant none of Einstein's work on general relativity had been seen. Frankly, though, no one would have likely been reading his papers anyway. The topic and the author were both rather obscure. Eddington's vague previous knowledge of Einstein was about as good as it got. De Sitter offered to write a paper summing up general relativity for one of the Royal Astronomical Society's publications. Eddington steered him toward the *Monthly Notices*, which had a simpler editorial process, so that de Sitter's paper would appear much more quickly. Eddington wanted more on relativity, as soon as possible: "I am immensely interested in what you tell me about Einstein's theory."

Over the summer of 1916 a handful of letters made their way back and forth between de Sitter and Eddington, with the former agreeing to write an additional short piece for the non-technical journal *The Observatory*. The annual meeting of the British Association for the Advancement of Science was coming up in September in Newcastle, and Eddington tried to get de Sitter invited as a speaker in a session dedicated to gravitation. Unfortunately he discovered that Newcastle was considered a "restricted area" and non-British citizens were not allowed. Instead, Eddington would present his own meager understanding of relativity: "So far as I can make out, no one in England has yet been able to see Einstein's paper and many are very curious to know the new theory. So I propose to give some account of it at the Meeting."

During that summer and fall Eddington struggled to make sense of relativity. He had to put aside the new research he had been doing on the internal structure of stars so he could focus on this strange new theory. When de Sitter first learned the theory, he had the advantage of personal contact with Einstein and others who already understood it. Eddington, on the other hand, had to rely on infrequent and often delayed correspondence. There was no one else on his island who knew anything about general relativity; there were no textbooks, no tutorial from Einstein. Eddington persevered, asking de Sitter for clarifications of fundamentals and interpretation. He constantly stumbled through the thicket of a secondhand theory, occasionally becoming stranded by a particularly difficult mathematical pit or philosophical obstacle. Eddington reassured his Dutch correspondent that these worries did not at all sour him on the theory as a whole: "I need scarcely say that this philosophical difficulty does not to my mind detract in the least from the remarkable practical implications of the paper."

The final sentence of this letter reveals that Eddington was concerned not only with Einstein's equations but also his role in the war: "I was interested to hear that so fine a thinker as Einstein is anti-Prussian." "Prussian" was common shorthand for the militarist German nationalism on which the British had blamed the war. Eddington saw a crucial opportunity. Einstein was not only a brilliant physicist but also opposed his own nation's worst excesses. As a peaceful German, he could be just what a Quaker scientist needed to convince his colleagues of the error of their own jingoistic ways. Relativity could show what was lost when science became consumed by wartime hatred.

De Sitter wrote to Einstein to tell him of relativity's spread. "Your theory still seems to be almost entirely unknown in England." Einstein was delighted with de Sitter's efforts to overcome the rifts within science: "It is a fine thing that you are throwing this bridge over the abyss of delusion." He liked what he heard about Eddington and was impressed with that astronomer's insights into relativity (Eddington caught a few mistakes in one of Einstein's early papers on general relativity from 1914). It seemed that Eddington might join the group with whom Einstein felt he could talk freely: "When peace has returned, I shall write to him."

To ENSURE DOMESTIC tranquility in Berlin, Einstein had to make a trip to Switzerland. In order to marry Elsa as she so fervently wanted, he needed to secure a divorce from Mileva, still living in Zurich with their children. He planned a trip for Easter 1916, assuming he could persuade both the central bureaucracy and the border guards to let him through (the former was no guarantee of the latter). His general absentmindedness did not help with the intensely ritualized crossing procedure, such as when a guard asked his name and he had to hesitate before he remembered it. At the border crossing in Lindau he underwent an inspection that was "very thorough . . . but entirely decent and polite. Jacket & vest off, shirt opened; even trousers down, collar off. Every single piece was searched through." He appreciated that the inspector was graceful, at least.

Einstein's plan to secure a divorce agreement was complicated by his resolution to never see Mileva again—even while he was in the same city he continued to converse with her only by letters. She was also increasingly ill, a state that their mutual friends blamed on Einstein's clumsy communications. Those friends acted as intermediaries and finally convinced Einstein to put off his demands for a formal divorce. While dealing with Mileva, he hoped to repair relations with his twelve-year-old son, Hans Albert, who would no longer even answer his letters. Despite all these family matters, Einstein's first order of business upon arrival was to contact Michele Besso so they could talk relativity. They went boating together. Even as the pair's lives continued to push them apart, Besso remained an important sounding board for all things relativistic—appropriate for the friend whom Einstein credited with the original inspiration for special relativity.

Einstein returned to an increasingly imperial and inhospitable Berlin. Intellectuals there continued to trumpet the close relationship of science and the war. The annual report of the Kaiser Wilhelm Society declared:

> But our enemies, by their surprise attack, achieved something quite unexpected and, for them, unwelcome: *they brought German science and military strength as close together as possible.* Of course we knew all along that these two pillars . . . , deep down, have a hidden connection; but we did not know that this connection is so immediate

> that military strength can be directly promoted by science. [emphasis original]

It was not just words. It was normal for soldiers to be seen marching down Unter den Linden past the Academy of Sciences. That fall the army moved through the city confiscating anything with a lens—telescopes, binoculars, and opera glasses were all seized and salvaged for the front. Public clocks were darkened at night; bells no longer tolled the hours. The soap shortage even made the Academy a smellier place. The government had begun restricting the buying of new clothes as well. Einstein assured his friends that he had plenty to wear and that they could regard "my aesthetic demands in this regard as minimal."

EINSTEIN CLAIMED TO his friends that thinking about the political situation was simply too much: "I shut my eyes as best I can to the insane goings-on in the world at large, having completely lost my social consciousness." His actions, even if modest, said otherwise. In an attempt to promote peace and internationalism in his profession, he took on the presidency of the German Physical Society. He put aside his general dislike of bureaucracy and titles to try to do some good. He was almost completely unsuccessful. Peace groups such as the BNV had come under increasing harassment from the military authorities: letters were censored, passports revoked, sometimes even imprisonment was threatened. Unable to continue its work, the BNV was forced to shut down.

A few months after, the art history professor Werner Weisbach wrote asking Einstein to join his new group, the Association of the Like-Minded. Einstein was intrigued by Weisbach's project. He replied to the invitation: "I am convinced the malady of our times is that moral ideals have almost lost their force." He worried that a German victory in the war would be seen worldwide as support for nationalism and aggression. But if they lost, "people will lose faith in the empty ideal of power and will extend the principles of justice willingly and fairly to the states. Then our hotly pursued goal of an organization of states eliminating war . . . will get implemented in a short time." Einstein said he would support Weisbach publicly, though characteristically was unwilling to join the organization itself. It might, he worried, cut into his time for physics. Just joining such an organization was dangerous—Germany continued to be governed by the wartime "state of siege" laws, which gave police control over any kind of political meeting.

JUST AS DE Sitter was expanding relativity's reach across the English Channel, the small community of Einstein supporters was dealt a sudden blow. Schwarzschild, whose mathematical skill provided the first solution to the field equations, fell ill in his post on the eastern front. He developed painful blisters, symptoms of a strange skin disease possibly induced by contact with chemical weapons. He was sent to a hospital in Berlin, where he died after two months of fruitless treatment. The German scientific community had lost an extraordinary mind at the modest age of forty-two. His death was a sober reminder that over the course of human warfare far more soldiers have died from disease than from enemy action.

Einstein was distressed by the loss of Schwarzschild. He wrote to Hilbert that "among the living there are probably only a few who know how to apply mathematics with such virtuosity as his." His grief was mixed, though. He could not forget that Schwarzschild had enthusiastically volunteered to fight and did everything he could to support a German victory. Einstein's June 8 public memorial address carefully avoided mentioning Schwarzschild's role in the war or the circumstances of his death. Instead, it praised his modesty and his "indefatigable theoretical creativity." In addition to describing his specific technical achievements, the lecture celebrated his ability to derive "artistic pleasure from devising finer mathematical systems of thought." Einstein's private ruminations, though, continued to struggle with reconciling Schwarzschild's scientific prowess with his moral culpability in the war: "He would have been a gem, had he been as decent as he was clever."

CHAPTER 8

The Borders of the Universe

"The lines of latitude and longitude pay no regard to national boundaries."

NEWS OF SCHWARZSCHILD's death eventually made its way to Britain. Eddington had known him well—they had both worked extensively on the statistics of stellar movements and the physics of stars. In August 1916 he wrote a moving obituary for his friend in the British astronomical journal *The Observatory*. Oddly, it had many references to how un-German Schwarzschild was: his "characteristics were not those which are usually associated with the scientists of his nation." Many people, Eddington wrote, complained about German scientists being overly thorough or plodding. Not Schwarzschild, though. He was described as clear and keen. Indeed, he was presented as shattering all the stereotypes of his people: "We would rather say that through him a new spirit was arising in German astronomy from within, raising, broadening and humanizing its outlook."

This unusual memorial makes more sense when we see that it was but one shot in a broader battle being fought that summer. British astronomers were in the middle of a ferocious debate about, essentially, whether Germans could even *do* science. Perhaps their conduct during the war had placed them outside the kind of civilized discourse needed for scientific work. These opinions had been simmering widely since the war started, but burst into *The Observatory* via the anonymous regular column "From an Oxford Note-Book." Although the column had no byline, everyone knew it was written by H. H. Turner, a distinguished professor of astronomy at Oxford. The column usually was a bit like the water cooler in an office—a place where Turner shared quips, gossip, and anecdotes about astronomy. Athletic and genial, Turner was well liked by everyone, and the column captured his avuncular role in the astronomical community.

THE MAY 1916 column started off with a dry discussion about whether there was a need for any new scientific journals. Then Turner suddenly declared that the Germans had made it impossible to return science to its prewar state. At the beginning of the war, he remembered, it seemed as though science could be "above all politics"—working in a rarefied realm untouched by such mundane things. Now he had changed his mind:

> We have seen how engagements and relationships, which we all thought were "above all politics," and safe to be respected even in time of war itself, have, nevertheless, been broken and tossed aside in a moment if Germany took the fancy that it could thereby benefit itself. Many of us do not see how, after such an exhibition, we can face the mockery of new understandings and undertakings with such a nation.

He acknowledged that tempers might calm after the war was over. Nonetheless, he wanted a decision made about whether to exclude German science now. And would time really make any difference? Hadn't the Germans already shown their true colors? "Is not the die really cast already?"

This anger—in a professional journal—was all the more notable for its source. Turner was once a pioneer of scientific internationalism. He had worked extensively with foreign scientists and oversaw many transnational projects, including the first standardized international star map. Eddington had even recommended him in 1914 to be the new foreign secretary of the RAS. In that letter he praised Turner as uniquely qualified through his extensive international connections and "intimate" relationships with scientists in other countries. Turner knew, and had worked closely with, astronomers in all parts of the world.

BUT NOW THE trauma of the war had erased decades of cooperation. Instead, Turner contended that the nature of the German character was fundamentally contradictory. It was "intellectual without being refined." Germany could "discipline its mind but cannot control its appetites." He quoted the official British investigation into wartime atrocities to state that Germany was essentially a "pre-Asiatic horde." "The dilemma is inexorable: we can readmit Germany to international society and lower our standard of international law to her level, or we can exclude her and raise it. There is no third course." Turner argued that scientists could simply no longer deal with their German and Austrian colleagues.

Further, he wrote, the war had shown that scientists could not hold themselves apart from politics. The integrity of an experiment, the persuasion of a mathematical proof, could only be as trustworthy as the civilization from which it had come. And the Germans had forfeited their right to that trust. Turner imagined a future in which their entire language would no longer be heard at a scientific conference.

The next issue of *The Observatory* contained a lonely response. Eddington wrote a letter, titled "The Future of International Science," that imagined a world where science could be free from nationalism and racism. Refusing to deal with scientists because of what country they came from would be a disaster, he wrote. He first pointed out the practical difficulties. Astronomers did not have the privilege of ignoring whole chunks of the planet—"the lines of latitude and longitude pay no regard to national boundaries." But his objection to Turner was based on more than just problems of measurement. It was based on what it meant to do science: "Above all, there is the conviction that the pursuit of truth, whether in the minute structure of the atom or in the vast system of the stars, is a bond transcending human differences—to use it as a barrier fortifying national feuds is a degradation of the fair name of science." Shallow and selfish patriotism, he warned, would destroy the progress of science.

Eddington reminded his readers that many German scientists regretted signing the Manifesto of 93, and some were even undertaking the dangerous task of helping British citizens interned behind enemy lines. The Berlin Academy of Sciences had twice refused to eject its foreign members; the German astronomical community continued to send publications to English members (as much as they were able).

Eddington's tactic was to humanize the enemy, a method he had learned as a Quaker. Vilifying an abstract nation was one thing. Thinking about individual people was something else. He wrote:

> Fortunately, most of us know fairly intimately some of the men with whom, it is suggested, we can no longer associate. Think, not of a symbolic German, but of your former friend Prof. X, for instance—call him Hun, pirate, baby-killer, and try to work up a little fury. The attempt breaks down ludicrously. . . . The worship of force, love of empire, a narrow patriotism, and the perversion of science have brought the world to disaster.

This was precisely the approach used by Quaker pacifists trying to bring the war to a close. An enemy who has a face, with whom you share values and history, is much harder to kill. Eddington brought his strategy into the realm of science. He simply asked everyone to view things from the standpoint of a German scientist, not the German government.

OTHER BRITISH SCIENTISTS were not receptive. Joseph Larmor, a fellow physicist at Cambridge, publicly responded to defend the exclusion of Germans from science. Larmor, of an older generation and also a Member of Parliament, lectured his apparently naïve young colleague.

He told Eddington that this exclusion was the most prudent move. National priorities needed to be considered differently from personal preference. He completely rejected the call to see things as the Germans do. He not-so-subtly reminded Eddington that the British government had given certain groups the right to sit out the war on grounds of conscience, and those groups should be content with that privilege. Translation: you Quakers don't have to fight, but that means you don't get to criticize the rest of us either.

Eddington anxiously responded to Larmor to clarify his position. He was frustrated at the difficulty of writing anything that would not be misinterpreted. He had tried to be moderate but was read as a radical: "It is really very difficult to know how to use the English language." He continued to try to shift Larmor's position toward his own. Viewing things from the German standpoint also meant hoping that the Germans would try to understand the British perspective, wouldn't it? Surely the Germans told outrageous stories of British atrocities, just as the British press manufactured its own tales? He signed off with a demurral that he did not want to bring politics into an astronomical journal, but that Turner forced his hand. After a hasty signature he added a postscript: "Naturally I do not regard the position taken by conscientious objectors as in any way connected with 'viewing things from a German's standpoint.'"

Before the next issue of *The Observatory* could appear, events on the Continent made reconciliation even more unlikely. The French fortresses at Verdun had suffered crushing German assaults since February 1916, and the British High Command planned an offensive along the Somme River to relieve pressure on their ally. Some 3 million artillery shells were stockpiled, and the preparatory barrage lasted a week (the final hour saw nearly a quarter-million shells fired). On July 1 at 7:30 a.m., officers blew their whistles, signaling soldiers to throw back a drink of rum and climb out of the trenches. Most of the British infantry were inexperienced troops (the so-called Pals battalions of friends who had enlisted together). The generals thought these green soldiers incapable of serious tactics, so the order was simply to advance shoulder to shoulder in straight lines.

The infantry did so, cheerfully confident that the massive shelling had completely subdued their enemy. They were wrong. The same finely crafted trenches that allowed Schwarzschild to do his calculations kept virtually all the enemy soldiers alive and ready to fight. Five hundred yards of uncut wire and undamaged German machine guns resulted in the worst slaughter of any day in the entire war. Twenty thousand soldiers died. Forty thousand were wounded. Battlefield surgeons had plenty of opportunity to use their new tools of antiseptic powders and blood transfusions. The British poet Siegfried Sassoon was present. He described "staring at a sunlit picture of Hell." Capt. W. P. Nevill famously brought four soccer balls for his troops to kick across no-man's-land as they advanced. He never came home. Raymond Asquith, the son of the prime minister, was shot in the chest. He casually lit a cigarette so the men under his command would continue to attack. He died within hours. J.R.R. Tolkien, a signals officer during the battle, described officers being killed "a dozen a minute." He found psychic shelter from the fighting by crafting intricate stories of elves, orcs, and dragons. He described writing "by candle light in bell-tents, even some down in dugouts under shell fire."

THE DEVASTATING LOSSES and pathetic gains of July's fighting—some 600,000 casualties for six miles of territory—hardened attitudes at home. Turner's response to Eddington's attempt at pacification channeled all this anger. Eddington had asked what stands in the way of international cooperation in science. Turner's brutal answer: "My reply is that the *facts* stand in the way—hard, horrible facts." Eddington, he wrote, "proposes to shut his eyes to these facts, and to test the situation by the play of our imaginations in connection with some individual." He jabbed at Eddington's "own shrinking from horrors"—how could a conscientious objector truly understand the nature of the war? The Quaker seemed to be ignoring reality:

> Is it not an actual fact that babies have been killed in ways almost inconceivably brutal, and not as a mere individual excess, but as a

> part of the deliberate and declared policy of the German army? Is it not a fact that the *Lusitania* was sunk with a national rejoicing that puts the cold-bloodedness of former pirates to shame? Is it not a fact that German men of science have gone out of their way to declare their adhesion to these things, and that one of them who ventures some excuse still boasts of a "quiet conscience"? If we cast our memories back before the war, it is easy to recall that we should have vowed these things incredible; but that does not alter facts.

His litany of German war crimes was hung around the necks of all German scientists. There seemed to be no going back. The atrocities (both real and imagined) were so horrible that it became impossible to remember old friends before they became villains.

The Battle of the Somme continued into the autumn, eventually seeing the British Army pioneer the first use of tanks in combat. About the same time those mechanical monsters debuted, an enormous German airship raid dropped bombs across England. With these technological horrors in the back of his mind, Eddington prepared for the British Association for the Advancement of Science's annual meeting in Newcastle. He presented a paper on general relativity. Or tried to, at any rate. Since he was still not fluent in the theory, his paper was based largely on de Sitter's letters and was incomplete and opaque. Indeed, he had just managed to find a copy of Einstein's review article from May. He tried to explain exactly what the principle of relativity was, and how the elevator thought experiment worked. He was actually more comfortable with the mathematics of the theory than the meaning, though he warned the audience that the equations were "highly complicated in form" and could not be easily written down. His inexperience with the technical issues went unnoticed by the sparse listeners in the audience. There were few people interested in an exotic theory from Berlin.

Eddington felt more and more out of place in British science. His solitude at the observatory was matched by his increasingly long solitary bicycle rides through the English countryside. While he composed Schwarzschild's obituary he took a break for a ninety-mile trip from Sibford to Cambridge. Thanks to his obsessive record-keeping, we know this was one of his longest bike rides ever. This gave him plenty of time to think about the twin problems of the fracturing of international science and the enigmas of relativity.

Eddington had hoped that his memorial of Schwarzschild would serve as a reminder to his colleagues that the Germans, too, were human and essential parts of the universal scientific quest. Yet Schwarzschild had been an enthusiastic German soldier, so was of limited value for this. The still-obscure Einstein and his mysterious theory might provide something better. World-changing science and a figure untainted by the German war machine. A symbol of how science could push beyond the pitfalls of nationalism. But Britain remained a hostile land; Eddington needed to plan relativity's invasion.

EINSTEIN DID NOT know much about his new ally across the English Channel, and looked for help closer to home. He had been eagerly digging into his own theory to see what other mysteries might be hidden there, and corresponded actively with those who might help. Supporters were still somewhat rare. One of these was Hermann Weyl, a Göttingen-trained mathematician. In a letter thanking Weyl for his interest, Einstein commented that while not everyone agreed with the theory, its advocates always seemed to be smarter than its critics. "This is objective evidence, of a sort, for the naturalness and rationality of the theory."

David Hilbert, Einstein's rival in the original formulation of the field equations, regularly reported his own progress. A late May 1916 letter from Hilbert commiserated about Schwarzschild's death and invited Einstein to come visit Göttingen again. In a sign of the dire living conditions of wartime Germany, he casually assured Einstein that if they couldn't find any food for their guest, they could go to a neighboring village that had better supplies.

Hilbert's research that spring had focused on the energy laws that emerged from the principles of general relativity. The law of conservation of energy was one of the cornerstones of physics, but it was not obvious whether it could be directly derived from relativity. Einstein had one solution to this; Hilbert had another. To help resolve the puzzle Hilbert brought in one of his students, Emmy Noether. Noether had overcome extensive sexism (both casual and structural) to attain her PhD in mathematics. Despite being one of the world's experts in differential invariant theory, she was unpaid as a lecturer at Göttingen—statutes prevented women from becoming formal instructors. Hilbert tried to persuade the establishment that the sex of a scholar was irrelevant: "After all, we are a university, not a bathing establishment." He recognized her genius and sought her assistance in a variety of projects. She was famous for speaking incredibly fast. One of her colleagues invited her for walks so she would become tired and speak more slowly.

Her expertise was particularly relevant to developing the mathematics of general relativity. Einstein was grateful regardless of her gender and did what he could to support her work: "Upon receiving the new work by Miss Noether, I again feel it is a great injustice that she be denied the *venia legendi* [the right to lecture officially]. I would very much support our taking an energetic step at the Ministry [to overturn this]." Eventually she would formulate what is now known as Noether's theorem. This established symmetry as a fundamental principle of modern physics, and is an indispensable tool for theoretical physics today. After the Nazis came to power she was expelled from her position at Göttingen, along with many others of Jewish descent. Like Einstein, she went to the United States as a refugee.

But in 1916, Einstein was still hard at work in Berlin. While Noether and Hilbert considered the relationship between energy and relativity, Einstein tackled a crucial problem: how does gravity get from place to place? And how fast does it get there? Right now the sun's gravity holds the Earth in its orbit. If the sun suddenly vanished, how long would it take for that disruption of gravity to be felt here? Instantaneously? The eight minutes it takes light to get here? Some speed unique to gravity? Newton never gave a satisfying answer for his theory of gravity, so if Einstein succeeded in this it would be a significant boost for his alternative.

Einstein modeled his approach to this problem on the analogous issue in one of the other fundamental forces of nature: electromagnetism. Back in the nineteenth century James Clerk Maxwell (one of Einstein's idols) had figured out that electromagnetic force traveled in waves. A wobbling electrical charge set up a jiggle in the surrounding electromagnetic field, which then traveled at the speed of light. Maxwell inferred that this meant light was itself an electromagnetic wave. This was confirmed in the laboratory by Heinrich Hertz, unifying light and electricity. Einstein hoped he could do the same for gravity.

So his task was to inspect the field equations and see if they could be arranged in a way that could be interpreted as a moving jiggle in space-time—a gravitational wave. He wrote to Lorentz, de Sitter, and—a few months before his death—Schwarzschild for help. The differences between electromagnetism and gravity meant he couldn't follow Maxwell's recipe exactly, but he kept trying. He slid back and forth from certainty that gravitational waves couldn't exist to certainty that they must, eventually settling on a lukewarm confidence. He found that his equations allowed for a mathematical entity that carried energy and moved at the speed of light—a pretty good candidate for a gravitational wave. But was it real? Unfortunately, gravitational waves make their presence known by absurdly small deformations in space-time as they travel. To see even the largest of these deformations a physicist would need to be able to measure a change in length of 1 in 10^{21}—vastly beyond the ability of Einstein's colleagues (they were not detected until September 14, 2015). Gravitational waves were not going to persuade anyone of the truth of general relativity.

Einstein needed to talk—there was only so much that could be done by letter. He decided he could not wait for peacetime. That summer he renewed his efforts to make a trip to the Netherlands to see Ehrenfest, Lorentz, and de Sitter. In August he went to the Foreign Office in Berlin to get permission. They were reluctant. His frequent trips to Switzerland—the most common point of entry for spies into Germany—had raised suspicions. Many peace activists were having their travel curtailed as well. He quickly fired off a letter to Ehrenfest saying that an official invitation from a Dutch university would be helpful. Ehrenfest complied immediately. Among other hoops he had to jump through, Einstein still had to get the original of his Swiss citizenship certificate, perhaps to demonstrate his good reasons for travel there. He complained: "A long chain of other still obscure obstacles awaits me. So don't be surprised if many more delays occur." Finally, on September 27, he boarded a train for Leiden.

Einstein arrived at the Ehrenfest home a grateful man. Once again he was with friends who understood both his physics and his politics. As he put it, he was pleased with "the concurrence in opinion on nonscientific matters." In addition to views on the war, this meant music. Einstein brought his violin and played duets with Ehrenfest on the piano, as they did every time they met. Usually they played Beethoven, but Einstein made a daring new suggestion—Bach. Ehrenfest never cared much for the Baroque composer and was skeptical. Nonetheless, Einstein converted his friend and Ehrenfest spent more time on Bach than physics for the next few months. His wife, Tatyana, herself a distinguished physicist, was not happy about this change.

There was plenty of science during the visit, though. Einstein bounced ideas off his friends, and they drowned him in questions. This personal interaction was particularly important for de Sitter, who had started learning general relativity well after Lorentz and Ehrenfest. Even though he and Einstein had a lively correspondence, when learning a theory there is no substitute for actually sitting down with an expert and talking. It is not a coincidence that Leiden and Göttingen were the best places to learn relativity and were also where Einstein made regular trips. The spread of relativity followed the trenches closely—the only person in Belgium who made any progress in grasping relativity was the physicist Théophile de Donder, who happened to live in a German-occupied area. This need for personal contact emphasizes the challenge Eddington had in teaching himself relativity.

Einstein was particularly pleased to be able to talk with Lorentz during his trip. The elder physicist often set the younger man up with difficult questions just for the joy of seeing him solve them. Ehrenfest described dinner at Lorentz's house: Einstein was given the best easy chair and a fine cigar before being presented with a tough problem.

Then, as Lorentz pushed harder: "Einstein began to puff less frequently on his cigar, and he sat up straighter and more intently in his armchair. . . . [Eventually] the cigar was out, and Einstein pensively twisted his finger in a lock of hair over his right ear." While Lorentz sat smiling, as though at a "beloved son," Einstein finally declared that he "had it"! The debate continued with "a bit of give and take, interrupting one another, a partial disagreement, very quick clarification and a complete mutual understanding, and then both men with beaming eyes skimming over the shining riches of the new theory."

Einstein reluctantly prepared to return to Berlin after two weeks. Elsa had asked him to buy some lard (a rare commodity in wartime Germany) while he was in the Netherlands. He had failed to find any and warned that she would "have to receive me lardless but with kindness all the same." Upon his return he wrote a gushing letter to Ehrenfest to thank him for his hospitality: "The reinvigorating days spent with you have melted into a beautiful dream which I relive tirelessly in my imagination." He even wrote to his friends in Switzerland to let them know how much he had enjoyed his time in the Netherlands. He reported that relativity had "already come very much alive there. Not only are Lorentz and the astronomer de Sitter working independently on the theory but a number of other young colleagues as well. The theory has also taken root in England."

In a scene familiar to many of us, Einstein had come home from vacation to piles of mail and work. One of his tasks was putting into motion a plan he had developed with Lorentz. The two of them had discussed creating a commission to investigate reported German atrocities in Belgium. Einstein tried to recruit Planck to this cause, but the senior scientist thought it would be impossible to get honest testimony from anyone involved. He decided that Planck was "skittish" about anything with "the faintest political flavor." He went on to press Wilhelm von Waldeyer-Hartz, secretary of the Prussian Academy, on the idea. Reportedly the secretary received it "very warmly." After his political overtures were made, Einstein returned to his normal routine. This included Elsa's companionship as well as mundane irritations—on November 17 he had to stay home, as he once again couldn't find his keys to the front door.

As winter came to Germany, conditions degenerated. The season was even worse than the previous year—in February 1917 the average temperature was –1 Fahrenheit (–18 Celsius). Further, the British blockade was working. Before the war, Germany imported nearly a third of its food, and famine was now widespread. The potato crop failed and food riots erupted in thirty cities. This was the so-called turnip winter, when that former animal feed replaced potatoes, which had themselves been the replacement for wheat. It is estimated that some 120,000 Germans died of malnutrition in 1916 and 1917. Food prices doubled, and then quadrupled. Anti-Semitic conspiracy theories emerged that blamed Jewish refugees for the shortages. Mass food protests were a daily occurrence in Berlin, occasionally bursting into full-scale rioting. Many farmers stopped bringing their produce into the city because of the danger. After taking control of the economy in 1916, the military began seizing foodstuffs directly from farms and distributing it themselves. It was hoped that some military efficiency would be brought to food distribution. In reality, everything was made worse as the military took first pick of food and monopolized the railways.

The combination of poor diet and concentrated work took its toll on Einstein. Soon after 1917 began, his stomach started bothering him more and more, and at Elsa's insistence, "contrary to my most ingrained principles," he went to the doctor. His doctor diagnosed him with inflammation of the stomach and recommended a special fat-rich diet that was essentially impossible to maintain in starving Berlin—milk was unknown and average meat consumption had dropped to one-eighth of prewar levels. Most sausage available there was of the ersatz (substitute) variety, of which there were officially 837 government-approved types. Ersatz sausage was often adulterated, and scams proliferated. Berlin teachers took students into the forest to gather materials for ersatz foodstuffs.

Einstein asked his friends in Switzerland if they could send the ten pounds of rice, five pounds of semolina, five pounds of macaroni, and a "most substantial quantity" of zwieback (a type of cracker) that he needed every four to six weeks. His relatives in southern Germany could no longer send any of it—they no longer had any. The food parcels began arriving in February—Einstein thanked his friends for "the chicken feed." He reported that one package never arrived, an unsurprising development given the unreliable state of the mail. Having food sent from elsewhere was a common practice in Berlin to get around food shortages, though the practice was soon made illegal. The situation in Berlin became so bad that German soldiers in occupied countries sent care packages home. Elsa came over to his apartment to prepare all Albert's meals without salt—so as not to excite the "wrath of the evil spirits"—and ministered to his needs while he tried to work.

Einstein's sickness got worse, and he lost nearly fifty pounds in two months. He had a "sickly appearance" and his hands were always cold. His doctor shifted the diagnosis from stomach to liver. Maybe gallstones? He wanted Albert to take some time at the spa in Tarasp, Switzerland, as a treatment. The patient declined, agreeing only to two glasses of Mergentheimer mineral water per day. At least the pain was better under the diet.

Bedridden, he still tried to push through the illness. He was finally making progress on problems that had been bedeviling him all year. These were, essentially, questions about the nature of the universe. Does the universe have a border? And if so, what is there? As Einstein described it, he was trying to uncover "the limiting conditions in the infinite." In technical terms, Einstein was exploring something called *boundary conditions*. Often scientists will have a differential equation with many possible solutions, and choosing the right ones is made much easier if you know a little bit about the state of the whole thing you are trying to understand. To use Einstein's own metaphor, imagine a piece of cloth suspended in the air. The possible twists and turns of the cloth are limited by the material it is made of, whether it is moving, and what is happening to the edges—is the cloth free to move around? Is it being held securely? If so, how tightly? These states of the edges are called the boundary conditions of the cloth, and help determine how it can behave. In this analogy, the cloth is like the space-time fabric of the universe. What, Einstein wondered, were the boundary conditions of the universe? Was space-time fixed in place? Could it move around? Was there a limit at all?

Einstein actually came to these questions indirectly. He was still trying to establish the validity of what he saw as one of the foundations of general relativity: Mach's principle. Even though the principle had led him astray in the hole argument, he still thought of it as essential to his ideas. As Einstein presented it, the principle suggested that inertia (that is, resistance to being pushed or pulled) was not an intrinsic property of mass but rather was generated only through gravitational interactions with other matter. This made mass and inertia relative, much like the other fundamental categories of space and time that were warped by relativity. If the principle was correct, then the inertia of a coffee mug would be due to the gravity of huge amounts of unseen, diffuse matter far away in the universe. Your coffee is hard to pick up in the morning because at vast, essentially infinite distances, matter was hiding and gently tugging on it. These implications stimulated Einstein to think about what was happening at the very edges of the universe. Einstein saw these investigations as a test of general relativity—could his theory handle being extended to infinity? Was there a limit to the domain of relativity?

He was thinking about these problems as early as May 1916, but it was not until that fall that he began focusing on them in earnest. An exchange of letters with de Sitter brought them into sharper focus. It was fine for Einstein the theoretical physicist to speculate on matter spread conveniently throughout the universe in just the right way that Mach's principle would apply. But de Sitter was an astronomer. To him, the structure of the universe was an astronomical problem—something you could see in a telescope—about which he and his colleagues knew a great deal. Einstein couldn't just imagine gigantic masses because it rescued his explanation for inertia! De Sitter wrote: "If I am to believe all of this, your theory will have lost much of its classical beauty for me. . . . I would prefer having *no* explanation for inertia to this one." He assured Einstein that he spoke so frankly only because he knew Einstein would not take it amiss.

Einstein took the critique in the spirit de Sitter had offered it. He was quick to reassure his friend that these were not very important questions: "I am sorry for having placed too much emphasis on the boundary conditions problem in our discussions. This is purely a matter of taste which will never gain scientific significance." Nonetheless, he stuck with his investigations. He came to the view that the problem was the always-sticky concept of the "infinite"—infinite distance meant possibly infinite mass, and Mach's principle would generate infinite inertia here on Earth, which made no sense. If he could get rid of the idea of an infinitely large universe, then those problems went away.

He consulted with several astronomers in Germany, chiefly the old reliable Freundlich. What was known about the actual structure of the universe? Consensus of the time was that the Milky Way—the group of billions of stars to which our sun belongs—was essentially everything that existed. Astronomers had a decent estimate of the total mass and size of the Milky Way (about 10,000 light-years across). There were notable dissenters to this view, but Einstein, taking this as true, was on firm ground in 1916. So he had a blob of matter in an otherwise empty void, a fairly straightforward situation to calculate with his equations. He cranked out the numbers and found that in such a system (essentially the Milky Way) space-time would loop back in on itself, creating a pocket from which one could not escape—physicists thus give it the nickname "closed."

This meant the universe would have the peculiar status of being *finite but unbounded*. This is so counter to our intuitions that it is worth a moment to think about. Finite means you can run out of space in the universe. There would be a limit to how many Starbucks we can have. Our everyday experience suggests that if this is true, there must be an edge, a border, a place where you can see no more Starbucks. But in four-dimensional space-time this is not true. Stepping out of one coffee shop will always take you into another. Eventually, though, you will be back in the one where you started. There is no escape. The surface of the Earth is a useful analogy here. Our planet's surface is finite (you could get enough paint to cover it) but unbounded (no matter how long you walk you will never get to the edge). Instead, you will eventually get back to where you started. Einstein's universe worked the same way.

There was an odd side effect, though. According to his equations this stellar blob's own gravity should make it collapse, which didn't seem to be happening. As he formally wrote, "The relative velocities of the stars are very small." In other words, we're not all crashing into one another. To explain why not, he did something extraordinary. He added an entirely new force to general relativity. It was the term λ (lambda), now called the cosmological constant, a placeholder for a mysterious repulsive force that would keep the universe from collapsing inward as a result of gravity. This was a difficult moment for Einstein. The whole point of general relativity was that it was supposed to emerge inevitably from a handful of universal principles, without any special fixes or adjustments. But this was the ultimate example of a special fix—the prediction of his equations (collapse) didn't match observations (a fairly stable universe). So he just stuck λ in, like duct tape over a leaky pipe. He confessed that although this worked, it was not truly "justified by our actual knowledge." He wasn't very happy about this, and the physicist George Gamow called it Einstein's greatest blunder.

Nonetheless, it gave Einstein a finite universe with no edges, which was what he wanted. The specific numbers did not quite work out perfectly. His calculations predicted a universe about 10 million light-years across, as opposed to the 10,000 light-years astronomers had measured. He had confidence that further work on stellar statistics could help fix the problem. As usual when he encountered challenges with astronomical observations, he tried to enlist the erstwhile Erwin Freundlich to resolve things. And as usual, Einstein failed to get him any time off to make it happen.

Einstein had created a new universe, with paper and ink and patient thought. This was not his goal—remember that he was just trying to see what the limits of his theory were. And writing to his friends, that was still what he emphasized: "It is at the very least proof that general relativity can lead to a system free of contradictions." Along the way he found that the universe needed to be structured a particular way. As usual, solving one problem introduced others—how big was our home? Was there really a mysterious cosmological force pushing apart the stars? He wondered idly whether "the eternal enigma-giver" would ever let him truly understand the physical world. Surely "Jehovah did not found the world on such a crazy basis."

Einstein presented this to the Prussian Academy on February 5, 1917. What started as a small exploration became a challenging project, especially once his liver ailments severely cut back on his ability to work. He actually crafted the final version of the paper from his sickbed. He didn't mind working from his bed—it cut down on visitors and allowed him to avoid the fine clothes required at the Academy. It did, however, mean he had to cancel his next trip to the Netherlands.

Writing to his friends, Einstein confessed that the work would seem "rather outlandish." To Paul Ehrenfest he wrote that it "exposes me a bit to the danger of being committed to a madhouse. I hope there are none over there in Leyden, so that I can visit you again safely." He reassured everyone that he didn't take these ideas about the whole cosmos too seriously even though "I argue as I do." He particularly worried that de Sitter's professional expertise would find this repellent: "From the standpoint of astronomy, of course, I have erected but a lofty castle in the air." It was just a side effect of trying to resolve problems within relativity; it was not about reality. He could now forget about the infinities: "Now I am no longer plagued with the problem, while previously it gave me no peace." Whether our universe was actually structured this way, he wrote, we could probably never know.

De Sitter eagerly received the results from his own sickbed. He had contracted tuberculosis and was trying to work from a sanatorium near Doorn. He was not impressed with Einstein's attempt at universe building. Simply juggling equations to determine the nature of the entire universe seemed a little much. He replied to Einstein: "Well, if you do not want to impose your conception on reality, then we are in agreement. I have nothing against it as a contradiction-free chain of reasoning, and I even admire it."

But within a week de Sitter had a change of heart and wondered what other universes were allowed by general relativity. Bizarrely, he found that a universe without any matter—completely empty—was perfectly compatible with Einstein's equations. This featureless universe was, perplexingly, somehow also in motion. If two intrepid astronauts ventured into de Sitter's cosmos, they would still have mass and would be driven apart. The nothingness was expanding. De Sitter pointed out that if general relativity allowed an empty universe in which individual objects can still have mass, then Mach's principle could not be part of the theory—there were no distant masses to provide inertia. De Sitter did not really understand what this meant, and he was content with that: "I do not concern myself with explanations." This was probably the end of Einstein's infatuation with Mach's principle, and general relativity went forward without it.

DE SITTER AND Einstein continued to spar over their different universes. Was it possible to tell the difference by observation? Were certain inexplicable features "real" or just mathematical illusions? They ended up agreeing to disagree; Einstein dismissed it as a simple "difference in creed." In other words, it was of no more significance than theological debates.

Others picked up the project, though, taking it somewhat more seriously. Einstein and de Sitter had created the first and second of what we now call *cosmological models*, and accidentally invented the foundations of the modern field of relativistic cosmology. The entire discipline is built around the idea that these models can describe the true state of the universe—static? moving? eternal? reincarnated?—and that observation could be used to decide between them. It is this field that eventually led us to our current models of a universe starting with a Big Bang and galaxies hurling apart. Quite different from the universe as Einstein and his contemporaries imagined it, our universe today is infinite and expanding, with our entire galaxy as no more than an insignificant speck. Einstein had no idea what he was setting in motion.

WHILE EINSTEIN AND de Sitter were creating universes without boundaries, borders back on Earth continued to matter. In March 1917, several months after his debate with Einstein, de Sitter sent Eddington a letter describing the two models. Mail remained unreliable across the Channel—de Sitter speculated that the reason he never received any copies of his paper from the *Monthly Notices* was that the boat carrying the packet had been "probably torpedoed." Unrestricted submarine warfare by German U-boats—the practice of torpedoing civilian ships without warning—had just resumed. After the sinking of the *Lusitania*, international outrage had brought a suspension of the practice. By early 1917, though, Germany's fortunes of war looked grim enough to justify a resumption. Wrecks and bodies began to fill the Atlantic. In one month nearly a million tons of Allied and neutral merchant shipping was sunk by the Germans. And it looked like it might work. British grain supplies were down to six weeks' worth.

Berlin felt increasingly surreal to Einstein. "When I speak with people, I sense the pathology of the general state of mind. The times recall the witch trials and other religious misjudgments." He felt as though the people who were kindest and most thoughtful in private were somehow also the most despicably patriotic in public. By the spring, he was venturing out occasionally, sometimes to take part in meetings of the Deutsche Friedensgesellschaft (or DFG, the German Peace Society). They met at the Café Austria. One of the other members was surprised to learn that Einstein was someone important in the scientific world. The DFG was no more welcomed by the government than the BNV was, and one meeting in April was broken up by the police.

Whom could he trust? His fellow pacifist Friedrich Wilhelm Foerster asked him to write something on general relativity for the popular reader. Foerster worried that the only thing most Germans knew about relativity was that it opposed British science by replacing Newton's gravity. "This agitation is probably connected to the almost psychopathic state of the current, widely spread sentiments among the populace."

It was not clear what the future would bring for Einstein and his theories. Relativity had managed to cross the trenches and find a foothold in enemy lands. But his hopes for confirmation of the theory—empirical support for his abstract speculations—remained as frustrated as ever by the war. Seeing the gravitational deflection of light seemed an almost idle fantasy. The war showed no signs of slowing or international science of resuming. His brief excursions to the outer reaches of the universe did not help him escape the brutal situation humanity had made for itself: "It is a pity that we do not live on Mars and just observe man's nasty antics by telescope. . . . Jehovah no longer needs to send down showers of ash and brimstone; he has modernized and has set this mechanism to run on automatic."

CHAPTER 9

The Resistance to Relativity

"The most remarkable publication during the war."

ELSA WAS TRYING to keep Albert alive. This became much easier to do once he moved into the apartment next door to her on Haberlandstrasse in the summer of 1917. One of her goals was to keep him to one cigar a day. Previously his office was known for being filled with a blue haze of smoke almost constantly. No more. Although her efforts to keep his lungs working were constantly sabotaged by his friends, who would smuggle him a second or third cigar when she wasn't looking.

Keeping him alive also meant taking care of him throughout his increasingly acute illness. This included cooking his special meals. Elsa had been trying to get him to move closer to make this easier but he resisted. So while he was away visiting family, she simply moved him into her building herself. She may have regretted this decision once she discovered that he snored "unbelievably loudly." He accepted the change with equanimity: "The grub is good and I rest a lot." She kept him eating healthy even when he was traveling—while in Frankfurt he sent a letter assuring her that he was still following the doctor's orders. The envelope also included her keys to the apartment, which he had thoughtlessly taken with him. In Zurich he excitedly wrote that he had found a bar of soap and a tube of toothpaste he would bring back for her. This was the last of his wartime trips. The military authorities decided that he was, in fact, a political danger (he was number 9 on a list of 31 pacifists) and they would soon restrict his movements to officially approved travel only.

Back in Berlin he continued to receive his food parcels from friends. He warned them not to seal any packages they sent because the military censors would open them anyway, or perhaps simply destroy them. The food was even more needed as supplies became tighter. As an unmarried man Einstein received only the completely inadequate "bachelor's packages" of rations. The middle of 1917 saw the introduction of "ersatz ersatz" food—when the substitutes themselves had to be replaced with even faker food (coffee became roasted barley with coal tar, which became chicory with sugar beet). Haber, Fischer, and Nernst worked with the government to improve the nutritional value of what food was available. The black market became an essential part of life. At one point, all news of the war was eclipsed by rumors about a missing jam shipment. When some forward-thinking revolutionaries were stockpiling weapons, they successfully kept the police from inspecting the crates by claiming they were actually smuggling fruit.

Einstein was starting to lose confidence in whether the work of he and his pacifist friends would make a difference. "All our exalted technological progress, civilization for that matter, is comparable to an axe in the hand of a pathological criminal." The ax was sometimes falling close to home: his friend Georg Nicolai, who drafted the internationalist manifesto he signed at the beginning of the war, had just been court-martialed. Nicolai's story was a bizarre one. A publicly avowed Socialist, he was a professor of physiology and private doctor in Berlin. A committed pacifist, he volunteered to run a cardiac clinic for the army, which provided enough political cover to continue practicing medicine and teaching.

For a little while anyway. As his internationalist views became better known, the military drafted him and began transferring him from one awful posting to another (including an infirmary in a Russian POW camp in the middle of a swamp). One of the finest doctors in the country, he was reduced to working as an orderly. This did, however, give him plenty of time to work on his intellectual projects. He struggled to get his book *The Biology of War*, his explanation for why the war broke out, published in Switzerland. He finally succeeded in 1917, though the text was incomplete, out of order, and a general mess (he never even saw the proofs). The military authorities were watching closely, though, and court-martialed him. He was fined 1,200 marks and made to write a public letter disclaiming the book as punishment. He was court-martialed again a few months later, this time for insolence.

Nicolai was unstoppable, though. He kept pressing Einstein to participate in a new publishing project, marshaling any argument that came to mind (including that, since Einstein hadn't explicitly said he *wouldn't* participate, he was really obligated to do so). Einstein couldn't take it anymore and blasted off a crushing response: "Thus I raise my voice with the force of a bullock just come of age and ceremoniously, fervently, and energetically call (bellow) it off herewith." He immediately regretted the outburst and sent an apology, saying his previous letter was merely "a coarse joke." He still didn't want anything to with Nicolai's crazy project, though.

Unfortunately for Albert, Nicolai was not his only politically complicated friend. The historian Peter Galison tells us the extraordinary story of Friedrich Adler. He had competed with Einstein for the Zurich professorship in 1908, and lived in the same apartment building. Their families were friendly. Albert and Friedrich would talk physics. When Einstein went to Prague he hoped that Adler would replace him. Instead, Adler became a full-time radical Socialist, even meeting Trotsky in 1914.

Like Einstein, Adler opposed the war. Unlike Einstein, he decided to do so by shooting the prime minister of Austria, Count Stürgkh. Shortly after disentangling himself from Nicolai, Einstein received a prison letter from Adler asking if he would be interested in discussing some aspects of relativity. Oh, and would Einstein mind being a character witness at his trial? Einstein quickly wrote a testimony praising Adler's trustworthiness and asking the emperor for clemency. Adler continued to send his thoughts on relativity, which amounted to an alternative theory that leaned more toward Lorentz than Einstein. Adler's father, desperate, tried to use his son's alternative physics as evidence for an insanity defense. In May 1917, Adler was found guilty and sentenced to death, though his execution was stayed. From solitary confinement, perhaps with not much else to do, he continued to try to convince his old friend that his theory was fatally flawed.

LIKE EINSTEIN, EDDINGTON had trouble getting tobacco. Neither his sister nor Trimble minded his smoking, but his nicotine fix often ended up on the bottom of the Atlantic thanks to the U-boats. Alcohol was in short supply too. That wasn't a problem for him—in 1917 he still kept to his mother's old-fashioned advice on drinking. He actually offered the use of the observatory for the making of a temperance propaganda film at one point. Later in life he was trapped on a sweltering boat where the only drink available was Champagne. Once he gave it a try, he decided alcohol wasn't so bad after all.

Along with those luxuries, chocolate began to disappear as well. Shopkeepers faced steep fines if they sold any. As the blockade intensified, it became common to have to stand in line for meat. There was a famous incident where some thousands of people queued to buy margarine (butter was essentially gone). There was never enough protein to go around. Sugar bowls were usually empty. And the worst shortage of all: tea. His pipe, Eddington could do without. But tea? Impossible.

Even with these difficulties, London and Cambridge didn't have the problems of Berlin. Bread was never rationed and the black market was quite small. In any case, the fortunes of the war were beginning to improve. It seemed that the sleeping giant, America, had finally been roused and would be entering the conflict on the side of the Allies. America was hardly champing at the bit for war. Instead, the shift came about because of some spectacularly poor choices on the part of the Germans. The reintroduction of unrestricted submarine warfare was one of the few actions that would almost certainly bring in the Americans. Even worse, someone in the kaiser's diplomatic corps decided that it would be a good idea to convince Mexico to attack the United States preemptively (in hopes of recapturing Texas, New Mexico, and Arizona), thus preventing their intervention in Europe.

Incredibly, the offer to Mexico (now known as the Zimmermann telegram) was sent via the American embassy in Berlin. Since the British cut their undersea cables, Germany had few options for fast international communication, and the Americans allowed the use of their lines as a favor. The telegram was in code, at least, though British intelligence deciphered it almost immediately. On April 6, the United States of America declared war on Germany. At that moment, the United States essentially had no army. No one thought their raw recruits were a match for the seasoned troops currently occupying the trenches. But there were a *lot* of them. Their entry into the war meant that Germany was essentially fighting an enemy with limitless reserves. They had no intention of giving up, though. The German High Command had made a daring strategic decision to pull some of their troops back from their farthest positions to a new, more defensible set of fortifications—the Hindenburg Line. This would be much more difficult for the Allies to breach. The Germans made it clear that they were in the war for the long term. They finished their withdrawal the day before the American declaration of war and awaited their new enemies.

Along with vast numbers of soldiers, America brought with it an enormous industrial capacity and technical expertise. When Churchill was asked what he most wanted from the United States he replied, "Send us chemists." The historian Roy MacLeod has documented how American scientists had been preparing for this moment for a while. George Ellery Hale had been trying to put US science on a war footing since the *Lusitania*, well before any combatant was asking for technical help. Hale wore wire-rimmed glasses under a high forehead and stubbornly parted hair, and always sported a high collar. He suffered from depression and insomnia while still making American astronomy the world leader in powerful telescopes. His expertise was solar astronomy, far from any war work, but he was unrivaled in his powers of scientific organization. His efforts resulted in the formation of the National Research Council within the National Academy of Sciences. Hale even visited Europe in August 1916 to see how best to organize the NRC for potential war work and see what was most needed. So when the country joined the war, Hale was able to present a "chemical reserve" ready to deploy thousands of scientists instantly.

Sir Horace Darwin (ninth child of Charles) wrote to Hale asking for a hundred or so scientists to come work on British military projects. He also warned Hale that the United States needed to avoid the same mistake Britain had made and not to conscript scientists as common soldiers. That, he said, had been a disaster. Cooperation among the Allied scientists was not easy. The British and French had made little progress. Hale had thought about it extensively, though the American scientists found it difficult to work within the constraints of classified communication and military priorities.

Despite the challenges, Hale and his colleagues worked hard to create a genuine sense of "Allied science" as a community. They had numerous conferences tackling chemical weapons, munitions production, and so on. Hale saw these meetings as models for what science would look like after the war: American, British, French scientists working closely (with maybe a couple of Belgians). The Germans would be unwelcome for a hundred years. The arrival of the Americans, ironically, had broken international science even further. The practice of science was pushed to become tied even closer to nationalism and partisanship.

THAT NATIONALISM MADE it difficult when Eddington went to his friend to ask a favor. A big favor. In some sense, when your friend is the Astronomer Royal—the most important scientist in the country—any favor that takes up his time will be fairly big. He had known Frank Dyson for years and they were close. But Eddington needed some help with enemy science—relativity.

It was not obvious that Dyson would be amenable to providing what Eddington needed. He was something of an internationalist, in that he helped set up the replacement system for the Kiel telegraph network. That system, where scientific messages were sent to neutral Copenhagen, was increasingly seen as being pro-German (in that the Germans were not explicitly excluded and that Strömgren was too friendly with them). H. H. Turner had stopped using it. French and American astronomers asked Dyson to switch the system's hub to an Allied country. One British astronomer supported this merely because it would inconvenience the Germans. Dyson capitulated and wrote to Strömgren officially withdrawing the Royal Observatory. He complimented him on the efficiency with which the system had been run: "I am very sensible of the good feelings towards Astronomers of the belligerent countries which prompted you to undertake the transmission of Astronomical telegrams during the war, instead of the Kiel central station. . . . Nevertheless I have reluctantly come to the conclusion that for the present at any rate the Greenwich Observatory should withdraw from this society." Dyson enclosed a payment for services rendered. We don't know precisely where Dyson's political views lay; he was certainly no fan of the Germans, though. He wrote to a friend, "I hardly know whether to be glad or sorry that our boys are too young to go [to war]." Near the end of the war he announced his pleasure that the "great evil with which the world was threatened is being overcome."

So Eddington was not likely to get much traction with his friend by using internationalist arguments. Nor was Dyson particularly interested in relativity itself. Even Eddington had to acknowledge that Dyson was "very skeptical about the theory." The best Eddington could do was get Dyson *interested*. The Astronomer Royal was gradually persuaded that relativity was scientifically important even if it might be wrong. Technical interest and personal friendship seem to have been enough for Dyson to join in on Eddington's project. He was willing to help.

Eddington didn't need help with the theory itself. The mathematics was far beyond Dyson, and Eddington had already spent months working through the physics and philosophy of relativity (he relaxed by reading *The Brothers Karamazov*). He had come to the same conclusion as Einstein, that even with a fully formed theory, what relativity needed was some kind of physical test—an observation that could support or wreck one of its unique predictions. Mercury was good but insufficient. It was technically a *retrodiction*, not a prediction, because Einstein knew what the answer should be before he developed his equations. Observations of the gravitational redshift were, at best, not supporting Einstein's prediction; at worst, they were strongly against it.

That left the gravitational deflection of light, visible only at a total solar eclipse. And the person in charge of British eclipse observations? Frank Dyson. The Astronomer Royal headed the Joint Permanent Eclipse Committee (JPEC), which handled British-sponsored expeditions to solar eclipses, so no test of relativity was going to happen without his approval. Once Dyson had agreed to support Eddington on this, his first move was the same as the one made by the astronomers in Germany—check photographs from old eclipses to see if the gravitational deflection was visible. The effect was small enough that if you weren't looking for it you could easily miss it. He and the staff at Greenwich searched through the records with no luck.

That meant a whole new expedition would need to be mounted to look for the deflection. Unlike Einstein and Freundlich, trapped in central Europe, the British ruled the waves (unless there were U-boats around) and at least had the possibility of going eclipse hunting. Total solar eclipses can be hard to come by; fortunately astronomers even then were very, very good at predicting where and when they would occur (down to the second). We do not know who did the calculations, but Eddington and Dyson quickly realized that there was a perfect opportunity coming up soon. On May 29, 1919, there would be a total eclipse in which the sun would be squarely in front of a prominent star cluster. The deflection of light would appear visually as stars near the edge of the eclipsed sun being slightly displaced from their actual location. So ideally an astronomer hoping for a clearer measurement would want to have several bright stars close together—the Hyades.

The path of the eclipse was not particularly convenient, stretching over Africa and South America. No one knew if the war would be over

How star images in the Hyades are deflected from their positions by the sun's gravity. Detail from the Illustrated London News, *November 22, 1919.* COURTESY OF THE AUTHOR

by 1919 and Dyson was unsure whether the JPEC would be able to mount such an expensive, complicated expedition. Eddington thought the difficulty of such a project was actually a good thing. He had been trying to convince his colleagues of the necessity of truly international science, and this expedition could be the perfect example. A German theory, British astronomers, travel across three continents—exactly the sort of international cooperation that Eddington had been saying was essential to the very spirit of science. Even better, it was an opportunity to bring a pacifist, brilliant "enemy" scientist to the world's attention. This expedition would not just be a scientific test, it would be a scientific demonstration—of international science as Eddington saw it. The stakes were high. In March 1917, even before he fully understood relativity, Eddington convinced Dyson to make a brief statement of support for sending an expedition in 1919. Everything was still uncertain, but it was the first step.

Even the Astronomer Royal couldn't make this happen on his own, though. Eddington knew he needed to persuade the British scientific community that relativity was important and interesting enough to spend scarce resources on. He was an apostle bringing good news to a strange land. What he needed was some scripture—a fundamental text that he could point to and say, "This is what relativity means." And, of course, it had to be in English. Einstein wasn't likely to write it, so that meant Eddington had to do it himself. Once he felt he had a genuine command of the theory, he sat down to begin writing.

As is often the case when trying to convert a land to a new belief system, the locals in Britain already had their own deity: Newton. Most scientists were perfectly happy with the Newtonian system and did not appreciate Eddington's missionary efforts. Chief among those was Sir Oliver Lodge, a walking, talking embodiment of the science of the Victorian age. He was an expert in electricity and magnetism at the University of Birmingham, developing many of the technologies that would eventually become radio (his patent dispute with Guglielmo Marconi over wireless telegraphy lasted for decades). He was also one of the great science writers of the time, with a few dozen books to his name. For a generation his name was synonymous with physics. He was famous enough to be caricatured in *Vanity Fair*. The drawing captured his lanky frame and the way his face-enveloping beard was balanced by an impressively smooth crown.

Lodge's science was completely intertwined with the ether. He saw radio as the ultimate evidence for the ether's reality. Maxwell had predicted that if there was ether, there would be electromagnetic waves. Hertz found those waves, and now Lodge was using them to talk across the sea. Surely that was definitive proof that the ether was real? However, Lodge's ether did more than just allow for radio waves. Tragically, his son Raymond had been killed on the battlefield near Ypres in 1915. Two weeks later, though, they were talking again—through a spiritualist medium. Lodge had been trying to study psychic phenomena for many years (a fairly respectable Victorian scientific project) and he had concluded that the ether not only carried electromagnetic waves but also the spirits of the dead. The ether, superfluous for Einstein, was the very foundation of reality for Lodge. It provided both physical laws and spiritual meaning. It surrounds us and binds us.

While the ether was technically from Maxwell's physics, Lodge thought of it as part of Newton's universe. Newton did not have much to say about electromagnetism, but to Lodge it was simply an extension of the *kind* of physics that Newton did. It was part of his universe's absolute space and time, objective knowledge, and clear concepts of force and mass. So to Newtonians like Lodge, relativity was dangerous not just because it denied the ether. Instead, it attacked everything that seemed essential to science, everything that had made science work since Newton published his *Principia Mathematica* in 1687. This strange German theory was not just an irritating enemy intrusion, it was a threat to the very foundations of human knowledge. Lodge stepped forward to defend Newton and all that he stood for.

Lodge was no dogmatist, though. Einstein's successful explanation for the wobble in Mercury's orbit was impressive. It needed to be met with equally impressive physics. Lodge used a fairly standard move in science. If an opponent pointed to a piece of observable evidence as support for their theory, you came up with your own theory that explained that piece of evidence in your own terms. If successful, both theories went back to the starting line and other arguments could be deployed. This kind of scientific judo—using your rival's successes for your own benefit—is an essential part of theoretical physics. In a sense, this is what Einstein had already done with Newton's entire theory of gravity. He had struggled to make sure general relativity was equivalent to Newton's theory in most circumstances and thus could take over all of the evidence associated with the earlier theory.

In summer 1917, Lodge decided to create an alternative theory that preserved Newtonian physics and the ether but could also explain Mercury. His idea was that perhaps as our solar system moved through the stationary ether, there was a kind of drag that would cause the wobble. It proved to be a challenge just understanding his competitor, though. His brother Alfred, a mathematician at Oxford, and Joseph Larmor helped. But there was only one person he could turn to who really understood relativity: Eddington.

So, amazingly, Eddington helped Lodge understand relativity well enough to attack it. Letters flew back and forth. By January, Lodge had developed a full-fledged neo-Newtonian theory of gravity that accounted for Mercury's wobble. Eddington gently pointed out that, unfortunately, Lodge's theory would throw off the orbits of all the other planets, so that was not much of a victory. One interesting side effect, though, was that Lodge's theory also predicted a deflection of light near the sun: 0.74 arc-seconds, roughly half of Einstein's predicted value. Two different predictions of the same effect, for different reasons. Eddington realized this could be of great value for his missionary campaign.

CHEMISTS SUCH AS Haber had also been thinking about how to achieve the same effect with different means. Chlorine proved to be fairly easy to defend against, once the shock value had worn off. Phosgene gas, with its distinct musty hay smell, was introduced as an alternative. In July 1917, just as Lodge was preparing to embark on his new ether theory, the Germans introduced a new weapon: mustard gas. Its yellow-brown clouds caused blisters on the skin and lungs, and induced temporary blindness. It rarely killed, though it created one of the definitive images of the Great War: a column of broken, sightless soldiers, gauze over their eyes, each with a hand on the man in front, slowly making their way back from the trenches.

That summer also saw air raids on England on a scale never before seen. The Germans set up their 3rd Bombing Squadron, consisting of the Gotha heavy bombers, around Ghent, in occupied Belgium. The first day of raids from that base caused 286 casualties and inaugurated what came to be known as the "Gotha Summer." This was three weeks of bombing that included a daring daylight raid on Liverpool Street Station that killed and wounded 594, which led to yet another round of anti-German rioting in London. The royal family, horrified that the formal title of their house—"Saxe-Coburg and Gotha"—shared a name with the German bomber, changed their appellation to the charmingly English "Windsor."

In the aftermath, Parliament gave the Home Secretary the power to revoke the naturalization of citizens of German descent and declare them to be enemy aliens. The leadership of the Royal Society took their equivalent move. Scientists resident in enemy countries still technically held membership in the Royal Society, and the group took action to finally expel them. The resolution read:

> In view of the war having continued . . . without any indication that the scientific men of Germany are unsympathetic towards the abominable malpractices of their Government and their fellow-countrymen, and having regard to the representative character of the Royal Society among British scientific bodies as recognized by the patronage of His Majesty the King, this Council forthwith take the steps necessary for removing all enemy aliens from the foreign membership of the Society .

The stated reason for throwing them out was that no German scientists had made any statements against the "abominable malpractices" of their government and military. There had not been any such statements, had there? Everyone of importance had surely signed the Manifesto of 93. If there had been resistance, someone in Great Britain would have heard about it, wouldn't they? No, German science was surely united in support of the war.

Beyond the lack of explicit protest, other reasons emerged to prevent the Germans from being part of scientific organizations. Oxford's H. H. Turner suggested that scientific communities rested on "the good faith of the contracting parties: can we accept in scientific matters assurances which are, by some of the parties, not considered binding in other connections?" If the Germans, as a people, did not respect the treaties protecting Belgium, how could one take their word on things such as experimental reports and mathematical analysis? Science relied on trust. And now, Turner argued, the Germans had revealed that their racial character made it impossible to trust them. "We have tried to think that the exaggerated and false claims made by Germans today were due to some purely temporary disease of quite recent growth . . . [but one wonders] whether the sad truth may not lie deeper." Perhaps the apparent success of German science in the past was merely due to "plagiarism and piracy" of more civilized nations.

Despite the hopefully imminent arrival of the Americans, late 1917 became a desolate time for the war in Britain. Apparently endless political unrest in Russia threatened to knock that ally out of the war just as the new one joined. Other allies were proving of dubious reliability. Some French units were refusing to go on the offensive—it was hard to keep morale up through endless suicidal attacks—though this was not widely known. Everyone heard about the Italian army routed at Caporetto by a combined Central Powers offensive intended to keep the wavering Austria-Hungary in the conflict. Rumors spread that many of the quarter-million Italian soldiers taken prisoner had surrendered eagerly.

Resources were directed south to support the Italians, just as they were desperately needed in a long-planned attack toward the Belgian village of Passchendaele. Formally known as the Third Battle of Ypres, it was British general Douglas Haig's plan to strike deep into occupied Belgium to seize German submarine bases. The U-boats were destroying shipping at an alarming rate. Eddington's lack of tea was a warning that Britain couldn't hold out much longer.

The aloof, polo-playing Haig was famously unconcerned with the likely casualties from his offensives. His reputation for being "stubborn, self-righteous, inflexible, intolerant" had not improved since the slaughter of the Somme. Haig was both devoutly Christian and as much a spiritualist as Lodge; there is some sense that he saw himself as having a divine role to play in the world.

Prime Minister David Lloyd George was deeply unimpressed with the plan Haig presented, but with no military experience of his own had no alternative to offer. Haig retained his confidence in massive barrages and huge attacks—more than 4 million shells were fired in preparation for the assault. Anticipating serious casualties, the

Royal Army Medical Corps set up the first blood bank on the western front.

Unfortunately the start date coincided with some of the worst weather Flanders had seen in decades. Massive rainfall turned the ground to mud, and the artillery barrage destroyed the local drainage systems. Tanks massed to penetrate German defenses quickly sank into the muck. Simply walking became nearly impossible. Boards had to be laid across the oozy ground so units could move forward—not a situation given to tactical flexibility. One Canadian soldier described his experience:

> Gas shells were bursting over me and I couldn't see where I was going. All of a sudden, my foot slipped on the slippery plank and I went right into a muddy hole. I was up to the neck in mud and couldn't get out and tore my nails trying . . .

Bodies floated in the water. The mud not only dragged down the advance, it swallowed shells before they exploded, making the preparatory barrage even more useless.

If there was one universal experience of the front, it was watching one's friends die: "A big shell had just burst and blown a group of lads to bits; there were bits of men all over the place, a terrible sight, men just blown to nothing. I just stood there. It was still and misty, and I could taste their blood in the air." The poets Francis Ledwidge and Hedd Wyn were killed the first day of the battle. Casualties mounted but the front barely shifted. Passchendaele Ridge was supposed to have been seized the first day, but the Germans still held it months in. Haig became obsessed with it and continued throwing more and more troops into the fighting. This battle was the distillation of the idea that the war was fought by lions, led by donkeys. The infantry nicknamed their commander "Butcher Haig."

Haig swapped subordinates to protect himself from responsibility. He was a master at political maneuvering, if not battlefield strategy. After three months, Allied forces finally captured Passchendaele and he declared the campaign over. The total gains were about five miles, but the submarine pens remained intact. The cost was about half a million dead and wounded on both sides, 4,000 per day or so.

Passchendaele became symbolic of not knowing where the war was going or whether it would ever end. The grinding combat of the western front aggravated the manpower crisis and there was growing anxiety at home about how to solve it. Conscription was extended to married men, many exemptions were rescinded, and it was made possible for tribunals to impose "finality" (meaning no appeals were allowed). Many men had to present themselves once more at their local tribunal to defend their exemption. During Passchendaele, conscription was extended to British subjects abroad and Allied citizens living in the UK.

Whatever the level, recruiting was never considered to be enough. The conscientious objectors who refused to fight made little difference in the recruiting numbers, though their resistance became increasingly frustrating for the government. They were made examples of to prevent any further disruption to the war effort. COs were subjected to military discipline, which had a variety of horrible means to crush someone's spirit. At Lyndhurst military prison Quakers like Eddington were "punched and pelted, knocked down and kicked and sneered at." Eddington's friend Ernest Ludlam disappeared into one of the camps. Eddington himself remained free to do science as his conscience allowed.

On November 10, Eddington attended the first meeting of the JPEC dedicated to discussing the 1919 eclipse. That was three days after the Bolsheviks seized the Winter Palace and two days after Lenin signed the Decree of Peace committing Russia to withdraw from the war. That JPEC meeting went smoothly; it was still high-level planning at this point. Nonetheless, Eddington had been sharpening up his debating skills. Fighting for Einstein was going to take a public presence. The shy boy from Weston-super-Mare would never be able to convert the British Isles. He was getting some good practice at the Royal

Astronomical Society, where he had been battling over the nature of the stars.

As much as we remember Eddington for his work on relativity, his lasting scientific reputation was built on being one of the first astrophysicists to understand why stars shine. He developed the equations that let us peer inside the sun. His great antagonist for this work was James Jeans, a doughy mathematical physicist who played the organ in his spare time. Eddington learned how to debate by clashing with Jeans, and their sparring became legendary (many scientists joined the RAS just to watch). Somehow this mild-mannered Quaker not only became an expert at scientific combat—he came to enjoy it. One of his students later recalled an opponent thinking they had dealt a fatal blow to one of Eddington's theories, but then "fire suddenly springs into Eddington's eyes and steel meets steel with sparks flying." In older years he was known for getting an "impish satisfaction" from anticipating a rival's entire presentation from their title, or performing their whole calculation before they had even taken the podium. Not everyone appreciated this side of him. The young Subrahmanyan Chandrasekhar, freshly arrived in Cambridge and eager to work with his astrophysical idol, was deeply upset by Eddington's typically rough-and-tumble treatment.

But at the end of 1917 he had to focus all his energy on Einstein. To convert the heathen Newtonians he needed to finish his scripture, what would eventually become his *Report on Relativity.* But he also knew that a scientific essay by itself wasn't going to have the impact he wanted. If the eclipse expedition was to restore international science, everyone needed to be watching. Everyone needed to be invested. He had to set the stage for a scientific event that could be seen even through the clouds and smoke of the war.

Papers given at specialist conferences wouldn't gather the attention he needed. Instead, he wrangled a spot onstage at the Royal Institution of Great Britain (usually known as the RI). The RI had been the public face of science in London for more than a century. Lecturers there spoke to packed audiences of people from all levels of education and all walks of life. It was where Michael Faraday gave his famous

Christmas Lectures, including the classic "Chemical History of a Candle." There was no better place for Eddington to start taking the case for Einstein directly to the people.

On February 1, 1918, the RI became Eddington's pulpit for relativity. His lecture fully embraced all the strangeness of Einstein's universe (perhaps he was inspired by his recent reading of H. G. Wells's *The War of the Worlds*). Listeners heard about wholly new views of space and time, mass, and energy. The speech ignited curiosity among both scientists and laypeople, priming the pump for the appearance of Eddington's *Report on the Relativity Theory of Gravitation* in April.

The *Report* was a remarkable document: less than a hundred pages to introduce an entirely new view of the cosmos (and to set up Eddington as an expert on it). It was the culmination of eighteen months of Eddington's work to understand, digest, and translate relativity for an audience that was actively hostile to German science. He had little of Einstein's actual work to model it on. Instead, the *Report* is distinctly Eddington's take on relativity. He had the same equations as Einstein, de Sitter, and Hilbert. But a theory is more than just the equations. It needs a framework: to be interpreted, given meaning, and connected to everyday life. Most of the world's first encounter with relativity would not be through Einstein's framework; it would be through Eddington's.

He explicitly wrote the document to make it as accessible as possible. The powerful but opaque Hamiltonian and Lagrangian methods, along with the complicated tensor mathematics, were exiled to a special section. Everything was laid out to get to the experimental consequences as quickly as possible. As fascinating as the theoretical aspects were, he knew that his audience needed to be persuaded that relativity was not mere speculation. It could, and would, be checked with that most powerful tool of science: looking closely at nature. That would be what determined whether, as Eddington put it a couple years later, "Albert Einstein has provoked a revolution of thought in physical science."

The *Report* began with the Michelson–Morley experiment. Its strange null results gave Eddington a chance to question precisely what it meant to *measure* time or space. Once the question was open, he then presented Einstein's positivist arguments for length contraction and time dilation. The *Report* was where Eddington first tried out the vivid illustrations that would push his later books to the top of the bestseller lists. In one of those, *Space, Time, and Gravitation*, he asked the reader to imagine someone moving near the speed of light. When she consulted a mirror on board her own ship, everything looked normal. But looking out on us on the street she saw "a strange race of men who have apparently gone through some flattening-out process; one man looks barely 10 inches across the shoulders." Even harder than accepting this strangeness, though, was the realization that those of us on the street saw her flattened in precisely the same way.

How to get his readers to understand this fundamental paradox? With *Gulliver's Travels*, of course:

> Gulliver regarded the Lilliputians as a race of dwarfs; and the Lilliputians regarded Gulliver as a giant. That is natural. If the Lilliputians had appeared dwarfs to Gulliver, and Gulliver had appeared a dwarf to the Lilliputians—but no! that is too absurd for fiction, and is an idea only to be found in the sober pages of science.

There was no confusion for Gulliver about who seemed big and who seemed small. In Einstein's terms he might try to declare himself to be a privileged reference frame. But under relativity, no observer could be privileged over any others. So the Lilliputians could come to the same conclusion as Gulliver about the other's strange size—each seemed absurdly small to the other. Length contraction was an inherently odd thing.

Similarly, Eddington said, time dilation meant that two people could disagree about how long a cigar would burn for (perhaps there was some bitterness over blockade shortages hiding in that example). Clocks ran slower and slower the closer one moved to the speed of light, so "if man wishes to achieve immortality and eternal youth, all he has to do is to cruise about space with the velocity of light. He will return to the earth after what seems to him an instant to find many centuries passed away."

The temptation upon hearing such outrageous claims was to try to dismiss them as, perhaps, not real. But they were fundamental to understanding relativity. Eddington appreciated that the frequency with which Einstein presented such ideas meant that "the relativist is sometimes suspected of an inordinate fondness for paradox." It was not mere fondness, though. Einstein's universe was genuinely different from our traditional one, and we needed to get used to it: there was no person, no place, no orientation more fundamental than any other. There was no Newtonian "super-observer" who was always right. It was only the laws of nature themselves that were absolute.

Once the reader had grasped (or at least accepted) this fundamental malleability of measurements, Eddington introduced them to the four dimensions of space-time. The basic unit of measurement in the universe was now the interval, that strange combination of space and time that all observers would agree on no matter what. The interval marked the 4-D distance between *events* (things interacting with each other in some way). These intervals could be warped and curved by the presence of large masses, and those curves were perceived by us as gravitational forces. He warned that while words like "curvature" were extremely helpful (they allowed us to avoid phrases like "differential invariant"), one needed to remember that they were only *analogies* to familiar three-dimensional space. When we say that gravity is like the puckering of a rubber sheet caused by a bowling ball, that's a helpful image—but remember that there is no giant rubber sheet out in space. It is merely an image to aid us in visualizing an inherently unvisualizable four-dimensional surface.

Eddington described how the surface of space-time stretched across the entire universe. He presented Einstein's and de Sitter's cosmological models, their attempts to describe mathematically the cosmos as a whole. He emphasized a particular detail about Einstein's closed universe (the one with finite Starbucks). Just as you, exploring that universe, would eventually come back to your starting point, light from a star would eventually bend around and return to itself. This meant that most of the lights in the night sky would not be actual stars but only light trapped in the curvature of space-time. He called these "anti-stars": "It suggests that only a certain proportion of the visible stars are material bodies; the remainder are ghosts of stars, haunting the places where stars used to be in a far-off past." De Sitter's model, on the other hand, had any and all objects flying apart. Eddington speculated that this might be related to then-recent observations that the spiral nebulae (what we now call galaxies) seemed to be hurtling away from one another. Indeed it was—he had spied the earliest evidence for what we now know as the Big Bang.

Eddington empathized with the reader at this point. He suspected they had a voice in the back of their head whispering that the fourth dimension was "nonsense"—he certainly did. But he pointed out that much of modern science could be seen as similarly absurd. "I fancy that voice must often have had a busy time in the past history of physics." Was it nonsense to say that this solid table is a collection of moving atoms, or that the air was trying to crush you, or that the Earth was moving even though you couldn't feel it? "Let us not be beguiled by this voice. It is discredited."

The reality of the fourth dimension could not be directly seen, but that wasn't a reason not to believe in it. Imagine that you are looking at a circular object with a flat portrait on it, and someone else on the other side sees a different flat image. A third observer sees only a thin rectangle. These disparate points of view can all be reconciled if the observers are all looking from different angles at the same three-dimensional object—a penny. No reasonable person could doubt that the penny is *real*, even if it looked different to different people. Doubting the fourth dimension was like doubting the penny.

The four-dimensional world had further strangeness in store. According to general relativity, all the objects in the universe, from cookies to badgers to every atom in your body, have a path through space-time called a worldline. Your worldline bounces from event to event—intersecting with the worldline of the coffee shop before intersecting with the worldline of your boss before intersecting with the worldline of your ride home before intersecting with the worldline of your bed. You ride your worldline from event to event, encountering each like a train coming into a station. We humans, limited three-dimensional creatures as we are, only experience those events one by one. But the full four-dimensional fabric of space-time doesn't—it "sees" all those events at once. A being who could perceive the true nature of space-time would see their future and past stretched out along their worldline. Past, present, and future would be only relative terms.

Now imagine, Eddington said, the worldlines for every particle in existence. The lines would be tangled and twisted, but this enormous skein would give us "a complete history of the configurations of the Universe for all time." General relativity presents us with a universe that is sometimes called *deterministic*. That is, the future is already set. We only see one little part of our worldline, so we think the future is not set. Our 4-D friend, though, can clearly see that the future already exists. A deterministic universe doesn't seem right to many people—surely I can decide what I will have for breakfast in the morning, and therefore change the trajectory of my worldline from intersecting with a doughnut to intersecting with oatmeal? Relativity says no. There is no free will. This sense that you can alter your future is an illusion, one caused by our incorrect perceptions. Eddington was not very happy with this conclusion—he certainly felt able to affect his own destiny—and he spent many years later in life trying to analyze the nature of free will and human sensation of the passage of time in his books such as *The Nature of the Physical World*.

Readers of the *Report* at this point would have been struggling with abstractions such as space-time and worldlines, so Eddington brought them back to more normal experience with Einstein's elevator thought experiment. Like Einstein, he used that mental image to show how gravity and acceleration were the same (or more properly, indistinguishable). Unlike Einstein, he used some classic science fiction to make sure his readers understood it. Eddington reminded them of Jules Verne's *Around the Moon*, in which some intrepid explorers build a gigantic gun with which to shoot themselves to the moon. In the book the proto-astronauts experience a moment with no gravity when the pull of the moon equaled that of the Earth (what Verne called the "dead point" and what astronomers call the Lagrange point). Eddington pointed out that Einstein had shown this to be wrong: the bullet riders would have felt intense acceleration/gravity while in the gun's barrel, but the equivalence principle says that once they were no longer being pushed they would have felt no gravity at all. They would have been floating inside their craft as it moved from the end of the gun barrel all the way to the moon. Once he finished the correction, Eddington immediately apologized. "Pedantic criticism of so delightful a book is detestable."

From the equivalence principle it was a short jump to the three classic tests of general relativity. Einstein's precise explanation of Mercury was presented as a powerful reason to take the theory seriously. The gravitational redshift was more complicated. Eddington noted that the best attempts to find the shift had returned nothing. He was willing to concede that it might be actually disproved. But the difficulty of the test was so great that "we may perhaps suspend judgment; but it would be idle to deny the seriousness of this apparent break-down of Einstein's theory." As Eddington saw it, the redshift was a necessary consequence of the equivalence principle. If it were never found, every part of relativity that depended on that principle would have to be discarded.

That left the gravitational deflection of light. This was the climax of the *Report*—the empirical test to which everything had been building. He told his readers that he hoped to look for this phenomenon at the solar eclipse in 1919. He claimed that what was really being asked was not an obscure question of theory—it was simply asking whether *light* had *weight*. Einstein said it did—specifically that energy had a mass equivalence. Eddington later joked that this meant the electric company could charge for power by the pound. He estimated that the rate would be something like £140,000,000 per pound of light (1918 currency values, of course). If you could catch all the light falling on the Earth each day, it would come to about 160 tons.

The nature of the deflection test was not quite that simple, though.

Without mentioning Lodge by name, Eddington commented that there were nonrelativistic theories that also predicted a deflection of a different value. The test would need to be precise enough to decide between the two theories, two versions of reality—Einstein's or Newton's. To find the truth about the universe would require the highest level of skill, dedication, and international cooperation. Could it be done? Could it even be considered, in these darkest days of the war?

The *Report* was carefully written to achieve exactly this goal—to capture the imagination of the English-speaking world, to evoke passion for a great clash of scientific ideas. He demurred that it did not even really matter which one triumphed. It was about investigating a new kind of science: "Whether the theory ultimately proves to be correct or not, it claims attention as one of the most beautiful examples of the power of general mathematical reasoning."

He closed the document with a brief suggestion of the philosophical implications of relativity. Relativity, he offered, had taught us a great deal but had given no "ultimate explanation" of forces like gravity. The theory's positivist emphasis on measurement and events suggested that a full picture of reality could not be achieved by science. Things beyond "the purview of physics" were needed for that. The human mind and spirit still had a role, Eddington hinted, in a world of non-Euclidean geometry and mass-energy equivalence.

Eddington's first epistle to the Newtonians was certainly a success. There were not many immediate converts, but that wasn't really what he needed. He needed people to pay attention; to wonder about the theory; to want to know the answers to the questions Einstein had raised. And by that standard it worked. The *Report* flew off the shelves (by the standards of a science paper anyway) and one reviewer called it "the most remarkable publication during the war." Most people were still cautious, though, and worried about relativity's complexity and seeming split from common sense. Regardless of skepticism, the *Report* allowed conversations about relativity in a way that had been impossible before.

We shouldn't overemphasize the public interest, though. Einstein was still obscure and German. The war commanded all the attention of the newspapers; the food shortages commanded the attention of every family. By the time the *Report* was published, food controls in Britain had become strict. Milk was the first item controlled. Sugar was the first to require ration cards. Children at boarding schools had to bring their ration cards with them. It was forbidden to throw rice at weddings or use starch in the laundry. It was an offense to feed stray dogs. Wheat flour became rare. "Flour" purchased at the store was adulterated with oatmeal, barley, and potato to stretch scarce reserves. Bread became mealy. The greatest offense, though, was the disappearance of the muffin and its teatime cousins. Virtually impossible to make with the ingredients available, the normally ubiquitous muffin's absence felt like a blow to national identity. On the other hand, one housewife reported, "We consoled ourselves for our muffinless, crumpetless state by owning gloomily that when we had neither butter nor 'marge' . . . of what use to us was the muffin?"

BY THAT POINT, Einstein had been without proper baked goods, much less anything to smear on top of them, for three years. With or without jam, though, the German academic bureaucracy marched onward. In October, Einstein was informed that it was time for him to receive a reward that he would have just as soon forgotten about. When he was first lured to Berlin, one of the promises Planck made was that Einstein would be made director of his own Kaiser Wilhelm Institute. It took four years for the Institute to actually be created, which was fine by Einstein. He was in no hurry to have administrative duties, though the extra money would be nice.

Initially his institute consisted of a board of trustees composed of Einstein and four of his friends . . . and that was it. It was run out of his apartment, and he was able to hire a secretary to handle his official correspondence (letters to him were addressed "Your Honor"). He hired Elsa's twenty-year-old daughter, Ilse, for the job. He was essentially clueless about how to handle even minor administrative tasks. Planck hovered constantly, helping him through. Control of the institute's meager funds, though, allowed him to—finally!—hire Freundlich to work full-time on relativity.

It is perhaps not a coincidence that shortly after he took on these new duties, Einstein suffered a slew of new stomach attacks. He was increasingly skeptical of his doctor's diagnoses. With typical cheek he complained, "I do not believe in the new medical magic with X-rays. I am at the point where I only trust post mortem diagnoses, nothing else." His friends wondered if there were better places than Berlin for finding treatment. By the end of 1917 he was bedridden again.

It was hard for him to do much science, though it gave him some time to think about what the postwar world might look like. He imagined a "pacifist union" with an international court and favored trade networks, restrictions on conscription, enforced democratic principles, and guarantees of territorial integrity for all members. He speculated that the economic benefits of such an arrangement might help bring in otherwise wavering nations. He tried to convince Romain Rolland, still sheltered in Switzerland, that this was worth striving for. Intellectuals in Germany, he wrote, have become subservient to "a kind of religion of might" and could only be "steered by hard facts."

Rolland thought Einstein was being rather politically naïve (he was not the first to make that accusation, nor would he be the last). Einstein seemed to think that only German intellectuals had been corrupted by the war, and that other countries would surely be more receptive to internationalism. Rolland, much more in tune with the wider European situation, warned that scholars in other lands were just as bad: "Evil spreads like a splotch of oil." Perhaps suggesting one opening, though, he asked if Einstein had seen any of Bertrand Russell's pacifist writings? Einstein had not.

Einstein was still confined to bed as 1918 started. He had been trying, unsuccessfully, to invite Gunnar Nordström (his earlier rival in gravitation theory) to visit. When Haber heard that the military authorities were making that difficult, he saw a perfect opportunity to help out his friend. Using his many contacts in the military, he tried to smooth out some of the bureaucratic difficulties. When Einstein heard about this, he exploded. He wanted nothing to do with the military, and he certainly did not want to benefit from Haber's morally compromised position! Haber apologized meekly after his friend's outburst. Einstein, as always valuing his close colleagues over almost everything else, apologized too.

Perhaps his stomach ailments were affecting all his personal relationships, because around the same time he also managed to make his home life even more complicated. He had been living with Elsa and her two adult daughters for several months (he liked to call them his "small harem") when he had some kind of romantic or sexual encounter with the twenty-year-old Ilse. We don't know the details of what happened, only that Ilse was suddenly unsure whether Einstein wanted to marry her or her mother. Einstein, astoundingly, told them that he was indifferent—the two women should sort it out among themselves and he would marry whomever. As one of Einstein's biographers has pointed out, he did not really connect sex and marriage (a lifetime of his mistresses could confirm). He did not consider the whole episode particularly important. Ilse quickly decided to step aside. We can only imagine how painful this incident must have been to Elsa. Very soon after, though, Einstein finally agreed to marry her. Perhaps this was his attempt to make up for that enormous embarrassment. The problem was that Einstein was still *technically* married to Mileva. For an official divorce, Einstein had to ask her for an exquisitely painful favor. He needed her to make a formal accusation of adultery against him. Then a divorce could proceed on those grounds. Mileva, finally having some leverage, was in no rush to make the process easy for him. Negotiations went back and forth—her asking for more money, him threatening not to send any more. Finally she decided she was willing to accept a kind of gamble on Einstein's success in science. If Einstein were to be awarded the Nobel Prize, she would receive the prize money. On that condition, the deal was made.

As WINTER RECEDED Einstein was feeling a little better and was allowed by his doctor to venture out for a half hour at a time. He immediately went to visit Planck for much longer than that. Elsa scolded Planck (who, of course, was blameless) for disrupting Albert's rest. Planck then chided Einstein for not telling him about the doctor's guidance. It seems the patient was beginning to chafe against his restrictions. He also tried to play the violin for an hour (he wasn't going out, so the doctor couldn't complain, right?). This immediately resulted in another painful stomach attack.

His condition was bad enough that he worried he might miss the scientific event of the season: Planck's sixtieth birthday. Despite the tumultuousness of the war years, Planck remained universally beloved in Germany, and a huge party was planned for April 26, 1918. Einstein rallied and he gave one of his finest speeches. It was a love letter not just to his mentor and friend, but also to their shared passion of theoretical physics.

Einstein presented his audience with a bold metaphor. Imagine, he said, the "temple of science." If we cast out those who pursued science just for personal ambition or for utilitarian application, so few would be left—and Planck would be foremost among them. He offered a meditation on what might motivate Planck and his ilk, if not some form of personal gain: "The state of mind which enables a man to do work of this kind is akin to that of a religious worshipper or the lover; the daily effort comes from no deliberate intention or program, but straight from the heart."

It is not hard to see Einstein describing himself here, evoking what he would later call the "cosmic religious feeling" that he saw as the core of both true science and true religion. This was not religion in the conventional sense—Einstein had no interest in organized faith or belief in a personal God—rather a kind of awe and reverence toward the laws of nature themselves. We can see more of this reverence as he explained what Planck's great contribution to science was:

> The supreme task of the physicist is to arrive at those universal elementary laws from which the cosmos can be built up by pure deduction. There is no logical path to these laws; only intuition, resting on sympathetic understanding of experience, can reach them.

Planck's gift, and clearly what Einstein was trying to achieve himself, was to find the "pre-established harmony" of the world. It was to find the fundamental principles on which the universe was organized. He wanted universal laws—like relativity. Theories were supposed to reach across the chaos of ordinary life and show the hidden rules that governed everything. They organized reality for us with invisible laws and abstract concepts, revealed only by patient thought and laborious investigations.

Einstein saw relativity as being precisely in this tradition, and as being one more contribution to the theoretical physics created by Newton, Maxwell, Lorentz, and Planck. He saw it as a natural extension of the work that had come before him, not a radical break with those traditions. As one of Einstein's biographers, Albrecht Fölsing, phrased it, he saw himself as perfecting classical physics, not overthrowing it. Ironically, it was Planck who was the first to talk about relativity as revolutionary. He emphasized its boldness and willingness to discard accepted notions.

Like Planck, Eddington wanted relativity to mark a new era, a new kind of science. He wanted a scientific revolution, a great moment that would draw the world's attention. He needed to present the theory as thrilling and exciting enough to capture attention even across the trenches. So the theory was both revolutionary and conservative, depending on whom you asked. It would remain to be seen which way relativity would be remembered.

CHAPTER 10

Angels of the Revolution

"The Director is the sole remaining member of the staff."

At Planck's birthday party in April 1918, everyone was happy. Einstein was happy to celebrate his mentor and friend. Those who opposed Einstein's politics were happy too. The party came just a few weeks after the beginning of the Kaiserschlacht—the Emperor's Battle. This massive offensive was war-weary Germany's last chance for victory before the Americans began flooding in. Nearly a million of the kaiser's troops had been freed up by Russia's collapse. Those fifty divisions were moved to the west for a decisive attack. By moving at night and staying far behind the trenches until the last moment, they achieved complete surprise. A brief (by Great War standards) barrage of a few hours was followed by infantry using the new "storm trooper" tactics. The attack fell on a section of the trenches only recently taken over by British troops, who had not yet had time to fortify properly. The defense collapsed. The long-awaited breakthrough had finally begun. Patriotic fervor was reignited across Germany in a way not seen since the early days of the war.

R. C. Sherriff's famous play *Journey's End* takes place in a bunker in the days leading up to the Kaiserschlacht. It captured the daily banality of poor food and pointless conversations about prewar rugby juxtaposed with the constant threat of overwhelming violence. George Bernard Shaw called it "a 'slice of life'—horribly abnormal life." One of the protagonists, Lieutenant Osborne, passed the time with *Alice's Adventures in Wonderland*, which was also one of Eddington's favorite books. Eddington used Wonderland to illustrate the strangeness of Einstein's space-time; Sherriff used it to show the otherworldliness of no-man's-land. Both were completely alien to ordinary life; both would lead to revolutions that would shake the world.

The shattering conclusion to *Journey's End* captured well the experience of the British troops on the receiving end of the onslaught. The breach of their lines completely changed the character of the war—from long periods of not even seeing the enemy to a terrifying war of movement. The Germans advanced twelve miles in two days (farther than four months of fighting had gained the British at the Somme). The British fell back to protect their supply lines connecting them to the coast. If the Germans managed to capture the strategic railways there, the British could be knocked completely out of the war. One soldier recalled the frantic retreat:

> What remains in my memory of this day is the constant taking up of new positions, followed by constant orders to retire, terrible blocks on the roads, inability to find anyone anywhere; by exceeding good luck almost complete freedom from shelling, a complete absence of food of any kind except what could be picked up from abandoned dumps.

If the British were outflanked, the French would have to pull back to protect Paris. The lines that had held since 1914 would be abandoned and the war perhaps lost.

Retreating troops were sometimes forced back into the fighting at gunpoint. Private William Hall remembered a staff officer riding up and telling his squad to dig into a nearby hillside and "stand firm." The deep divisions between the command staff and the ordinary soldiers erupted and Hall's friends "didn't take any notice whatsoever, they began to stampede." They told the officer, "We've no chance, Sir, we've got no chance whatsoever, the Germans are coming."

Less than three weeks after the British were first pushed back, the Germans unleashed a second attack on the French lines. They had similar success there, and it looked like they might be on the verge of victory. Haig issued his famous "backs to the wall" order demanding that every position be held to the last man. This did little for morale either at the front or back home. One newspaper recommended that everyone "write encouragingly to friends at the front. . . . Don't repeat foolish gossip. Don't listen to idle rumours. Don't think you know better than Haig."

The German successes were not to last, though. They outran their supply lines and were unable to maintain their momentum. One serious problem was soldiers stopping to plunder the liquor and luxuries that they had been denied by years of the blockade. The attack ground to a halt. They inflicted about as many casualties as they took. The Allies, however, had reinforcements on the way from America—some quarter million per month. The Germans had no way to replace any of their losses. If the war returned to a state of attrition, they would surely lose.

Berlin felt barely in the war—400,000 workers had called a general strike, accompanied by a million people protesting the food crisis. Martial law was declared and the strikers were prodded back to work at bayonet point. There was an important change, though. The protesters were no longer demanding only an end to the war but also universal suffrage, the monarchy's end, and a dismantling of the system that had led to the conflict.

DISMANTLING THE SYSTEM was exactly what Einstein was hoping for. He imagined rallying scholars for an anti–Manifesto of 93, a new call for internationalism in both politics and science. The result would be a series of short essays on internationalism written by "men of science and the arts" published as a book. Perhaps it could even include work from neutral and enemy countries. He wanted no one who had lent their name to any nationalistic declaration, which narrowed his options considerably. His first stop was David Hilbert, whose politics he knew and whose academic reputation had survived the controversies of the war.

Hilbert thought this was a terrible idea. It would essentially be self-denunciation. "The very word 'international' is like a red cloth to a bull for our colleagues." It would only hurt the cause of international science—it would be firing off the gunpowder at the wrong time. Hilbert wasn't alone. Another friend of Einstein's warned that the idea had "grave problems." Unable to find any supporters, he gave up on the project. He was deeply disappointed that his colleagues were not as committed to internationalism as he. He had wanted to make a public statement that he was a man of culture first and a German only second.

To most in the country, that was an impossible inversion. Adolf Kneser, a Breslau mathematician, included general relativity in a lecture about gravitation he gave to help celebrate Wilhelm II's birthday. He used relativity as an example of the outstanding "German work" that showed the strength of the nation's intellectual life during the war. When Einstein heard that his work had been used for patriotic purposes, he was furious. He told Kneser that this was the "*one* thing" that truly hurt him: "I suffer when my name and my work is abused for chauvinist propaganda." Even beyond that abuse, Einstein pointed out that the claim that relativity was German science was actually wrong. He was a Jew, a Swiss citizen, and "by way of thinking a human being and *only* a human, without special favor toward any state or national entity. I wish I could have said this before you gave your talk. Certainly, you would have taken into consideration my feelings and not made such utterances."

Kneser replied that, regardless, Einstein's work had only been made possible by the German nation and its resources. It is unlikely that Kneser knew anything about Einstein's politics before his lecture (which tells us something about how successful Einstein had been in promoting his views). To Kneser, it seemed safe to assume that Einstein was just like any other Berlin professor at the Prussian Academy: properly patriotic and conservative. He just happened to have picked the anomaly.

One of Einstein's fellow anomalous professors, Georg Nicolai, also had become frustrated with being part of the patriotic system. After his courts-martial Nicolai tried to escape to Switzerland. He was caught, and, incredibly, released. He immediately began planning another attempt. He made contact with the Spartacists, a revolutionary group within Germany. In a James Bond–worthy escapade, he and a Spartacist cell stole two biplanes from the German military and flew to neutral Denmark. The incident was celebrated by the Allied press as a humiliating blow to the enemy. Ilse Einstein, seized by the romanticism of the adventure, composed a ballad in Nicolai's honor.

When not being inspired by Nicolai, though, life for the Einsteins only became harsher. By summer 1918, Einstein was consuming half the calories of Eddington. Hoping to spare Albert some embarrassment, Elsa had been secretly communicating with their friends in Switzerland to have more milk sent. When one acquaintance sent apples, Einstein replied with profuse thanks. The fruit had made him "the envy of all Berlin." Too sick to do much science, and having too little fuel to heat his office, he passed the time reading Rousseau's *Confessions* and the Bible. He also made his way through Dostoevsky's Siberian prison camp novel *The House of the Dead*, which probably did not help his feelings of isolation and hopelessness. It seemed like the war would go on forever. Everyone around him was "becoming increasingly inflexible and unpleasant." Some part of him just wanted to go to sleep and hibernate until it was all over.

Einstein's stomach problems were verging on life-threatening and had brought on serious jaundice. In June 1918 he went to the Baltic for a few weeks of recuperative quiet and sunbathing. Given his dislike of socks, we should perhaps not be surprised at his delight at being able to go barefoot every day. He joked to Max Born that they should introduce the practice in Berlin.

An awkward opportunity appeared that month too. He received a job offer in Zurich that was perfectly tailored to his needs. The German mark was plummeting in value and the Swiss franc was much more stable. It was tempting but he again declined. He claimed that only in Berlin would his theories get their proper support: "If moreover you remember that my papers have become effective only through the understanding they have encountered here, then you will understand that I cannot make up my mind to turn my back on this place." This was at best an odd claim—other than Planck, who there was excited about relativity? A more likely explanation was that Elsa wanted to stay and Zurich would mean again being nearby to Mileva. Not an exciting prospect. He proposed a compromise where he would come to Zurich twice a year to lecture but would have no official professorship. Instead he would remain in Germany, even with the danger of his science being unwillingly drafted into the cause of German nationalism.

BRITISH SCIENTISTS CONTINUED to debate whether German science could even be separated from that nationalism. A paper in *Proceedings of the Zoological Society of London* by Sir George Hampson proudly announced, "No quotations from German authors published since August, 1914, are included. 'Hostes humani generis.' [*sic*]" That trailer, an attempt at the Latin for "enemy of mankind," was an old legal term justifying military actions against pirates. Pirates, it was thought, had committed such egregious crimes that they should no longer be treated as part of civilization.

In a letter to *Nature,* Lord Walsingham, entomologist and former Member of Parliament, wholeheartedly endorsed the application of this principle to German scientists. He recalled the meeting of the International Zoological Congress in 1913, where the German delegates tried to dominate the discussion and enforce their rules of nomenclature. A German butterfly catalogue, he told the reader, had "improperly but deliberately" assigned German names in preference to French ones.

Walsingham told this story to explain how it could be that "before the war every man, woman, and child in Germany, with scarcely an exception, was intent upon war"—and this included the "highly educated and scientific classes." The scientists could not be separated from the atrocities of their leaders. As a further example of why German scholars could not be trusted he described an incident in which a "certain learned professor" lecturing in London and Dublin turned out to be a German spy. He concluded that British science could have no dealings with Germany for at least twenty years. It was not emotional vengeance, it was justly considered punishment. Not using their scientific terms would be even further punishment. And any "honestly well disposed" German wouldn't mind using English or French terms.

W. J. Holland, a Pittsburgh zoologist, offered a slight corrective from the other side of the pond. Germans had done *some* useful work in science. The problem was the "Teutonic megalomania" that had "ridiculously overestimated" those contributions. So it was no great worry if they were removed from science. Hopefully, though, Prussianism would disappear and Germans could return to making modest offerings to human knowledge.

British, American, and French scientists gathered in the fall of 1918 to institutionalize these views. The Royal Society hosted the Inter-Allied Conference on International Scientific Organizations in London. The first act of the conference was to declare that Germany's conduct during the war had made it impossible to maintain scientific relations. No one, the announcement read, could "doubt the necessity of this conclusion." Further, Allied scientists were expected to refuse to even meet personally with Germans. The conference argued that such meetings would actually delay rather than speed the resumption of friendly relations because of the "bitter arguments that would certainly occur." This meeting laid the groundwork for what would become the International Research Council (IRC)—an entirely new organization to oversee the international workings of science. It was composed of subgroups for each discipline (Eddington's work would come under the International Astronomical Union). The politics of the IRC were not tangential. The war was built directly into the foundations of the organization. Not only were Germany and Austria not allowed to join, IRC Statute I.1.b explicitly required member states to withdraw from any scientific organization in which Germany or Austria held membership.

We do not know if Eddington attended the Inter-Allied Conference. Certainly he had other matters to worry about—two very different challenges to his attempts to evangelize relativity. The first was a wave of attacks on Einstein's theory as being mere "metaphysics." This was (and still is) a deep insult to level at a scientific theory. Scientists feel very strongly that science is different from, and superior to, philosophy. The difference is supposed to be that philosophy is only abstract reasoning and speculation, whereas science is grounded in empirical facts, experiments, and mathematical rigor. And metaphysics, as the most speculative and abstract part of philosophy, is supposed to be most distant from science (recall that Mach's positivism was aimed at getting rid of metaphysics). These distinctions are wildly unfair—Einstein himself credited his philosophical reading as essential to his work on relativity. Nonetheless, calling a scientific theory "metaphysics" is a powerful way to attack its status. Here in the twenty-first century the accusation is often leveled at string theory to mark it as too speculative. In 1918, it was one more way to attack a theory that was already suspicious for coming from an enemy country.

Eddington admitted that any theory that talked about time and space might appear to be "an admixture of metaphysics." Those were traditionally metaphysical concepts, so should relativity be placed in that category? Eddington said no. It was important, he said, to distinguish between transcendental, philosophical space and space as it was discussed by Einstein. Under relativity, "space" was simply a way to articulate our methods of measurement and how they are affected by the physical world. There was nothing metaphysical about using a ruler to measure length or a clock to measure time. Even the strangest claim of general relativity, the warping of space-time, was still scientific:

> There is nothing metaphysical in the statement that under certain circumstances the measured circumference of a circle is less than pi times the measured diameter; it is purely a matter for experiment. . . . We certainly ought not to be accused of metaphysical speculation, since we confine ourselves to the geometry of measures which are strictly practical, if not strictly practicable.

Eddington contended that, if anything, relativity was the opposite of speculation. Its grounding in the processes of actual measurement was as "matter-of-fact" as anyone could want. Just because philosophers talked about space, this did not make space a metaphysical concept. If scientific reasoning had something to say about space, so be it. Einstein's notions of space were both mathematically sophisticated and, hopefully, able to be checked by experiment. What more could one want from science?

Eddington worked hard to portray relativity as firmly grounded in the real world, and he was very good at making connections with the outlook of his audience. At one lecture he explained the difference between mass and weight with reference to a pound of sugar and the deflection of light with reference to the trajectory of a bullet. Two items unavoidably part of the everyday experience of the war, whether through rationing or the threat to a loved one. That lecture happened to be given to the British Astronomical Association, an amateur collective far from the mathematically skilled elite of the Royal Astronomical Society. If he could persuade that hands-on, practical group that the abstractions of relativity were worth thinking about, he could persuade anyone. Maybe.

THE SECOND CHALLENGE Eddington was dealing with threatened his ability to perform exactly that experimental check on relativity. The success of the German spring offensive had again put renewed pressure on recruitment for the British Army. The government revoked all occupational exemptions and the age limit was raised to fifty-one. There was talk of eliminating all conscientious objections as well, though that plan was shelved.

The Cambridge Tribunal went through all its exemptions, checking to see which could be revoked under the new rules. Eddington's certificate had been granted on the grounds that his work at the observatory was of national importance. Occupational exemptions were no longer valid, though. Eddington was informed that his exemption would be terminated on April 30, 1918.

The university panicked. They desperately did not want another embarrassing Bertrand Russell case. They again tried to persuade the tribunal that Eddington should not be conscripted, this time on the grounds that both his assistants had been killed in the war:

> In making application for the exemption of the Director, Prof. A. S. Eddington, it should be stated that in consequence of the death of the First Assistant [of the Observatory] in the explosion of the *Vanguard*, and of the death of the Second Assistant in action in France, the Director is the sole remaining member of the Staff.

The tribunal was sympathetic and extended Eddington's deferment for another three months. This was immediately appealed by the military representative on the tribunal, Lieutenant Ollard. The university's national-importance claim would no longer hold. If Eddington was to continue to avoid conscription, he would have to stand in front of the tribunal and refuse to fight on the grounds of his Quaker beliefs.

We do not know what his reaction was to this news. Frustration, certainly. If the hearing did not go well (they often did not) he would be unable to make the eclipse observations in May to test relativity. Indeed, he would be unable to do any more scientific work. Personal fear, perhaps. One would have to be foolish not to be afraid after seeing the casualty lists. Despair, likely. His deepest moral beliefs were now under question. The full power of the modern state would be deployed to force him to fight his fellow humans, violating the Peace Testimony he had held since childhood. And even if he could persuade the government to allow him an exemption based on his Quaker beliefs, he might still go to prison, where starvation and torture might await him.

Along with these horrors, though, perhaps he felt some relief. Years of watching his friends punished for the same beliefs he held, while he was able to work on science more or less undisturbed, must have gnawed at him. As a modern Quaker he was not supposed to merely not fight. He was supposed to demonstrate his pacifism to the world, to remind people of how war violated the most fundamental of Jesus's commands: love thy neighbor. Now it was his turn.

Declaring conscientious objection in front of a tribunal was hardly straightforward. Many tribunals felt their role was to judge applicants to see if their beliefs were "genuine." The hearings were often dominated by military service representatives looking to uncover shirkers, not affirm religious belief. Their presence was powerful: the Middlesex Appeals Tribunal denied 406 out of 577 conscience cases that came before them.

The *Cambridge Daily News* suggested that in addition to questioning whether the beliefs were genuine, they should also question whether the beliefs were *valid* (that is, could pacifism really be part of Christianity?). As a result, the hearings were often more like interrogations. Applicants had their beliefs aggressively attacked: "What would you do if a German attacked your mother?" If the applicant responded that he would defend her, he should be sent to the front. If he responded that he would not, he was clearly lying, and labeled "not genuine." Insults and slurs were frequent. One applicant was told, "You are exploiting God to save your own skin. You are nothing but a shivering mass of unwholesome fat."

This treatment was not restricted to the hearing chambers. COs were despised across the nation—they were seen as "unpatriotic, a slacker, a weakling." Eddington's local newspaper printed a letter reporting what those in the trenches thought of COs:

> You have presumably heard of the proposal that conscientious objectors should be made to wear white armlets with a large red "C" on them. They themselves can take "C" to stand for "conscience"; decent folk will consider it to stand for something else. There is a rumour that they are to be collected in a corps which will put up barbed wire between our trenches and those of the Huns. I have done that job myself, and they may have joy of it. They will not have much joy if they meet any of this battalion in days of peace.

The Earl of Malmesbury commented that COs were "sailing dangerously near the very ugly word traitor." This mind-set drove verbal, and sometimes physical, abuse directed at Quakers and other pacifists. Local Anglican clergy were unsparing in their attacks on those who would not fight. One declared, "Liberals, Socialists and Pacifists [are] worse than Jews."

Even if a CO was accepted as "genuine," they would only be exempted from combat duties. They were often given the opportunity to join the Non Combatant Corps, in which they would find themselves digging ditches and leveling roads in Flanders, or working in a munitions factory. Some refused even these assignments—they were still helping the British Army kill people—and were then arrested, fined £2, and given over to the military authorities.

From there they could be sent to a military prison or even the front. The COs were technically conscripted members of the army and therefore could be treated as soldiers refusing to follow orders. Military discipline could vary from solitary confinement to beatings. One Quaker prisoner wrote home:

> Things are coming near the end this morning. I was taken up to a quiet place and simply "pasted" until I couldn't stand and then they took me to the hospital and forcibly fed me. . . . The colonel was standing near me and thundered up and shouted "What! You won't obey me?" I quietly answered "I must obey the commands of my God, Sir." "Damn your God!"

Field punishments included "Crucifixion," in which the prisoner was tied, standing, to some object or structure and left exposed to the elements. On one occasion thirty-four COs were taken to the front and given an order. When they refused, a court-martial was immediately set up and they were sentenced to be shot—permissible for refusing an order in a combat situation. They were told they would be shot at dawn. Come morning, a rifle was loaded in their presence and a soldier ordered to fire. Only then were they told that the sentences had been commuted to ten years' hard labor. Prime Minister Lloyd George, when asked about the treatment of COs, promised only to make "the lot of that class a very hard one."

These were not abstract fears for Eddington. He had watched closely as members of his Quaker Meeting went through exactly these rituals.

Ernest Ludlam was a chemist at the Cavendish Laboratory for whom the university had secured an exemption for work of national importance. He had hoped that his work would be applied to agricultural problems but quickly became convinced that it would be used for military purposes. He quit the Cavendish to join the Quaker Emergency Committee doing relief work for refugees. No longer protected by his scientific labors, he was arrested. Then, refusing to fight, he was sentenced to hard labor and sent to the infamous Wormwood Scrubs Prison. Once he finished his sentence he was arrested again—these were the "cat and mouse" tactics the authorities had developed to keep suffragettes behind bars. After his second sentence to hard labor, his wife, Olive, wrote a letter to Eddington's Quaker Meeting thanking them for their support: "Ernest went to prison with a light heart feeling it a privilege to suffer in such a righteous cause." Eddington personally transcribed that letter into the Meeting's records.

Eddington was sure he could no longer count on support from the university. He had a letter from Dyson, which was a powerful asset. More would be better, though, and Eddington tried to find other allies within the scientific community. Thinking through his options, he made an interesting choice: Sir Oliver Lodge. Would Lodge, the great champion of Newton and British science, help defend a shirker so he could test a German theory? Their correspondence had always been collegial, though, and Eddington had helped him develop his alternative to relativity. It was an audacious strategy. He asked Lodge to write a letter of support for the tribunal:

> I should explain first that I am a conscientious objector. (No doubt you will deplore that, but I can only say that it is a matter of lifelong conviction as a member of the Society of Friends from birth, and I have always taken a fairly active part in the affairs of the Society.) . . .
>
> My position is that I should be willing to do work of that kind (not war-work) if ordered; but I find it difficult to believe that that would really be for the benefit of the world even from the most narrow point of view. . . .
>
> One feels reluctant to make much fuss about a particular case

> like this, when so many obviously far harder cases are being ruthlessly dismissed every day in order to supply the army. Still I think I ought to make the attempt to continue my work, provided that it is in the national interest.
>
> I shall quite understand if you think it best in your position not to appear to be mixed up with a conscientious objector's case; and you may be sure that I should not take a refusal amiss.

In the letter we can hear some fatalism—Eddington seemed to expect being forced into some kind of alternative work, and certainly did not seem optimistic about Lodge helping him. We do not have a reply from Lodge, nor did a letter from him appear at the tribunal. There are several possible explanations—Lodge supported the war, not objectors to it; he had no love for an Einsteinian; he did not feel scientists should become involved in political matters; or perhaps he did write a letter that is lost to history. We will never know, but we can see the challenges Eddington faced in gathering support.

Eddington appeared in front of the Appeal Tribunal, chaired by Maj. S. G. Howard, on June 14. We have little information about who else was present (wounded soldiers were sometimes present to shame applicants). Because this was an appeal, Major Howard first focused on the original reason for the exemption—whether Eddington's scientific work was of national importance. He suggested that the government could certainly find more important work for him than relativity. It was unlikely, he said, that Eddington would be taken as an ordinary soldier.

But Eddington had made up his mind to publicly declare his faith. He stood, rejecting Howard's line of thinking, to declare, "I am a conscientious objector." Howard shut this down immediately: "That question is not before us." The topic at hand was Eddington's science, not his religion. The astronomer himself refused to accept that distinction: he was a pacifist Quaker, and he thought his science was of national (if not global) importance. He reminded the board that he had filed an application for CO status years before but it had never been considered because of his occupational exemption. Puzzled, the tribunal retired from the room to consider the case behind closed doors. They returned and announced that Professor Eddington's skills would be better used by the government than the observatory. He now had no protection against conscription. The chairman briefly mentioned that they had not considered the issue of conscientious objection because the appeal was based on the question of the value of his science. To the authorities, it made no sense for a scientist to claim religious belief. Surely those were opposites? To Eddington, they were anything but—he had placed himself in peril by demanding to be heard on the grounds of both science and religion.

If he wanted to pursue a case based on his Quakerism he now needed to return to the Cambridge Tribunal and start anew. Despite the university's effort to keep events quiet, his case came to the attention of the media. The *Cambridge Daily News* blared the headline PROFESSOR OF ASTRONOMY AS CO. The story played up the contrast of the shameful status of a pacifist against all of Eddington's honors: "the Plumian Professor of Astronomy at the University, Director of the Observatory, and hon. Secretary of the Royal Society." At this June 27 hearing, finally, Eddington was able to make a statement of his pacifism:

> My objection to war is based on religious grounds. I cannot believe that God is calling me to go out and slaughter men, many of whom are animated by the same values of patriotism and supposed religious duty that have sent my countrymen into the field. To assert that it is our religious duty to cast off the moral progress of centuries and take part in the passions and barbarity of war is to contradict my whole conception of what the Christian religion means. Even if the abstention of conscientious objectors were to make the difference between victory and defeat, we cannot truly benefit the nation by willful disobedience to the divine will.

Representative Miller, speaking for the government, denied that Eddington could hold this position. He had earlier accepted a national-importance claim for his science. This meant he had by default given up his right to claim religious objection. Eddington maintained that his science was of great importance, *and* that he had a religious objection to the war—he was both an astronomer and a Quaker. Miller, in a tone like an exasperated adult explaining to a child, asked Eddington to withdraw his religious claim, because it was clearly not consistent with his earlier claim of being a scientist. Eddington refused. The tribunal threw out his religious claim "until required." They then deliberated behind closed doors, and apparently decided against him based on his national-importance claim instead of his conscientious one. The *Cambridge Daily News* reported that the tribunal "considered the case a very hard one—hard on Prof. Eddington." Unusually, they gave him until July 11 to get a cabinet-level intervention, perhaps an indication of the maneuvering being conducted by Dyson and the university behind the scenes.

One of the challenges Eddington faced was his claiming he should be exempt for two reasons. This caused trouble for many Quakers: a schoolteacher asked for exemption on grounds of both importance and conscience; the tribunal ruled these claims canceled each other out. Even John Maynard Keynes, one of the most influential economists of the twentieth century, found himself trapped by this paradox. As a Treasury employee he was exempt for work of national importance. Nonetheless, he wanted to be recognized as a conscientious objector as well (he was not a Quaker). The conscription apparatus was simply not set up for people with depth of character: citizens were supposed to fit neatly into bureaucratic categories. Those who did not, suffered for it.

Exactly how to think about conscientious objection was complicated. Ebenezer Cunningham, the other relativist in Cambridge, had been given CO status while teaching. The National Service representative objected that, as a CO, he was not a fit person to be around children. The tribunal agreed and he was sent to work in agriculture or minesweeping. This strange pair of possibilities both tells us something about the hazards of farming at the time and reveals a new government policy. Work given to COs was now officially supposed to be of "deterrent character." Menial, degrading, and often of no value. Sometimes COs simply stacked rocks from dawn to dusk. Some COs asked to be returned to prison.

The University of Cambridge, specifically Professors Joseph Larmor and H. F. Newall, were working furiously behind the scenes to secure Eddington an exemption free of the embarrassing letters "CO." Through unknown means, they were able to do so. It must have been a national-importance exemption, though the details are lost. Eddington received a letter containing the exemption; he needed only to sign and return it. He signed it—adding a postscript that while he accepted that his work was of national importance, he would pursue conscientious objector status regardless. He refused to let them split his religious identity from his scientific work. This invalidated the exemption. Larmor and Newall must have been furious. But Eddington could see no reason for them to be. He had been nothing but honest and transparent in his loyalty to the Society of Friends. He would not abandon them now in their moment of greatest crisis.

Eddington was called before the tribunal again in July. In a notable departure from procedure, the Ministry of National Service had allowed his case to be heard again. This time the hearing had the feel of a well-oiled machine designed for a particular purpose. The proceedings began with Eddington presenting a letter from Dyson. One does not become Astronomer Royal without some political acumen, and his letter amply demonstrated this. It was keenly designed to appeal to the patriotism of the tribunal members. Dyson underlined the importance of keeping prestigious British scientists working to counter the dominance of enemy science:

> I should like to bring to the notice of the Tribunal the great value of Prof. Eddington's researches in astronomy, which are, in my opinion, to be ranked as highly as the work of his predecessors at Cambridge—Darwin, Ball, and Adams. They maintain the high position and traditions of British science at a time when it is very desirable that they be upheld, particularly in view of a widely spread but erroneous notion that the most important scientific researches are carried out in Germany. . . . I hope very strongly that the decision of the Tribunal will permit that important work to be continued.

After establishing Eddington's importance for British intellectual life, Dyson invoked a spectacular opportunity for astronomy that only Eddington could carry out:

> There is another point to which I would like to draw attention. The Joint Permanent Eclipse Committee, of which I am Chairman, has received a grant of £1000 for the observation of a total eclipse of the sun in May of next year, on account of exceptional importance. Under present conditions the eclipse will be observed by very few people. Prof. Eddington is peculiarly qualified to make these observations, and I hope the Tribunal will give him permission to undertake this task.

After this argument for the importance of his scientific work, Eddington again described his conscientious objection to the war. He said he had been a member of the Society of Friends since birth (often the key criterion for being seen as a genuine objector). When asked whether he would accept alternative service, Eddington took the most common position of COs. He refused to accept service under military auspices ("he did not think the War Office could guarantee that he be employed solely in saving life"), but expressed willingness to work in other contexts: service in the Friends Ambulance Unit, the Red Cross, or helping with the harvest, "if it was thought he could be of more use to the nation in that way." The tribunal was not particularly interested in the details of his conscience, however, and wanted to know more about the eclipse. They interrogated him about whether this eclipse was of particular importance and Eddington assured them that it would allow observations that could not be made again for centuries. Without retiring to discuss his case, the tribunal announced that his work was of national importance, and "therefore gave Prof. Eddington, in order to cover [the period of the eclipse], 12 months' exemption, on condition that he continued in his present work." They also, startlingly, said they were convinced he was a genuine CO. This was immediately forgotten, however, and the actual exemption was solely for his science and made no mention of his conscientious objection. Eddington then left the hall, free to pursue his work.

The ease of Eddington's final hearing was clearly due to Dyson's intervention through his contacts in the Admiralty. The exemption for national importance was no different than if Eddington had accepted the deal arranged by the university. Why, then, did he accept the one arranged by Dyson and not the other? The difference was that the exemption received with Dyson's and his Admiralty contacts' help allowed Eddington to maintain a protest against the war. He was unwilling to accept the exemption for his scientific work without also announcing his religious identity. So there must have been some element in the final exemption that fulfilled his need to work for peace. This was the 1919 eclipse to test relativity.

Eddington saw the eclipse expedition as a kind of pacifist statement—it was a way to demonstrate to the world of science that pacifism and internationalism was superior to patriotism and war. It was therefore in some sense *equivalent* to conscientious objection for him. What was important for him was to express his pacifism along with his work on relativity. It was a matter of *conscience*, and of Eddington holding true to his own connection to his God.

THE QUESTION OF conscription was, in the language of the time, one of compulsion. The British state had traditionally worked hard to avoid any kind of compulsion. Even compulsory vaccination for conscripted soldiers was seen as worrisome. Nonetheless, British parents listened to their doctors and the army achieved an over 90 percent vaccination rate. The rate of typhoid dropped to one-fifteenth that of previous wars.

Even with the era's enormous advances in medicine, disease remained a constant companion of war, as it has throughout history. About 2 in 3 British soldiers were sick at some point in the war. Half of the medical treatments given were for disease (of the wounds representing the other half, 80 percent were caused by artillery). There were new kinds of wounds too—those of the mind. In February 1915 the physician Charles Samuel Myers published in *The Lancet* about "shell shock," what we today call post-traumatic stress disorder (PTSD). Whether it should be considered an injury or a sickness would be debated, even as psychology and psychiatry mobilized to meet this new threat.

The year 1918 saw an eruption of war-borne disease unlike any before seen. Spanish flu, as it was known, raged across the world in three waves. The first, in the spring, was fairly mild. American doctors had noticed influenza at Camp Funston in Kansas, through which huge numbers of troops passed. It arrived in Europe in April. It first appeared in Spain (thus the name). Then France, Britain, and Germany in May. The second wave ran from August through November. It marched through Africa, Asia, and Australia. Historians have pointed out that it appeared in essentially every inhabited part of the Earth.

The incubation period was short, a handful of days. Victims would fall ill in a matter of hours with fever, cough, nausea, rashes, and intense pain in muscle, nerve, and bone. Military doctors tried to treat or prevent it with tea, brandy, quinine, and gargling menthol (none helped). One soldier wrote home describing how 90 percent of his unit was sick in bed with the "fashionable illness"—like the latest cut of clothing, it was everywhere. One soldier returned safely from the front to find his family dead from flu: "There was no doubt of the existence of a God: only the Supreme Being could contrive so brilliant an afterpiece to four years of unprecedented suffering and devastation."

The war probably did not cause the outbreak, but wartime conditions turned it into a pandemic. The movement of soldiers made it impossible to quarantine; the Americans refused to turn back infected troop ships. Men were crammed into the trenches in the least sanitary conditions imaginable. Starving populations in Germany had little resistance and proved to be perfect incubators. Western civilization had barely needed to invent machine guns and barbed wire; 50 million people were killed without so much as a single bullet.

In one month Einstein's Berlin saw some 30,000 cases of the flu and 1,500 deaths. Cholera followed. It was perhaps a blessing that Einstein was already confined to bed by his liver ailments—with little human contact he was much less likely to pick up these new diseases. Elsa had been taking lozenges to keep the flu at bay, but they were inducing heart seizures (adulterated medicine was an enormous problem by this point in the war). The newspapers were full of reports of illness and the meat ration being completely replaced by potatoes. The government now recommended gathering berries from trees.

There was important news from the front but the papers were under heavy censorship. Berliners thought the spring advances were continuing and almost everyone thought the war was near victory. What few people knew was that by July the German offensive had completely run out of steam. Attempts to push the Allies back again failed.

Finally, the Allies launched their counteroffensive against the overextended Germans. The British attacked near Amiens on August 8, immediately breaking enemy positions. With the help of strategic surprise (using techniques they learned from the Germans) and nearly six hundred massed tanks, they advanced seven miles the first day. For the first time large numbers of demoralized German soldiers surrendered. A British soldier wrote excitedly in his diary after capturing a German officer and finding a fruit cake, cigarettes, cigars, biscuits, and sweets. Erich Ludendorff, the German commander, called it the "black day" of the German Army. This was only the beginning of one hundred days of Allied attacks all along the line.

The German retreat never turned into a rout, though, and the Allies had to pay for every mile with blood. From firsthand accounts it is sometimes difficult to tell the victories from the defeats. From one successful battle:

> I saw Wylie [killed] instantly alongside of me by a machine-gun bullet in front of Harbonnieres at about 5 or 6 o'clock in the evening. We had gone over that day and had reached our objective and were lying and crawling about in a shallow sunken road and Wylie lifted his head to look at a machine-gun position opposite when he was hit right in the throat. Within a few minutes Wylie, a man named O'Mara (shot through spine and killed instantly), Davies (through back) and Curly Hendry (through head instantly) were killed and

> Male was also mortally wounded. . . . They were buried at Harbonnieres. . . . Wylie was a short chap, slightly bow-legged. I think he came from Scotland. He had his leave there a short while before. He was a good soldier and a decent little chap.

It ended up being the most costly year of the war for both sides. About 2 million died as the kaiser's troops were pushed eastward. A series of "Second Battle of X" began as Allied forces reclaimed captured territory. On September 26 the Hindenburg Line was broken. Germany was desperately seeking a diplomatic conclusion to the war. The other members of the Central Powers were collapsing or making separate peace, and by the first week of November, Germany was standing alone.

The resolution of the war began neither on the front lines nor in the capital cities. Instead it erupted in Kiel, the German port that had formerly hosted the international scientific telegraph system. The German High Seas Fleet had spent most of the war sheltered there, not daring to confront the might of the Royal Navy. As the German military realized that defeat was inevitable, though, the fleet was ordered to sea for one last glorious, hopeless battle. On October 30 the sailors refused and mutinied. They quickly seized not only their own ships but the entire city. The admiral in charge of the port—Wilhelm II's brother—was forced to flee in disguise as the sailors called for revolution. The mutiny spread to other units, many of whom displayed red soviet-style flags. As it became clear military discipline was breaking down, opposition politicians in the Reichstag called for the kaiser's resignation.

Wilhelm himself had already left his palace for occupied Belgium, sensing (with rare political acumen) that the army might be safer than his actual capital. In Berlin the military government handed over power to a joint military-civilian group that would conduct any armistice negotiations. A charitable interpretation of this action is that it was intended to make sure responsibility was shared by all of the leadership of the nation. A less charitable one was that the military was planning on blaming the civilian leadership for the war's loss.

By November 7, Berlin was boiling over with barely contained revolution. Rosa Luxemburg and other Bolsheviks were organizing huge crowds calling for a socialist state. Lenin sent agents into the city with 12 million marks of funding (about 40 million of today's dollars). Wilhelm pondered whether he could order his army to put down the demonstrators. His generals knew that the rank and file wanted nothing but peace, and on November 9 he was persuaded with some difficulty that abdication was necessary. The next day he was on a train to shelter in the neutral Netherlands. There was no word about how Einstein's friends there felt about their new role as guardians of the emperor.

In the last weeks before abdication, as the government was collapsing, Einstein and his fellow pacifists felt increasingly emboldened. The BNV, disbanded by the military government in 1916, reconstituted itself. On October 19 they issued a new declaration calling for an investigation into war guilt, an introduction of civil liberties, and a new democratic assembly (constructed by universal suffrage, including women). Einstein probably wasn't at the meeting, given his health, but he immediately began disseminating the declaration. In a move as daring as Eddington's request of Lodge, Einstein wrote to Planck seeking the elder scientist's support for the document.

Planck's reply was emotional and earnest. He thought the declaration might backfire and make peace even more difficult. Further, after the Manifesto of 93, Planck had refused to make any more public statements about the war. On the crucial question of democracy or monarchy, he was restrained by loyalty to higher principles. He thought it would be "extremely fortunate" if Wilhelm II abdicated. Nonetheless, he could not actually ask the kaiser to do so: "Think of the oath I took. . . . I feel something that you admittedly will not be able to understand at all . . . namely, a reverence for and an unshatterable solidarity with the State to which I belong, about which I am proud—and especially so in its misfortune—and which is embodied in the person of the monarch." Planck, with whom Einstein felt he shared so much, could not give up what one historian calls his reverence for the state.

In the wake of Wilhelm's abdication, republics were declared and revolutionary councils formed across Germany. On November 11, representatives of the German government—though exactly what that meant was unclear—signed an armistice agreement in Marshal Foch's private railway car in Compiègne, France. The fighting officially ceased at eleven o'clock on the eleventh day of the eleventh month (Paris time). On that same day, British forces advanced to Mons—the battlefield on which they had first joined the war in 1914. The Germans agreed to complete demilitarization and other humiliating terms. The British blockade would remain in place until a formal peace agreement was signed.

Einstein was supposed to teach a course on relativity. His diary noted that class was "canceled because of revolution." Excitedly, he wrote to his sister, "The great event has taken place! . . . The greatest public experience conceivable. . . . That I could live to see this!" He celebrated gleefully the collapse of not just the tyrants he had been battling but their whole world view: "Militarism and the privy-councilor stupor has been thoroughly obliterated." He sent celebratory postcards to everyone he knew. He reassured his mother that everything was going smoothly and he was safe. That was only partially true. His friends at the BNV organized a mass demonstration at the Reichstag, which was dispersed by machine-gun fire. Einstein had likely been too sick to attend.

The streets of Berlin were filled with crowds, and competing revolutionary groups began maneuvering for influence. The revolutionary example to the east was on everyone's minds—would Germany go the way of Soviet Russia? The sudden shift of politics leftward was a great boon for Einstein; he was not a Communist, but he was a socialist, and that was suddenly a badge of honor. His academic colleagues, with their monarchist politics, hoped that he could help defend them against the new provisional government. "I am enjoying the reputation of an irreproachable socialist; as a consequence, yesterday's heros are coming fawningly to me in the opinion that I could break their fall into emptiness. Funny world!"

In fact, Einstein ended up being a critical go-between for the young socialists and the older establishment. Soon after the revolution, a student council at the university took prisoner some professors and the chancellor and issued new university regulations establishing an ideological framework. Einstein, as a reputable socialist, was asked to negotiate with the students and free the academics. He agreed and made his way into the chaos of central Berlin.

He didn't much care about the freedom of the university chancellor, who had been rabidly patriotic during the war, but he had a great deal to say about the students' new ideas for running the university. He took them to task for threatening academic freedom, what he called "the most precious asset of the German university. . . . I would regret if the old liberties were to end." He delivered a speech warning against tyranny from both left and right. Calling himself an "old-time believer in democracy," he urged the formation of a national assembly that would subordinate its will to that of the people. He warned, "Force breeds only bitterness, hatred and reaction." This was not a popular position even among Einstein's friends, some of whom had tried to get him to support a new intellectual aristocracy that would run the nation. He declined. Democracy, he thought, was the only way forward.

Einstein also met with Friedrich Ebert, the nominal head of the provisional government. We do not know exactly how much influence Einstein had on either side, but a few days later the professors and staff at the university declared their support for the new republic and repudiated the Student Council for its Communist leanings. Einstein signed the statement: "We do not keep to what was ruined, but side with what is coming. Without reservation we are at the disposition of the people, its will and its representatives. According to our abilities and where we are needed we will serve in the shaping of the future." This happy compromise allowed the reopening of the university.

On November 16 he attended the founding meeting of the Democratic People's Union, a new organization dedicated to immediate elections. It was promoted by Walther Rathenau, an important wartime logistics official now turned liberal statesman. Rathenau would eventually become the foreign minister for the Weimar Republic. He would go on to be assassinated by anti-Semitic terrorists in an early indication of what would become the absurd "stab in the back" idea—the right-wing conspiracy theory that Jews and leftists had caused Germany to lose the war. Rathenau's murder helped galvanize Einstein's support for Zionism and his own Jewish identity, which changed his life forever.

But in 1918, Einstein was still hopeful that a peaceful democracy could be formed from the ruin of Wilhelm's Germany. The BNV reorganized itself with Einstein on the steering board. They called for a socialist government, a controlled economy, the elimination of the old aristocracy, the end of compulsory military service, a democratic assembly, and reconciliation among formerly enemy countries. This matched nicely with Einstein's political fantasies during the war, and he actively tried to recruit politicians to the BNV's goals.

Einstein was delighted to be on the "winner's side," though Berlin was hardly a triumphant city. Skirmishes in the streets between Leninists and moderate leftists were common—Ilse was briefly caught in a crossfire. Things did not improve as 2 million soldiers began returning home with no food or jobs. In a letter to his sons, Einstein described the "cheerful stir" that accompanied their arrival, and then warned the children not to play soldier games. It became common for armed men to wander the streets. On December 7 there was a general strike that was suppressed with bullets.

Einstein supported the new government though he worried that those in charge were too interested in following Russia. There was great potential but a vacuum of authority remained: "The military religion has vanished. . . . Nothing has taken its place, of course." Arnold Sommerfeld, writing to Einstein in December about quantum statistics, was incredulous to hear that he supported the new government. Sommerfeld's rightward leanings were clear: "I find everything unspeakably dire and imbecilic. Our enemies are the biggest liars and scoundrels, we the biggest morons. Not God but money rules the world." Einstein struck back, declaring that he was "of the firm conviction that culture-loving Germans will soon again be able to be as proud as ever of their Fatherland—with more reason than *before* 1914. I do not believe the current disorganization will leave permanent damage."

His views soured as little progress was made toward a stable government or even tamping down the fighting in the streets. Many of his fellow leftists became aggravated as they realized that the Allies had little intent of following Woodrow Wilson's generous principles for peace. Among other insults, blockade-induced starvation continued. The BNV held a meeting at the Berlin Opera House calling for a peace treaty based on law and justice rather than revenge. At the end of the year, Einstein traveled to Zurich to deliver his contractual lectures and, incidentally, finish up the legal paperwork for his divorce.

As in Berlin, the armistice brought vast crowds into the streets of London. These, though, were celebrating their king rather than deposing him. About a million people joined the celebrations. Huge bonfires were lit under Nelson's Column—for the first time in years, no one worried that nighttime lights would provide guidance for enemy bombers.

One might imagine that the end of the fighting brought a breath of relief to Eddington and his fellow pacifists. Instead, it spurred them to renewed action. Eddington's Quaker Meeting made a statement of purpose for the postwar world:

> Suffering and unrest are by no means over, nor is the conflict, tho' it may be removed to another plane of action. And while we could take no part in the late warfare we feel that the inevitable struggle now before the world, social political or economic, is one to which we may rightly have some thing to give, indeed we feel that it is precisely here that our place may lie. The value and power of this contribution depends on one thing, the ability in which our Church and each one of us as individual members of it can lay hold of the power of the Spirit. Nothing else can avail in this great moment.

The Friends Emergency Committee began planning its own invasion of Germany to help ease the suffering there caused by the continued blockade. Sir Eric Geddes called for "squeezing Germany until the pips squeak." British and French leaders hoped to use starvation as leverage to wring a favorable formal peace treaty from Germany in the months to come.

Some Quakers had to wait to join this effort. Ludlam and others were not immediately released from prison with the armistice. In the end, they were not released but rather discharged, since they were technically members of the army. The grounds for discharge were listed as "misconduct." About seventy conscientious objectors died as a result of their treatment, and all COs were denied the vote for five years after the war.

Once free, Ludlam joined the Friends Emergency Committee to purchase food and medicine and ship them to Germany, directly defying government orders. These Quakers who crossed the Channel worked in difficult and dangerous conditions and became emblematic of what modern pacifism was supposed to look like. They journeyed into far and foreign lands as a duty of conscience.

And Eddington planned to do the same—but with physics and astronomy. The eclipse expedition was his chance to repair international relationships on an intellectual level. This kind of work was a recognized part of the Quaker relief efforts. Healing social and scholarly relationships was seen as being as important as material suffering:

> While chemists are testing out the deadliest types of poison gas for future wars . . . it is well that there should also be some notable attempts made to conquer the hearts of men by kindness and to demonstrate that one person who heads an expedition to heal the wounds and desolation of war is stronger than a battalion of men under arms.

This was the Quaker war, only able to be waged once peace had come.

This emphasis on intellectual connections meant the German educational system was a particular target for the Quaker relief workers. Ludlam's group undertook to feed thousands of students in Berlin. Connections with Berlin teachers were particularly important for the relief work at schools. To this end, Eddington suggested that his friend contact one particular faculty member for assistance: the very busy Albert Einstein, who was reported as being "closely connected" to the efforts.

CHAPTER 11

The Test

"The most favourable day of the year for weighing light is May 29."

USUALLY, WHEN SCIENTISTS test a theory, they get everything nicely under control. Laboratories are carefully designed, protected, and isolated. The whole point of an experiment is to create a space where there is no interference, no confusion—a locale where nothing unexpected can interrupt. These delicate places are what allow, for example, the predictions of quantum electrodynamics to be tested to a precision of 1 part in 10 billion. Accuracy and reliability of that sort can only be achieved when you can have everything just how you want it.

Eddington did not have that luxury. He was going to test Einstein's theory at a solar eclipse thousands of miles from the nearest precision laboratory. This was not easy. The young Eddington once described the difficulty of eclipse expeditions: "In journeying to observe a total eclipse of the Sun, the astronomer quits the usually staid course of his work and indulges in a heavy gamble with fortune." Weather and war made true control impossible. Instead, he had to perform a test that was persuasive in spite of the chaotic conditions. He had to make an expedition as scientific as an experiment. He was hardly the first scientist to have this problem, and he had a whole arsenal of strategies to draw on.

Instruments and devices were not the only things that needed to be handled carefully. As much as he needed to create a favorable physical environment for his tools, he needed a favorable social and political environment in which his results could be received. Experiments and observations do not speak for themselves. People must be willing to be persuaded; they need to understand how the data connect to the theory; they must be prepared to accept that the results mean what you say they mean. Scientists must create spaces—sometimes tangible, sometimes intangible—in which science can be done. Eddington had to do both.

EINSTEIN'S SPACES AT the beginning of 1919 were, at best, unstable. Berlin, his scientific space, was increasingly messy. His lectures on relativity were postponed because the university lacked coal to heat the lecture hall. He was frustrated both by the fecklessness of the new government (he called them "excessively dishonest") and the viciousness of the Allies in victory (compared to the kaiser, they were "*only slightly* the lesser evil"). Luckily, he was absent from the city during the bloodiest week of the revolution, when the Spartacists (the group that helped Nicolai escape in a biplane) tried to seize power.

He was away from Berlin because he was trying to get his personal spaces into better order. On February 14—Valentine's Day—a Zurich court ended his marriage to Mileva on the grounds of adultery (he admitted his affair with Elsa) and "character incompatibility." Custody of the boys was given to their mother, and Einstein was ordered to pay 8,000 francs annually, drawn from the hoped-for Nobel Prize money. Finally, he was ordered not to remarry for another two years.

Nominally he was in Switzerland to deliver some lectures and earn some money in a currency other than the plummeting German mark. He was stunned by the strangeness of being around "well-fed citizens who have nothing to fear." Back in Berlin for his fortieth birthday, he still couldn't get much science done. He gave a lecture on relativity whose admission fee went to support the Socialist Students' Union.

Unimpressed with the state of Germany, he returned to Zurich right away to deliver more lectures. There wasn't much interest—only fifteen students registered to hear Einstein speak about relativity—and the university canceled the event.

In Berlin it would have been hard to know that the war was over (ironically, the clearest sign was large numbers of armed men in the streets). Food and fuel were still short. All this was because technically speaking the war was only on pause. There had been an *armistice* but there was no *peace*. The latter would come only after a formal treaty was signed among the warring countries. Almost nothing to this end had been done in the months since the armistice.

The peacemakers began to gather in Paris in January 1919. Delegates on the Champs-Élysées strolled past captured German cannon. The conference officially opened January 18, which (surely by coincidence) happened to be the anniversary of Wilhelm I's coronation in 1871. The negotiations involved setting up the League of Nations as well as dividing Africa and the Middle East into new colonial possessions. The victorious empires gobbled up ever more of the world.

THOSE NEW IMPERIAL boundaries were of huge importance to astronomers, particularly ones who, say, were planning solar-eclipse expeditions for May 1919. This was not unusual. There is a long history connecting imperialism to eclipse expeditions—remember the infrastructure that made Eddington's overseas work possible back when he was chief assistant at the Royal Observatory. Traditionally governments put forward funds and resources for eclipse expeditions because those projects brought tremendous national prestige—who but a great power could perform science at any spot in the world?

Eddington's plans for his expedition were quite different. Rather than affirm national pride, he wanted this journey to shatter patriotism and celebrate what could be done across borders. He wanted to use the tools of empire to fight for internationalism; Dyson's support meant this was just possible. If everything went well.

Preparations for a solar-eclipse expedition were elaborate. The first step was simply to figure out where and when the eclipse would be visible. The zone of totality—the place from which the moon completely blocks the sun—is typically some miles wide, but the eclipse is visible only for minutes (if one is lucky). The shadow of the moon hurtles across the surface of the Earth at more than a thousand miles per hour, and astronomers need to be in the right place at the right time with their telescopes and cameras.

The path of totality was an arc across the Southern Hemisphere from Africa to South America.

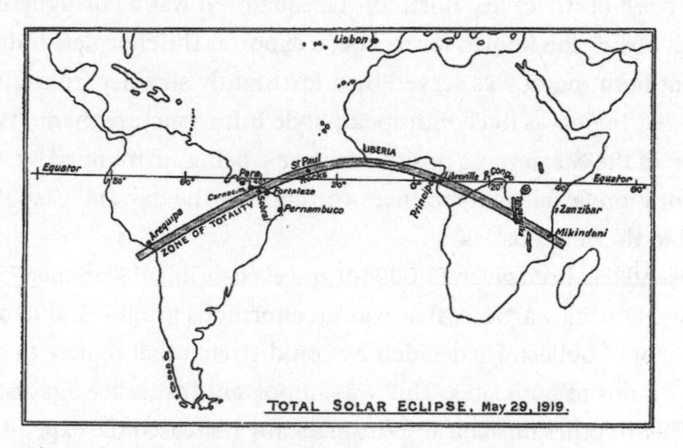

The path of the May 1919 eclipse, as illustrated in Andrew Crommelin's February 1919 article in Nature

COURTESY OF THE AUTHOR

Many factors entered into the choice of where to make the observations: Did the location have a reputation for good weather? How humid was it? How low in the sky would the eclipse be? Were there nearby steamship and railway networks to carry the astronomers and their heavy equipment? Did the schedules of those networks intersect with the eclipse dates? Was there a telegraph station nearby? Could food and water be procured at reasonable prices? Was there a friendly local government or colonial administration to help support the expedition? Answering these questions often relied on travelers' reports or expats of dubious trustworthiness.

Dyson and Eddington decided that there were two locales—each would have about five minutes of totality—that best answered all these questions, one on each side of the Atlantic. Sobral, eighty miles inland in Brazil, was on the rail lines. It was not quite in the center of the path, so totality would be a few seconds shorter. But the logistical advantages more than made up for that. Word was that the rainy season would be over by May (no one was quite sure). Much of the information was provided by a circular sent by Dr. Henrique Morize, the head of the Rio de Janeiro Observatory.

The other observation site was Principe, an island 110 miles off the west coast of Africa just north of the equator. It was a Portuguese imperial possession known for its cocoa exports. The chocolate industry meant both that it was served by a fortnightly steamer from Lisbon and that there was likely European-style infrastructure there. Its isolation in the ocean was a positive feature—being surrounded by water meant more stable temperatures throughout the day and easy sightlines to the horizon.

Dyson had been given £1,000 for travel costs in 1918 (about $75,000 today). During wartime, that was an enormous grant—£1,000 could buy a lot of bullets. He decided he could stretch that money to cover expeditions to both sites. This was important insurance against bad weather or other mishap, and dramatically increased the expeditions' chances of success. Eddington would go to Principe, accompanied by Edwin T. Cottingham, a clockmaker who had worked for years with both Dyson and Eddington maintaining the timepieces at their observatories. The observations in Brazil would be conducted by Andrew Crommelin, an assistant at the Royal Observatory, and Fr. Aloysius Cortie, a Jesuit astronomer, of the Stonyhurst College Observatory in Lancashire. Cortie was known both for his cheerful nature and for regularly delivering science-themed sermons at meetings of the BA. At the last moment, however, he could not participate and was replaced by Charles Davidson. Davidson had been overseeing the preparation of the eclipse equipment and had a reputation as an absolute wizard with mechanical devices and scientific instruments. Dyson trusted him implicitly to make any mechanism work properly.

The equipment that Davidson had been preparing included three carefully chosen telescopes. Eddington needed crisp images of stars, which is usually not something eclipse observers are looking for. So the teams decided to use astrographic telescopes—specially designed to capture precise, faint images. Dyson tried to secure two telescopes of this sort that had been used at previous eclipses where they accidentally captured good star images. One currently mounted in Greenwich was easily acquired. The other was at the Oxford observatory overseen by H. H. Turner, the most vocal anti-German astronomer in the country. We do not know how Dyson persuaded Turner to contribute this valuable instrument to Einstein-centric expeditions, but somehow he succeeded. Father Cortie also suggested that they take a smaller four-inch telescope to Brazil as a backup. It had captured good fields of stars at other eclipses, and would not add much logistical stress.

No one would be looking through these telescopes—a camera would be used. The telescope would focus the image of the stars on a photographic plate. The plate would be exposed briefly (five to ten seconds), and then the astronomers would swap it for a fresh one without disturbing the delicate equipment. There was a complicating factor to this procedure, thanks to Copernicus. Because the Earth rotates, the eclipsed sun and the stars appear to move across the sky. Even over the course of just a few seconds this apparent motion will blur the images on the photograph. One solution to this problem is to mount the telescope on a pivot and slowly turn it to match the Earth's movement. This is not a very good solution for an expedition, though—telescopes are heavy and large, very difficult to move smoothly without shaking or bending that would ruin the image.

The traditional answer was to use a coelostat, the same sort of clockwork mirror Eddington had used in his 1912 expedition. The telescope would be laid horizontally, nicely stable. The lens of the telescope would be pointed at the coelostat mirror, which would then be adjusted so the image of the sun would fall in the middle of the camera. Then the mirror could be smoothly turned during the eclipse to keep the image centered without blurring.

Greenwich had a set of these coelostats that had been used for many previous expeditions. Unfortunately they had been used for *many* previous expeditions and were old and unreliable. Normally, overhauling them would be a straightforward, if tedious, process, but the early preparations for the expeditions were happening during wartime, and a "priority certificate" from the Ministry of Munitions was required to get any precision work done. That was impossible while the war was on. Once the armistice arrived, Cottingham did the best he could to get the coelostats running smoothly. Mountains of fine, delicate work had to be compressed if the expeditions were to leave on time in the early spring.

Science is more than good equipment. Eddington and Dyson needed to make sure other scientists were ready to think about—to understand—the results captured by that equipment. The expeditions were not passive attempts to just look for something interesting during the eclipse. Their goal was to test a specific prediction of Einstein's theory of relativity. And Eddington had his own pacifist goals for the expeditions. Both the scientific and the political goals required groundwork to be laid ahead of time.

The first part was understanding Einstein's prediction. Einstein said, let's look at a star that appeared to be just at the edge of the sun's disk (the star was actually trillions of miles away, it just happened to line up with the edge). The image of that star is being carried to us by a ray of light. As that light passes by the sun, the curvature of spacetime there (in other words, the gravity) will bend that ray of light. To an observer on Earth looking at the star's image, the bending means the image will be shifted slightly from its original location. General relativity predicted the exact angle between where the star should be when the sun's gravity was not in the way and where it appeared to be when the sun's gravity was. That angle was measured in arc-seconds (one-sixtieth of one-sixtieth of a degree). Einstein said the change should be 1.75 arc-seconds. On the photographic plates Eddington would be using, that would translate to about one-sixtieth of a millimeter. Some scientists at the time objected that this was a very small size to measure precisely. To astronomers, though, this was no great challenge—they measured effects that small every day. Eddington reassured everyone that “this in itself calls for no extravagant precautions of accuracy.”

Astronomers were able to make these precise measurements because they took *everything* into account. The photographs taken during the eclipse needed to be compared to *check plates*—photographs taken of the same field of stars when the eclipsed sun is not in front of them. It is the *change* of position of the star that matters—they had to have an exact reference for that change. It can take months for the sun to move far enough across the sky that the images would be undistorted by its gravity. That means the check plates needed to be taken either months before or after the eclipse itself. Further, they had to be taken with exactly the same lens and photographic setup—every lens is a little bit different, and it was essential to make sure that an apparent change in the star's location was not really due to an imperfection introduced by a different lens. So photographs of the stars they would measure were taken from England with the lenses they planned to use in the field. Ideally additional check plates would be taken in the field to account for any atmospheric strangeness unique to that location. Calculation of the full results would take months of work back in Britain, but Eddington hoped that they could make preliminary measurements in the field. That required special tools, as well as research into how to develop the photographic plates in tropical conditions—every photographic manufacturer (and even individual production lines) required slightly different techniques. Hoping to get those preliminary results home as soon as possible, he and Dyson even arranged a special telegraphic code.

Before his departure, Eddington wrote an article presenting all this information to his colleagues so they would know how to interpret the results as they came back. Eddington declared that there were three possibilities: no deflection; 1.75 arc-seconds, the Einstein prediction; or 0.87 arc-seconds, sometimes referred to as the half-deflection.* The

* One sometimes sees the full deflection calculated to be 1.74, or the half-deflection calculated anywhere between 0.83 and 0.87. This can vary according to the constants used in the calculation, or how one chooses to round the results. The overall advice would be that one should not worry too much about the final digit in a calculation of this sort.

half-deflection probably came from Lodge's alternative theory based on Newtonian gravity. Thus Eddington was able to present only three possibilities: null, Einstein, or Newton.

Modern scholars have pointed out that Eddington made a shrewd choice in framing the possible results this way. This was a "false trichotomy"—there were certainly more than three prospects. There were other alternative theories of gravity (remember Nordström, Einstein's early rival). There were other possible effects that could mimic gravitational light deflection, such as ether condensation or refraction in the solar atmosphere. Deciding to leave those possibilities out of the "official" predictions created a scenario known as a *crucial experiment*, where a single measurement could immediately decide between two rival theories. The null result not being very interesting, the test suddenly became a direct struggle between Einstein and Newton—a single moment in which this upstart German could dethrone the greatest thinker in history. There are always many possible explanations for any experimental result, and the crucial experiment (ironically, Newton's favorite setup) narrows the options. This is helpful epistemologically and practically—one can never account for every possibility—but Eddington was probably more interested in the *narrative* value. It created a thrilling background against which to present the expeditions' results. One could hardly hope for a better title card—two geniuses enter the ring, one leaves—to gain attention in the world of science.

During preparations, Dyson found himself explaining these possibilities to Cottingham, the clockmaker going to Africa with Eddington. He was a hands-on worker with little interest in the mathematics of relativity; he just needed to know the measurement they were looking for. As Eddington described Dyson's lesson, Cottingham "gathered the main idea that the bigger the result, the more exciting it would be." Cottingham asked, "What will it mean if we get double the deflection?" (3.5, twice the Einstein prediction). "Then," said Dyson, "Eddington will go mad, and you will have to come home alone."

Even if the numerical results were not those predicted, Eddington could still have high hopes for the expeditions' value for international science. For that, they needed publicity beyond reports in technical journals such as *Nature*. Eddington and Dyson were in contact with reporters from the *Times* at least by January 13, 1919, when it published its first article about the upcoming expeditions. In those days before press releases, it was common to talk directly to reporters or editors about important news. The historian Alistair Sponsel has documented the extensive efforts Eddington and Dyson made to get regular press coverage of the expeditions even before they found any results. Articles about the project appeared every couple of months. There was usually no author listed—we know some were written by Dyson, some by an assistant of his. A "devout reader" of the *Times* would have, over the course of 1919, become familiar with the expeditions, Einstein, and relativity. By summer they would have been eagerly awaiting the results—just as Eddington wanted.

EINSTEIN DID NOT have a subscription to the *Times*. But after the armistice scientific publications from outside Germany were beginning to trickle in. Einstein's friend Arnold Berliner, a fellow liberal who edited the journal *Naturwissenschaften*, managed to get his hands on an issue of *Nature* that contained a description of the eclipse expeditions. He wrote to Einstein excitedly, telling him about the British efforts to test relativity. If Einstein was interested, he would happily translate the rest of the article into German. In any case, Berliner would publish the entire report in the next issue of *Naturwissenschaften*. He was as pleased by its scientific merits as the opportunity to tweak the "Anglophagists" like Lenard and Stark who had so viciously tried to eliminate English science from Germany.

Berliner admitted that finally having access to foreign journals was "a mixed pleasure," though. On one hand, he heard about exciting news like this. On the other, he could see an advertisement in *Science* in which a company declared, "Not one item in [our] catalog is made in Germany." This was hailed as an opportunity to "free your laboratory of German products." The upcoming expeditions were still an anomaly in the world of Allied science.

Planck was no doubt happy to hear about the British efforts to support his friend's theory. He had been working hard to keep German science active even in this "wretched time." He did not much care either for republics as a form of government or the actual new administration. By 1919, though, politics was barely a distraction. He was living in "gnawing pain" after wartime had devastated his family. His two daughters died in childbirth and his son Karl was killed at Verdun. His son Erwin was taken prisoner (many years later he would be executed by the Nazis for plotting against Hitler). Planck was not unusual for the tragedies that tore apart his family, only for the stoicism with which he bore them.

Some scientists wondered if this meant Planck could be lured away from Germany, perhaps to Switzerland. Einstein dispelled any such ideas. He said it was "totally inconceivable" that Planck would leave Germany. "He is rooted to his native land with every fiber, like no one else." Einstein was certainly not but showed no inclination to leave. He grew more and more disgusted with the steady stream of political upheavals, though: "The country is like someone with a badly upset stomach who hasn't yet thrown up enough." Still, he made no move to depart.

EDDINGTON, ON THE other hand, was in a serious hurry to get out of the country. All the preparations for the expeditions had happened at the "eleventh hour" because they had to wait for wartime restrictions to loosen—and the eclipse would not wait for them. At the beginning of March, Eddington rushed out the front door of the observatory and tossed his luggage into a waiting taxi. The expedition teams took a train to Liverpool, on which they were charged extra handling fees for the delicacy of the equipment. On March 8 they boarded the RMS *Anselm*, a decommissioned troop ship that was part of the Booth Line.

Of Booth's thirty ships, eleven had been requisitioned for war use and nine had been sunk by submarines. This was some of the first post-armistice commercial travel, though there were many reminders of the war—a held-over regulation meant that passengers were not allowed to know the boat's location or course at any given time.

The *Anselm* was a roomy boat, and the teams stayed in first-class cabins. Scientists sat alongside tourists. Food on the boat was exempt from rationing and Eddington commented on how strange it was: "unlimited sugar, and large slices of meat, puddings with pre-war quantity of raisins & currants in them, new white rolls, and so on." Crommelin and Cottingham were unable to enjoy that, though, as they spent most of the time seasick.

The ship stopped in Lisbon, where a local astronomer, Dr. Frederico Tomás Oom, took them on a tour by motorcar. Dr. Oom had been in contact with the JPEC helping to make arrangements in Principe, a Portuguese colony. He was instrumental in maneuvering through the political turbulence—Portugal had only become a republic in 1910, and just months before Eddington arrived the monarchy had been briefly reinstalled. That overthrow had momentarily halted all sailings to Lisbon, imperiling both expeditions.

After that brief stopover they landed on the island of Madeira on March 15. Arriving by sea was an unpleasant reminder of the war: "Three ships were torpedoed by submarine in Madeira harbour during the war, and one sees the masts of two of them sticking up out of the water. The town was also bombarded and there are a few traces visible." From there Davidson and Crommelin caught a steamer for Brazil. Eddington and Cottingham had to wait for the next ship going to Principe. Contrary to the information on which they had planned the trip, the sailing schedules were completely undependable.

Madeira was not a bad place to have to spend a few weeks, though. Eddington found the mountainous terrain perfect for hiking, and he sent home letters describing the vistas. Cottingham tried to join in but could not keep up. The precipitous slopes were a serious challenge—Eddington had to buy a walking stick. After climbing one mountain, Eddington descended via a four-mile toboggan run. He was disappointed to hear that swimming was out of the question—too many sharks. Food had been scarce during the war, except for locally grown sugar and fruit. Finally free of rationing, Eddington found himself eating a dozen bananas a day. Nipper, a dog at their hotel, often came along on his adventures. Eddington was typically excited to befriend local dogs during his travels, but he did not encourage Nipper at all, "as he was neither beautiful nor free from fleas."

One of the main attractions on the island was the local casino. Eddington's letters to his very conservative mother assured her that he went there only because that was the only place to get proper tea. In a letter to his sister, though, he confessed: "I expect Mother sends on my letters to some of our relatives, so I did not mention in them, that I played roulette, of course not seriously, but enough to get a good idea of it and experience the ups and downs of fortune."

By the second week of April, Eddington and Cottingham were on board the *Portugal* on their way to Principe. The other passengers were mostly Portuguese, so conversation was often limited (he had been reading *The Vicar of Wakefield* in Portuguese to learn some). The language barrier still allowed for games of musical chairs and an egg-and-spoon race on Good Friday. He enjoyed trying the exotic cuisine, although since the milk on the ship was not good he had to take his tea black.

After a journey of nearly five thousand miles, Eddington arrived off the coast of Africa on April 26. Principe is about four miles wide by ten miles long, about one-seventh the size of its nearest neighbor, São Tomé. It is thickly wooded and dominated by a mountain in the center. This was the spot that Eddington had chosen as the crux for his campaign to restore the international world of science, to show the strength of peace in the face of war. But when the inhabitants of the island thought about Quakers, they probably did not think about pacifism but rather some tragic history.

Principe was covered in cocoa plantations, which provided the raw materials with which the Cadbury family—a prominent Quaker clan—made their fortune. Quakers had been involved in the chocolate business for a long time (it was seen as a healthy alternative to alcohol).

Many of those companies, and the Cadburys in particular, were known for advocating workers' rights and improving living conditions. So William Cadbury was horrified to discover in 1901 that the cocoa workers were essentially slaves—brought from Angola, forbidden from leaving, and sold along with livestock and equipment. He launched a very public investigation that led to tensions between the British and Portuguese Empires, with Quaker commercial interests in the middle. The Principe authorities were pressured to radically improve conditions, and by 1916 the British Foreign Office reported that there had been marked progress. Workers were well paid and could return home at will. But those men—Eddington commented that there were virtually no women on the island—would certainly have remembered their earlier treatment. Whether they blamed the Quakers for that, we do not know.

Cadbury's investigation was carried out by the young Joseph Burtt. Burtt, a gifted writer, described his first view of Principe, which would have been unchanged by 1919:

> And as the light brightened the sea became sapphire blue over the rocks, and turquoise in the sandy shallows, while here and there beneath grey precipitous cliffs it lay in pools of deep translucent green that seemed too radiant for mortal eyes to look upon. Beyond this, and as white as silk, the tiny breakers foamed against the line of yellow sand where careless cocoa palms flung up their sloping stems and tossed their plumes in the fresh morning air. . . . Further still, and higher, the vast dome of Papagaio stood out against the pale blue sky, veiling one purple side in diaphanous clouds that rolled and rose like incense to the mountain gods.

Eddington and Burtt were both hopeful when they landed on the island. The former looked to repair the damage caused by four years of war. The latter sought to ameliorate the suffering caused by greed and empire. They each saw Principe as a pivot point, a place where singular effort could change the world. Burtt succeeded. Eddington's test was yet to come.

Davidson and Crommelin's arrival in Brazil went smoothly. The British Consulate helped speed their delicate instruments through customs (the British and Portuguese Empires could cooperate when it suited them). The astronomers had a leisurely trip up the Amazon, then a combination of trains and local steamers to get them to Sobral on April 30. They were met by both civil and religious authorities—Father Cortie had used his Jesuit connections to ensure their warm welcome.

The astronomers were given use of the house of Col. Vicente Saboya, the deputy of Sobral. In addition to comfortable quarters, the house provided a ready supply of cool water necessary for developing the photographs. They set up their two telescopes and the coelostats at a nearby racecourse. This was a very good place to observe the eclipse, with a large clear area near a shaded grandstand. Police patrolled to keep gawking locals from entering the racecourse, making it more like a controlled scientific laboratory and less like a sporting field.

Eddington and Cottingham, too, were greeted warmly by the establishment at their destination. Imagine the far-flung colonial outpost where the local dignitaries struggled mightily to re-create all the trappings of European civilization—that was Principe. Eddington had a letter of introduction from the colonial officials in Madeira that helped smooth his arrival. As Eddington put it, they were "in clover." Crucial to their work were Mr. Wright and Mr. Lewis, two men from Sierra Leone who ran the local telegraph station. Their English was excellent and they often translated for the expedition. Most evenings the group sat on the local judge's balcony overlooking the sea, listening to the governor's record collection (grand opera, mainly).

Eddington and Cottingham stayed in the port of Santo António for about a week as they scouted the island for the best observation sites. The locals' intimate knowledge of the jungled, mountainous terrain was crucial. They finally decided on the Roça Sundy Plantation on the northwest corner of the island, away from the cloud-gathering mountains, on a plateau overlooking a bay five hundred feet below. This provided shelter from the winds but still allowed excellent views. There was also a luxurious plantation house conveniently nearby. The plantation owner, Sr. Carneiro, put his workers at the astronomers' disposal to build the foundations and huts that would support and protect the observation equipment. They also carried that equipment by hand through the thick forest for about a kilometer. We will never know those workers' names; they join the legion of anonymous laborers who have made science possible.

The astronomers had some chances to relax on Sundays, including monkey hunting (they didn't catch any) and swimming (a plantation worker went with Eddington to keep him away from sharks). They took a trip to visit a particularly fruitful plantation where the trees bent under the weight of cocoa. As Eddington described it to his mother: "It was a very fine sight to see the large golden pods in such numbers—almost as though the forest had been hung with Chinese lanterns."

On May 5, Eddington started writing a letter home to his sister. He commented that he had just received their mother's letter dated March 14, and he had little idea what life was like at home: "Indeed I do not know what has been happening in the world in general—whether peace has been signed or any important events have occurred." He was deeply isolated and cut off. He had no sense of the current state of science or politics; he could only continue with the expedition as had been planned. He could not help but contrast his experience of tropical plenty with that of wartime Britain:

> I wonder if you are still rationed. It seemed funny on the boat at starting to see full sugar-basins, unlimited butter, and to eat in a day about as much meat as would have been a week's ration. We have had no scarcity of anything since we started.

He asked after Punch, their dog, and wished the pooch a happy birthday.

By May 16 the telescope was set up under one of the waterproof huts the plantation workers had constructed. Eddington and Cottingham began taking the check photographs that would provide the references for measuring the Einstein deflection. These check plates also provided valuable practice developing the photographs in tropical conditions—no one at Kodak had expected their products to be used in the jungle.

The final days leading up to the May 29 eclipse were nerve-racking, as they often were on eclipse expeditions. Years of planning, months of journeying, weeks of physically and mentally grueling preparation, all without knowing whether the sky would be clear at the critical moment. The day before an eclipse Arthur Schuster, one of Eddington's college professors, actually broke down crying.

Eddington, with his literary flair, pointed out that this methodical preparation made eclipse observations feel somewhat ritualistic. He suggested that this eclipse in particular had a bit of a divinatory aspect in its particularly fortunate circumstances. Years later he wrote:

> In a superstitious age a natural philosopher wishing to perform an important experiment would consult an astrologer to ascertain an auspicious moment for the trial. With better reason, an astronomer to-day consulting the stars would announce that the most favourable day of the year for weighing light is May 29.

This was because on that day the eclipse would take place right in front of the Hyades, a handful of bright stars perfect for measuring the Einstein deflection. He wanted bright stars so they could be easily seen on the photograph. He wanted more than one so he could see how the deflection changed the farther away from the sun they were: a star right at the edge of the sun should show the 1.75 arc-second deflection; a star slightly farther away would show less; a star well away would show almost none. Einstein predicted not only a deflection but also a specific way the deflection would change with distance from the sun's edge. Multiple stars meant this aspect of the prediction could be tested as well. A past or future astronomer might have to wait centuries or millennia for a background as auspicious as the Hyades.

The Hyades are found in the constellation Taurus. They are the bull's head, right by the blazing-red star Aldebaran. They were named after five nymphs, the daughters of Atlas. Weeping over the death of their brother, they were placed in the heavens just out of Orion's lustful reach. Their tears made the constellation traditionally associated with the arrival of the rainy season. Eddington probably grew up calling them the "April Rainers" (perhaps not a propitious omen for the eclipse observations).

As one of the brightest clusters in the sky they are visible to the naked eye and have been watched since antiquity. They appear in the *Iliad* when Hephaestus crafts a shield for the hero Achilles. On the shield is depicted the entire world and cosmos, from sheepherders to wars to the imperishable stars. The Hyades are among the constellations placed on the shield, along with Orion and Ursa Major. They were part of the ancient links between the heavens and the Earth, carrying meaning from the celestial realm to the terrestrial. Eddington had no shield on which to catch these stars, only a telescope with which to look for their message.

To see if the light from those stars was bent, he had to point that telescope into the darkness of a total eclipse. The experience of a total eclipse is unlike anything else. A 99 percent eclipse seems to be a cloudy afternoon—100 percent plunges you into sudden, awful darkness. The temperature drops, birds stop singing, and (crucially for Einstein) the stars become visible. If one is not ready for it, the experience is disorienting and unsettling. George Airy, an Astronomer Royal from the nineteenth century, warned that during totality "the most perfect discipline will fail." Astronomers since then had developed routines and rituals to keep their focus during an eclipse, to keep their attention on their observations and their science at a moment when the world seemed to have ended.

In Sobral, May 29 was cloudy. The previous few days had been unusually rainy, dampening everyone's spirits. The local community had been preparing to make the eclipse a public event, and festivities were ready to go. A small observatory near the edge of the eclipse path sold tickets to look through a telescope.

The clouds were still thick at the beginning of the eclipse. When the leading edge of the moon touched the solar disk (a moment called "first contact"), Crommelin estimated that 90 percent of the sky was clouds. But they rapidly diminished and the sun sat in a large clear patch as totality began. The landscape was plunged into surreal darkness and the astronomers began their work. Their focus was all on the instruments. One of the Brazilians watched a clock and called out the passage of seconds for timing the photographs. Nineteen photos were exposed with the large astrographic telescope and eight with the small four-inch lens. The clear sky held for the entire eclipse; everything had gone smoothly. They cabled home immediately: "Eclipse splendid."

On the other side of the Atlantic, the Principe dignitaries came to visit Roça Sundy the morning of the eclipse. They were immediately greeted by a tremendous rainstorm, the heaviest the British visitors had seen, and quite unusual for that time of year. It ended around noon, with a couple hours to go before the eclipse. Eddington watched as there "were a few gleams of sunshine after the rain, but it soon clouded over again." The clouds, he said, "almost took away all hope."

At first contact the sun was invisible behind the clouds. It was not until 1:55 p.m. that the astronomers began to get glimpses of the sun, shaped into a crescent by the moon's inexorable creep. It slipped in and out of cloud from moment to moment. Even in good conditions the last few seconds before totality have been described as "almost painful." We can only imagine what this kind of knife-edge waiting would have felt like. Totality was calculated to begin five seconds after 2:13 p.m. At that moment the astronomers became machines, carrying out the planned procedures regardless of what they could see with the naked eye. Machines, though, driven by hope and anticipation. As Eddington described it: "We had to carry out our programme of photographs in faith."

The telescope took all of their attention. Cottingham kept the coelostat mechanism running and handed Eddington fresh plates; Eddington removed the exposed plates and slid in the new ones. He had to pause for a delicate second after each swap, lest the motion cause some tiny tremor that would ruin the image.

Other than a quick look to make sure the eclipse had begun, Eddington did not get to see the celestial event on which he had gambled so much. There was no time to watch the sublime theatre going on:

> There is a marvelous spectacle above, and, as the photographs afterwards revealed, a wonderful prominence-flame is poised a hundred thousand miles above the surface of the sun. . . . We are conscious only of the weird half-light of the landscape and the hush of nature, broken by the calls of the observers, and the beat of the metronome ticking out the 302 seconds of totality.

When totality ended, the world returned to normal, with no lasting marker of the disruption of the natural order that had just taken place. Within a few minutes the sky was perfectly clear, perhaps due to the temperature change of the eclipse itself.

At Roça Sundy, sixteen glass plates sat covered, holding the secrets of the stars until they could be scrutinized. Great efforts would be required until they gave up their message. But the event itself was over. Eddington could take a moment to breathe. His telegram to Dyson was succinct: "Through cloud. Hopeful."

THE DECISION HAD been made to develop the photographs on-site in Brazil and Principe, and for reasons "not entirely from impatience." The glass plates were delicate and could easily be damaged on the long journey home. Developing them in place and making preliminary measurements would at least guarantee some results, even if they were gathered in imperfect conditions.

Davidson and Crommelin in Sobral developed four of the astrographic photographs the next night. They were shocked to see that the star images were ever so slightly distorted, as though the focus on the telescope had been changed.

> May 30, 3 am . . . It was found that there had been a serious change of focus, so that, while the stars were shown, the definition was spoilt. This change of focus can only be attributed to the unequal expansion of the mirror through the sun's heat. The readings of the focusing scale were checked next day, but were found unaltered at 11.0 mm. It seems doubtful whether much can be got from these plates.

The coelostat mirror that reflected the sun's image into the astrographic telescope was a thin sheet of metal. The portions of the mirror where the sun's light was brightest would be slightly heated, and when metal is heated it expands. This expansion would be different from the unheated parts of the mirror and cause the surface to warp and bend. For normal eclipse observations this effect would be negligible. But the Einstein deflection was such a small effect that it could easily be swamped by such a phenomenon.

The images from the four-inch telescope, brought along as an afterthought, looked much better. So not all hope was lost. In any case, the pair of astronomers had a long wait ahead of them. They needed to stay in Brazil until July to take check photographs of the Hyades once the sun had moved out of the way.

Eddington was not in a mood to wait. While there were good technical reasons to examine the photographs right away, it seems his incentive may have been more personal. For the six nights after the eclipse he and Cottingham developed two plates each night. They were not quite what he wanted:

> We took 16 photographs (of which 4 are not yet developed). They are all good pictures of the sun, showing a very remarkable prominence; but the cloud has interfered very much with the star-images. The first 10 photographs show practically no stars. The last 6 show a few images which I hope will give us what we need; but it is very disappointing. Everything shows that our arrangements were quite satisfactory, and with a little clearer weather we should have had splendid results.

Eddington then spent each day hunched over the photos with his micrometer making fine measurements. The effect he was looking for was "large as astronomical measures go" but still tiny by any ordinary consideration.

Even the measurements themselves were not enough to make a decision about the result of the Einstein test. The numbers had to be "reduced"—interference accounted for, optical effects eliminated, and so forth—before they became real data. Even with Eddington's

One of Eddington's photographs from the eclipse at Principe
ROYAL ASTRONOMICAL SOCIETY

legendary mathematical speed it still took him three days of feverish work. It was more complicated than he expected because the cloudy images forced him to use different methods from those planned. Explaining this to his mother, he wrote:

> . . . consequently I have not been able to make any preliminary announcement of the result. But the one good plate that I measured gave a result agreeing with Einstein and I think I have got a little confirmation from a second plate.

So at some point in the first week of June 1919, Eddington put down the pen he had been using for his calculations. Perhaps he rested his head in his hands. Three years after he'd received de Sitter's first letter, a year after he'd walked freely out of the Cambridge Tribunal, Eddington had his answer: "I knew that Einstein's theory had stood the test and the new outlook of scientific thought must prevail." He called this the greatest moment of his life.

Despite that solemnity, Eddington could not let the opportunity slip—he turned to Cottingham, recalling Dyson's warning, and quipped, "Cottingham, you won't have to go home alone." This moment was just a matter of Eddington persuading himself, though. A personal confirmation of faith. His preliminary calculations were not nearly enough to convince everyone back home. That would take months' more measurement and calculation (and he didn't even know what the plates from Brazil might look like). Not to mention making sure the results had the political and social impact he was hoping for. A great deal of work remained.

Eddington had hoped to stay on Principe to complete some of that work but his plan was disrupted by labor issues with the local steamship line. If he did not depart immediately he might be stranded for an unknown length of time. The governor of Principe commandeered space for him and Cottingham on the last ship leaving that summer (the SS *Zaire*). The boat was extremely crowded. Aggravating the situation was the fever that Eddington had been struck with. It got bad enough that he passed out at least once, though the sea air on the ship set him to rights. He wrote one last letter to his mother from the ship on June 21. He would likely arrive home before the letter did, hopefully in time to have the English strawberries that were, he said, better than anything he could get in the tropics.

EINSTEIN PROBABLY DID not have any strawberries that June, though they would have been appropriately celebratory: he and Elsa got married. On the second day of the month, they went to the Berlin registration office and made it official (despite the Swiss court's order that he wait). She had successfully pulled Einstein into a respectable middle-class life, much to the amazement of his friends.

Not much changed in their lives, since he had already been living in her apartment. The family took over the two attics above Elsa's flat and renovated them as workspace for Albert. That study, known as the "turret room," was distinguished by shelves of books and portraits of Newton and Faraday on the walls. No one was allowed to clean the study, save for light dusting. He sometimes received visitors there, but it was generally a place for solitary labor. He would often work for hours, come downstairs, strike a few chords on the piano, and then return to his study. In the weeks following the marriage he did not get much original work done (his illness was coming and going) but he was waiting anxiously for news from the eclipse. He had heard via the Dutch that Eddington had returned to Britain, and he hoped to hear results within six weeks or so.

Virtually no one else in Berlin was waiting for them, mostly because all of Germany was outraged by the recent news of the peace negotiations. They had been hoping that the final treaty would take Wilson's Fourteen Points seriously. Instead, the country would lose 13 percent of its land, 10 percent of its population, all its overseas colonies, and Alsace-Lorraine. The Rhineland would be occupied. Their army would be shrunk to the point where it was unclear if they could maintain order inside the country.

Most insulting of all was the clause that established Germany's sole guilt for the war. This became the justification for brutal monetary reparations. Prime Minister Lloyd George had won an election on the promise to squeeze the Germans: "We will search their pockets for it." John Maynard Keynes was part of the diplomatic team that arranged the terms, and he was immensely frustrated by the "empty and arid intrigue" going on. Disgusted, he resigned from the negotiations and went home to write his prescient *The Economic Consequences of the Peace* that warned of the long-term damage of a vengeful treaty. Total reparations were eventually set at 132 billion marks (about $330 billion in modern money).

Two weeks after the Einstein wedding, Germany was informed they had three days to approve the treaty, or else (they were eventually given another week). The unstable German coalition government was deeply divided over whether to sign the treaty, and there was a great deal of popular resistance. Berlin had an official week of mourning in response to the terms. In the end, though, they had little choice and the treaty was signed at Versailles on June 28. Only then, more than half a year after the end of the fighting, was the British blockade finally lifted.

Einstein was irate over the Allies' greed. "Just as well that we do not have to sell our brains or make an emergency sacrifice of them to the state." He didn't think all the provisions would be enforced, though. He was more worried about how the cruel terms had given a huge boost to right-wing politics in Berlin. The revolution's initial promise of an inclusive socialist democracy looked more and more empty. "Here the political wave has subsided. There is no energy left for grand emotions; rather, one is more or less passive. The populace perceives the end of the war as a liberation, the reappearance of vegetables as a relief." He continued to try to organize a commission to investigate wartime atrocities, with little support. Einstein had hoped that defeat would drive the Germans away from militarism. Now he worried that it would do just the opposite.

THE VICTORS, THOUGH, were generally pleased with how things were going. Along with the finalization of the peace treaty and the formation of the League of Nations came a rush of efforts to build a new international order: economic, military, and scientific. Part of the last was a group of American astronomers on its way to Brussels to help establish the new International Research Council (IRC). Led by W. W. Campbell of the Lick Observatory, they stopped in London en route to visit the Royal Astronomical Society. The RAS welcomed their visitors, arranging their first July meeting in sixty years. H. H. Turner, in the chair, explicitly greeted them as wartime allies: "We do not forget what we owe to our American friends for their help towards ending the war so successfully." He commented that he had last seen many of these astronomers at a scientific meeting in Germany just before the war. "We look back at that meeting with somewhat mixed feelings."

Campbell chose an interesting topic for his lecture—his observatory's attempts to measure the gravitational deflection of light at the eclipse of the previous year, June 8, 1918. Eddington was still on his way home from Principe, so Britain was eager for news on the Einstein problem. The Lick expedition had been led by Campbell's deputy, Heber Curtis. Their specialized eclipse telescopes were actually still in Russia, where they had hoped to observe the 1914 eclipse, when they were chased away by the start of the war. Without that equipment they had to borrow some lenses, and then Curtis was called to do war work. For some time. They did finally manage to observe the 1918 eclipse, though. Curtis measured the plates . . . and found no deflection.

Campbell was apologetic. He admitted that the telescopes used for the measurement were not the right type for this kind of work, and the star images were just too faint for much confidence in Curtis's result. Nonetheless, he did not think the plates could be reconciled with the full Einstein deflection; perhaps with the half-deflection.

With relativity's chief defender absent, Dyson was called to speak for the British expeditions. He had been reporting what little news there had been from Brazil and Africa. With no firm results yet, there was not much he could say beyond a mild statement that relativity was "an extremely difficult question to settle." He had received a letter from Eddington two days before in which he said he had "some evidence" for deflection, but nothing had been fully determined.

Campbell's closing message for the meeting was the American vision for the new international scientific community that the IRC embodied. He said the exclusion of the Germans and Austrians was a way of protesting the "over-development of militarism." If that meant breaking some scientific bonds, so be it: "It chances that certain astronomers may be criticized and cut off somewhat from others. We feel, however, that it is clearly our duty to be men first and astronomers later."

The Americans went directly from London to Brussels to inaugurate the new organization and its subsidiaries such as the International Astronomical Union. They saw themselves as inaugurating a new era for science of morality and responsibility, just as the founders of the League of Nations felt they were doing for politics. Of course, there was no chance that the recently enemy countries would be included in the new international scientific groups. The question was whether neutral countries would be allowed. Was the price of admission active engagement in the war, or just not being on the wrong side? In the end, thirteen countries that had been neutral during the war were allowed to join. A special rule was put in place to prevent them from using their numbers to vote the Germans in. Exactly one month after the signing of the Treaty of Versailles, the IRC statutes were set in stone for twelve years. Some prominent scientists, including Eddington's Dutch friend Jacobus Kapteyn, protested the German exclusion. Kapteyn circulated a public letter reminding the IRC, "Our grand governess, nature, mocks our petty hostilities."

EDDINGTON CAME HOME to this new world of science. "International" science was now officially defined as "everyone except Germany and Austria." But he had a trunk full of photographs intimately tied to a theory substantially developed in Berlin and had no intention of suspending his work on relativity. The next step was to turn those photographs into rigorous data. Scientific observations do not speak for themselves; they do not give up their secrets easily. Bringing the world around to his conclusion that Einstein was right would take months of tedious measurement and calculation.

Dyson and Eddington apparently decided to keep the expeditions separate even during the process of analyzing the data. Perhaps it was thought that independent measurements would be seen as more reliable. The Principe photographs would be analyzed in Cambridge, the Sobral ones in Greenwich. Eddington probably did the measurements and calculations for the former himself. Davidson worked with Royal Observatory staff on theirs.

The Brazil team had the slightly easier task. Because they were able to take check plates on-site, they could just directly compare them to the eclipse photographs. Since both were taken in the same place with the same telescope, they could just measure how far the image of a certain star appeared to move when the sun's gravity was present. This was not a matter of slapping down a ruler and lining up by eye, though. Small measurements were made with a complicated device called a micrometer that could assess much tinier distances than the human hand. These measurements required a great deal of training and patience, but were a standard part of an astronomer's toolkit.

Eddington needed an extra step. He had been unable to take check plates from Principe, so he could not make a direct measurement. He had to compare the image of the Hyades he took during the eclipse with the image of the Hyades taken in Oxford with the same telescope. But he had to account for the possibility that there was some subtle difference between Oxford and Principe that changed the image. So he had taken an image of a different star field in both locations, and by comparing those two photographs he could see what differences there were. Armed with that information, he could then account for that in his final measurements. It is very rare that a measurement in science has no interference or error. Rather, the trick is to understand those problems and correct for them.

The Principe observations produced sixteen photographs, though thanks to the cloud only seven had useful images of stars. Fortunately all seven had the two stars with the highest predicted deflection. A reliable measurement required five stars for cross-reference, though, and only two of the plates had that many. Those two were consistent, at least, and gave an average deflection of 1.61 arc-seconds, ±0.30. That uncertainty was not superb, but it was adequate. Einstein's predicted deflection was 1.75. For the first measurement of a completely unknown physical phenomenon, Eddington thought that was pretty good.

There were two sets of photos from Sobral, one for the astrographic telescope and one for the four-inch. The astrographic images had looked discouraging even in the field, apparently due to the heating of the coelostat mirror. It was clear to everyone involved that these would not give trustworthy information. Nonetheless, Davidson and Turner analyzed them thoroughly and they showed a deflection of 0.93 arc-seconds (the poor image quality made it difficult to estimate error). This was much closer to the Newtonian half-deflection than the full Einstein value.

Dyson was unsure how to handle this. The pattern of the image distortions suggested a systematic effect that could be accounted for mathematically. If they fixed the measurements with that systematic effect in mind, the numbers became 1.52 arc-seconds, much closer to the Einstein prediction. It was hard to be sure about the systematic effect, though. Dyson decided to leave the results uncorrected at 0.93.

The four-inch telescope, only brought along as a backup at the last moment, saved the day. Seven of the eight plates taken with it had excellent images of all seven hoped-for stars. Measuring those provided much better results than those from Principe: 1.98 arc-seconds, ±0.12. Significantly more precise, and still supporting Einstein.

By October, Eddington and Dyson had sent their results to each other by mail. Dyson sent the astrographic results first, causing Eddington a great deal of worry—they were simply not compatible with his own results. When the four-inch data arrived, he wrote back in relief: "I am glad the [4-inch] plates give the full deflection not only because of theory, but because I had been worrying over the Principe plates and could not see any possible way of reconciling them with the half-deflection." He was also pleased that the four-inch plates had enough stars to graph how the deflection changed with distance from the edge of the sun (stars farther away were deflected less). This more detailed, internal consistency of the data made it even more persuasive: "It seemed to me rather interesting and deals with a point that ought not to be overlooked and brings out the really remarkable agreement of individual stars at Sobral."

So they had three results: 1.61, 1.98, and 0.93. The standard thing to do would be to average the three (according to arcane rules for weighting some of the numbers). This gave a mean measurement of 1.64 arcseconds, arguably a solid confirmation for Einstein. As they began writing the report, though, Eddington had second thoughts:

> I do not like the combination of the astrographic with the other Sobral results—particularly because it makes the mean come so near the truth. I do not think it can be justified; the probable errors of both are I think below 0.1" so they are manifestly discordant. . . . It seems arbitrary to combine a result which definitely disagrees with a result which agrees and so obtain still better agreement.

It was true that this standard mathematical approach gave a number close to the prediction. But two of the numbers agreed, and one did not. Averaging them hid that. That might look suspicious, and it also gave an incorrect sense of the results. It would be like coming upon two whole pies and one half pie. On average, you still had something like four-fifths of a pie, and that's not bad. You might want to know, however, that two of the pies were pristine and one had already been half eaten. Sometimes quality is as important as quantity. Eddington and Dyson decided to present all three results separately, letting each stand on its own.

While Eddington and Dyson were furiously measuring and calculating, they somehow still made time to set the stage for the eventual presentation of the results. Dyson was interviewed by a *Times* reporter about the measuring process. When asked what the data revealed, he cagily responded, "They disclose something, but what it is I am not prepared to state yet. It is a very curious position—but only one thing among many other things."

About the same time, Eddington ventured to the annual meeting of the BA in Bournemouth. This was his first public appearance since he left for the expedition in March, and the audience was eager. He described the theoretical background and how the observations had actually fared in the field. He declined to give any results, saying that the calculations were ongoing. Oliver Lodge commented that he hoped the results would support the Newtonian deflection. Other scientists proposed alternative explanations for the deflection, should it be found.

Almost immediately after he and Dyson agreed on the results in late October, Eddington decided it was time for a test run. As a site he chose the ∇^2 V Club (pronounced "del-squared vee") in Cambridge, an informal organization of students and professors interested in physics and astronomy. This was the first meeting of the club in three and a half years, surely organized by Eddington for just this purpose. The president of the club was Ebenezer Cunningham, a fellow conscientious objector and relativity enthusiast. Fifteen other people were there, most of whom were highly skilled in both experimental and theoretical physics. So Eddington had chosen an audience that was extremely qualified to judge the results and still somewhat sympathetic. Perhaps most important was that it was a private meeting. If they shot everything down, it would not be a public disaster.

It seems his conclusion was accepted, if not without contest—conversation went until midnight. But it went well enough that Eddington was ready to go public. The day after the meeting Dyson asked the Royal Society Council to schedule a special meeting on November 6, at which the results would be formally presented. There was no turning back now.

WE DON'T KNOW who leaked the news. Balthasar van der Pol, a colleague of H. A. Lorentz in Leiden, was at the Bournemouth meeting where Eddington spoke about the expeditions. He reported back to Lorentz that the results were in favor of Einstein. But Eddington did *not* formally announce that at the meeting. Perhaps he spoke personally to van der Pol and whispered the news? This would not be a particular surprise, even if it bent some rules. He was concerned about making sure the news crossed the wartime borders. In the wake of the formation of the IRC, perhaps he wanted to take any victory for scientific internationalism that he could. It was still impossible to send a message directly to Berlin, so this was the next best thing.

Lorentz immediately sent a telegram on to Einstein, urgent and brief: "Eddington found stellar shift at solar limb, tentative value between nine-tenths of a second [of a degree] and twice that." Unfortunately we have no eyewitness account of Einstein first receiving the news. Fortunately, he then showed the telegram to anyone who came into his apartment, so we can see it through other eyes.

Ilse Rosenthal-Schneider, a young physics student, was sitting with Einstein at his desk going through a book full of objections to relativity. Einstein suddenly interrupted their reading to reach for a document on the windowsill. He coolly remarked, "This may interest you," and handed her Lorentz's telegram. There are a few different versions of the story from this point. The most complete goes:

> Full of enthusiasm, I exclaimed, "How wonderful! This is almost the value you calculated!" Quite unperturbed, he remarked, "I knew that the theory is correct. Did you doubt it?" I answered, "No, of course not. But what would you have said if there had been no confirmation like this?" He replied, "I would have to pity our dear God. The theory is correct all the same."

Around the same time, two friends were visiting while he was ill in bed. In his pajamas, with his socks showing, he presented the telegram, saying, "I knew I was right." In another version of the story he went to sleep even knowing that the results were on their way, confident in the outcome, while Planck waited up anxiously. This is a complete fabrication—he did not know the results were on their way.

Einstein's confident manner in these stories makes for great anecdotes, though it certainly does not mesh with his years of efforts trying to get precisely this confirmation. We can see more of his relief in a letter to his mother ("Today some happy news. H. A. Lorentz telegraphed me that the English expeditions have really verified the deflection of light by the Sun.") and another to Planck:

> I cannot postpone telling you . . . how deeply and how heartily pleased I was about the news contained in Lorentz's telegram. Thus the intimate union between the beautiful, the true, and the real has once again proved operative. You have already said many times that you personally never doubted the result; but it is beneficial, nonetheless, if now this fact is indubitably established for others as well.

He had been hoping for precisely this for years. And he was in no mood to be shy about spreading the word.

His friends in the Netherlands were overjoyed, particularly at their role in making it happen. De Sitter wrote to Einstein to "congratulate you heartily on the fine success of the English eclipse expeditions. The agreement is really *very* good, much better than I had expected, and the whole thing is very convincing." Lorentz was baffled that no British journal had published the results yet: "This is certainly one of the finest results that science has ever accomplished, and we may be very pleased about it." Ehrenfest hoped to get Einstein and Eddington both in Leiden at the same time so they could finally meet.

Einstein immediately sent a note reporting the results to Arnold Berliner at *Naturwissenschaften*, who published it on October 10. He also wrote to his friends in Switzerland. The physics group in Zurich sent Einstein a poem:

> All doubts have now been spent
> At last it has been found:
> Light naturally is bent
> To Einstein's great renown!

Einstein's responding verse was less mellifluous. One friend wrote, "So all is going better for you, even light has been bending a few million years to please you." He wondered if Einstein could get the stars to do any other tricks.

On October 23, Einstein was in the Netherlands for a physics colloquium. Ehrenfest had arranged for a visiting professorship so he would have an excuse to be there a few weeks each year (and to provide some hard currency—the German mark continued to plummet and Einstein was having trouble paying alimony). Eddington, busy with the final stages of data analysis, could not be there, but he did have time to write a letter to the astronomer Ejnar Hertzsprung reporting the state of the results. Given the timing, Eddington must have sent the letter right before his test presentation to the $\nabla^2 V$ Club—essentially as soon as he had results from both expeditions. Einstein was delighted to see the more detailed version of the news:

> This evening at the colloquium Hertzsprung showed me a letter by Eddington, according to which the precise measurement of the plates has furnished the exact theoretical value for the light deflection. It was gracious destiny that I was allowed to witness this.

As always, Einstein had a fabulous time with his Dutch friends, going on long walks and playing music in the evenings. He thanked Ehrenfest for "keeping this rickety trunk in good spirits" and for the gift of a thermos of soup to make the unheated train ride home bearable. He rhapsodized about his joy in spending two weeks deep in thinking about physics: "How hard this magnificent creation of ours seems to have to struggle just for that little bit of live awareness of its beauty! This beauty is so subtle and strange that one can't quite shake off religious notions." But the news from the eclipse dominated all his thoughts and he rushed to share it. Writing to his mother: "The result is now definite and signifies a *perfect verification of my theory*." And to Elsa: "My theory has been verified exactly with the greatest precision conceivable. Eddington reported it here. Now no rational person can doubt the validity of my theory anymore."

THAT WAS THE attitude Eddington was hoping to instill in his British colleagues. Dyson had requested that a joint meeting of the Royal Astronomical Society and the Royal Society take place November 6. JPEC regulations stipulated that such a meeting should be held as soon as practical after an expedition, though this one attracted attention like none before. On that Thursday the meeting was held at the Royal Society's rooms at Burlington House in Piccadilly. The audience was seated in pews, with an overflow crowd standing among the columns lining the sides. Alistair Sponsel estimates that between 100 and 150 people crammed into the room that day.

One of the attendees was Alfred North Whitehead, the distinguished philosopher-mathematician. He reported the excitement in the air:

> The whole atmosphere of tense interest was exactly like that of the Greek drama. . . . There was a dramatic quality in the very staging:– the traditional ceremonial, and in the background the picture of Newton to remind us that the greatest of scientific generalizations was now, after more than two centuries, to receive its first modification. Nor was the personal interest wanting: a great adventure in thought had at length come safe to shore.

Dyson was scheduled to present the overall results, Crommelin to speak for the Brazil observers, and Eddington for Principe. Dyson carefully described the methods used, perhaps trying to emphasize their caution, perhaps to heighten the tension, before announcing, "After a careful study of the plates I am prepared to say that there can be no doubt that they confirm Einstein's prediction. A very definite result has been obtained that light is deflected in accordance with Einstein's law of gravitation."

There was no doubt a stir in the room as Dyson then sat to allow Crommelin to describe the Sobral results in detail. He was effusive in thanking the various locals who supported the work there. He explained the problems with the results from the astrographic and how they concluded that the coelostat mirror was the source. He presented those results even though everyone involved accepted that they were not reliable and should not be given much credence. The excellent data from the four-inch telescope took up most of his time.

Eddington took the final part of the presentation so he could both describe the Principe expedition and place all the results in the wider scientific context of relativity. Eddington emphasized that, as Dyson said, the deflection was real and it was close to Einstein's prediction. He acknowledged an important gap, though—the other remaining test of general relativity, the gravitational redshift, had still not been observed. This meant one could say that Einstein's *law* (that is, the mathematical formulas) had been proved, but the *theory* itself (that is, the ideas from which the formulas came) could still be questioned. This gave him some wiggle room in which he was able to say "Einstein is right" without having to grapple with thorny issues such as the nature of space-time or the curvature of the universe.

The chair of the meeting, J. J. Thomson (Nobel Prize winner for the discovery of the electron), then took over. He said that the deflection was not an isolated fact, rather it was "part of a whole continent of scientific ideas affecting the most fundamental concepts of physics." His commentary was extraordinary:

> This is the most important result obtained in connection with the theory of gravitation since Newton's day, and it is fitting that it should be announced at a meeting of the Society so closely connected with him. . . . If it is sustained that Einstein's reasoning holds good—and it has survived two very severe tests in connection with the perihelion of Mercury and the present eclipse—then it is the result of one of the highest achievements in human thought.

Thomson had been no Einstein devotee like Eddington, and he had almost certainly had the opportunity to inspect the results closely before the meeting. For him to place Einstein on the level of Newton, to call relativity one of the highest achievements in human thought, was an astonishing embrace of a theory that just a year before had been barricaded behind battleships and barbed wire. He did, however, close his statement with a complaint about the difficult mathematics Einstein had used.

Alfred Fowler, president of the RAS, was then given the floor. Fowler was not quite as impressed as Thomson. With a slightly backhanded compliment he suggested that perhaps this should not be the final word: "The conclusion is so important that no effort should be spared in seeking confirmation in other ways."

Thomson opened the floor to questions and comments from the audience, and scientists leaped to their feet. Interestingly, everyone there seemed to accept the accuracy of the photographs and the reality of the deflection. No one was bothered by the discarding of the Sobral astrographic results, even though they seemed close to the Newtonian prediction. The astronomers and physicists gathered there trusted that the observers in the field had made the right call. No one knew better than they whether a measurement was reliable.

Instead, the objections were about the nature of relativity itself, and whether the deflection should be considered evidence for it. F. A. Lindemann, who had tried to measure the deflection with daytime photography of stars, followed up on Thomson's complaint about the mathematics. He was sure it was "elegant" in some sense, though he could not believe "that a profound physical truth cannot be clothed in simpler language." He asked if Eddington could please translate the theory into such a form. H. F. Newall acknowledged that deflection had been seen but suggested that it was in fact due to refraction from the solar corona. Dyson replied that the location of the corona made that unlikely; Eddington provided detailed calculations refuting that interpretation too.

Ludwik Silberstein, who saw himself as an expert on relativity, objected that the deflection should not be connected to the wider ideas of relativity, and that Einstein's theory had not been confirmed. This was surely the objection Eddington had anticipated with his law/theory distinction (and, indeed, Silberstein would go on to propose alternatives to general relativity for decades to come). Silberstein insisted that it was "unscientific" to say that the deflection confirmed relativity without the gravitational redshift as well:

> The discovery made at the eclipse expedition, beautiful though it is, does not, in these circumstances, prove Einstein's theory. We owe it to that great man [pointing to Newton's portrait] to proceed very carefully in modifying or retouching his Law of Gravitation; this is by no means defending blind conservatism.

It was noted that Oliver Lodge, the great defender of Newton, left the meeting halfway through without posing any questions. This departure was read as perhaps an insult or a de facto rejection of the results; Lodge reassured everyone that he simply had to catch a train.

As the discussion settled and the meeting began to disperse, Silberstein came up to Eddington and reportedly quipped, "Professor Eddington, you must be one of three persons in the world [meaning Einstein, Eddington, and Silberstein himself] who understands general relativity." As the story goes, Eddington demurred, whereupon Silberstein responded, "Don't be so modest, Eddington," and Eddington replied, "On the contrary, I am trying to think who the third person is."

THE NEXT DAY, the *Times* of London presented the greatest scientific headline in history: REVOLUTION IN SCIENCE. Sharing the page was a reminder of the first Armistice Day observance (THE GLORIOUS DEAD). The proclamation of revolution headed a breathless article describing the joint meeting, reporting that "the greatest possible interest had been aroused in scientific circles." The discovery was attributed to "the famous physician Einstein" (he was neither).

This article was the culmination of nearly a year of Dyson and Eddington's careful tending of the press, preparing the newspapers and their readers for this moment of scientific drama. It was written by Peter Chalmers Mitchell, a scientist turned journalist interested in physics and sympathetic to German science (he had studied in Leipzig and Berlin). We do not know if he and Eddington had met, though they were both Fellows of the Royal Society at the time. He could hardly have asked for a better spark to light public interest in relativity.

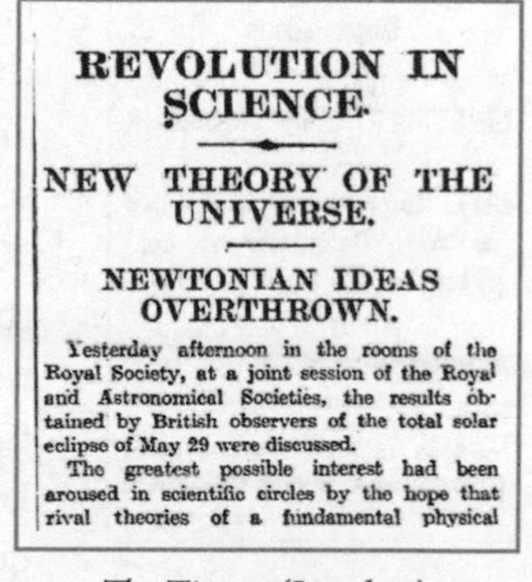
REVOLUTION IN SCIENCE.

NEW THEORY OF THE UNIVERSE.

NEWTONIAN IDEAS OVERTHROWN.

Yesterday afternoon in the rooms of the Royal Society, at a joint session of the Royal and Astronomical Societies, the results obtained by British observers of the total solar eclipse of May 29 were discussed.

The greatest possible interest had been aroused in scientific circles by the hope that rival theories of a fundamental physical

The Times *(London), November 7, 1919*
COURTESY OF THE AUTHOR

On Saturday there was a follow-up article with the same title, with the addition "EINSTEIN V. NEWTON." This was the general public's first introduction to Einstein, and he appeared exactly as Eddington wanted to present him: a peaceful genius who repudiated all the wartime stereotypes of the militaristic German. He was described as a Swiss Jew who had only taken a position in Berlin due to the large salary. Further, "During the war, as a man of liberal tendencies, he was one of the signatories to the protest against the German manifesto of the men of science who declared themselves in favour of Germany's part in the war."

The excitement jumped the Atlantic and on November 10, 1919, the *New York Times* blared, LIGHTS ALL ASKEW IN THE HEAVENS, along with "Men of science more or less agog over results of eclipse observations" (no word on how to determine one's level of agogness). This article leaned heavily on the ones from London. The writer knew nothing about relativity and felt free to add embellishments such as "A book for 12 wise men—no more in all the world could comprehend it, said Einstein when his daring publishers accepted it." Of course, Einstein said no such thing, although one still occasionally hears today that only twelve people "really" understand relativity. It is important to look back and remember that this was virtually the first mention of Einstein in the *New York Times*—he was a person of little consequence until this moment.

LIGHTS ALL ASKEW IN THE HEAVENS

Men of Science More or Less Agog Over Results of Eclipse Observations.

EINSTEIN THEORY TRIUMPHS

Stars Not Where They Seemed or Were Calculated to be, but Nobody Need Worry.

A BOOK FOR 12 WISE MEN

No More in All the World Could Comprehend It, Said Einstein When His Daring Publishers Accepted It.

The New York Times, *November 10, 1919*
COURTESY OF THE AUTHOR

In England, the story of Newton's dethroning spread rapidly. Relativity was reported to be "a lively topic of conversation in the House of Commons." Joseph Larmor said he had been "besieged by inquiries as to whether Newton had been cast down and Cambridge 'done in.'" There was a genuine sense of wounded national pride that heightened attention. As usual, it was the satirists who were the real judges of what people were interested in. *Punch* gave this lovely quatrain:

> A patriot fiddler-composer of Luton
> Wrote a funeral march which he played with the mute on,
> To record, as he said, that a Jewish-Swiss-Teuton
> Had partially scrapped the *Principia* of NEWTON.

The *Daily Mail* headline, LIGHT CAUGHT BENDING, placed a salacious spin on the results.

Scientists were stunned at the broad public interest in Einstein and relativity. At least some of the credit for this goes to Eddington's tireless efforts to promote exactly that interest. He held a public lecture in Cambridge where "hundreds were turned away unable to get near the room." He gave interviews and wrote articles. And every time, it was a story of a scientific revolution made possible by scientific internationalism. From one piece: "The theoretical researches of Prof. Albert Einstein, of Berlin, now so strikingly confirmed by the British eclipse expeditions, involve a broadening of our views of external nature, comparable with, or perhaps, exceeding the advances associated with Copernicus, Newton and Darwin."

These were often adventure stories, too, with Eddington and his compatriots venturing to the edges of the Earth in the cause of science. Someone offered some new verses for "The Astronomer's Drinking Song," punning on the meaning of "Einstein" in German ("one stone"):

> *We cheered the Eclipse Observers' start.*
> *We welcomed them returned, Sir;*
> *Right gallantly they played their part,*
> *And much from them we've learned, Sir.*
> *No pains nor toil they thought too great,*
> *Nor left ein stein unturned, Sir,*
> *Right heartily we asseverate*
> *Their bottle a day they've earned, Sir.*

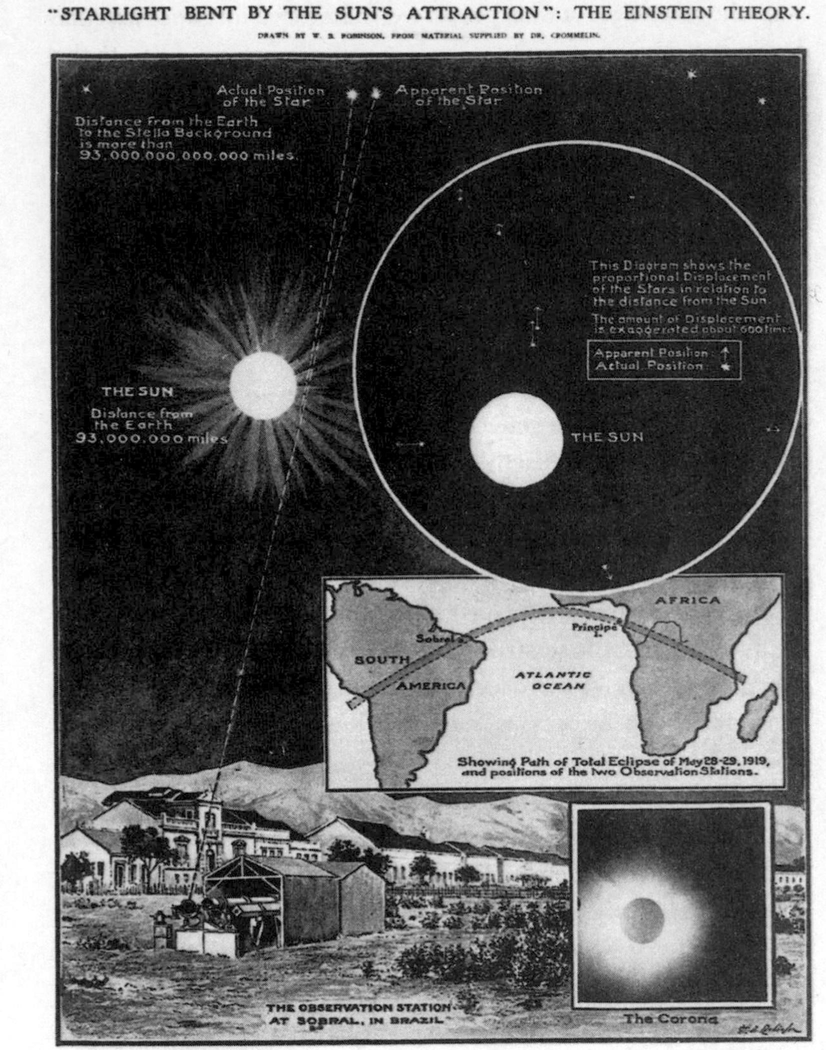

"STARLIGHT BENT BY THE SUN'S ATTRACTION": THE EINSTEIN THEORY.

DRAWN BY W. B. ROBINSON, FROM MATERIAL SUPPLIED BY DR. CROMMELIN.

THE CURVATURE OF LIGHT: EVIDENCE FROM BRITISH OBSERVERS' PHOTOGRAPHS AT THE ECLIPSE OF THE SUN.

The results obtained by the British expeditions to observe the total eclipse of the sun last May verified Professor Einstein's theory that light is subject to gravitation. Writing in our issue of November 15, Dr. A. C. Crommelin, one of the British observers, said: "The eclipse was specially favourable for the purpose, there being no fewer than twelve fairly bright stars near the limb of the sun. The process of observation consisted in taking photographs of these stars during totality, and comparing them with other plates of the same region taken when the sun was not in the neighbourhood. Then if the starlight is bent by the sun's attraction, the stars on the eclipse plates would seem to be pushed outward compared with those on the other plates. . . . The second Sobral camera and the one used at Principe agree in supporting Einstein's theory. . . . It is of profound philosophical interest. Straight lines in Einstein's space cannot exist; they are parts of gigantic curves." (Drawing Copyrighted in the United States and Canada.)

The Illustrated London News, *November 22, 1919*

COURTESY OF THE AUTHOR

There was a long tradition of newspapers reporting on eclipse expeditions with this sort of "astronomer-adventurer" frame. Now they could add wartime drama—science heals international rifts—to that as well.

People in Britain flocked to Team Einstein, including those who had been rabidly anti-German during the war. Even H. H. Turner, who had called for the utter abandonment of German science, took the first Armistice Day anniversary as an opportunity to meditate on the extraordinary victory of relativity: "Who could have anticipated, at the date of the actual Armistice, such a wonderful outcome? . . . Has there ever been so unexpectedly and completely successful an Eclipse Expedition? . . . A positive answer to a definite question is surely something new?" He hoped someone had placed a "modest wager" on the outcome. He did, however, call for many future tests of "The Great Result."

Turner also noticed an important theme of the press coverage and popular interest: the incomprehensibility of relativity. "The vain attempts of the reporters to apprehend exactly in what the revolution consists have been amusing, and would have been more so but for our own similar difficulties." He recounted a story of uncertain origin and accuracy but of great entertainment value. A newspaper reporter allegedly went to the Royal Society and asked for an explanation of relativity: "The Secretary rubbed a hand over a dome-like brow, and frankly admitted he was beaten. The theory is down in black and white, with plenty of $x=0$, but compared with it the Rosetta stone in the British Museum is a child's rag alphabet. . . . A distinguished scientist was next seen. 'I don't understand it at all,' he said, wearily. 'Don't mention my name.'" The reporter then went to the library, read through the theory three times, "and was led out sobbing."

Fictitious stories like these helped cement the reputation of relativity as being beyond understanding. It wasn't (and isn't), but it looked scary, dealt with traditionally overwhelming philosophical ideas like time and space, and was distinctly foreign. We can see all of those in the tale of Turner's reporter, who closed with this: "It is said that Professor Eddington, of Cambridge, claims to understand the theory, so, until he consents to put it in schoolroom prose—*Gott strafe Einstein.*"

The final words were an inversion of the German wartime slogan "*Gott strafe England*" (May God punish England), a not-so-gentle reminder that Einstein was, until recently, the enemy.

THIS EXPLOSION OF interest around relativity finally made possible an important milestone. For the first time, Eddington and Einstein were able to write directly to each other. On December 1, Eddington sent his distant colleague a letter. He expressed all his hopes for what relativity and the expeditions might achieve:

> All England has been talking about your theory. . . . There is no mistaking the genuine enthusiasm in scientific circles and perhaps more particularly in this University. It is the best possible thing that could have happened for scientific relations between England and Germany. I do not anticipate rapid progress toward official reunion, but there is a big advance toward a more reasonable frame of mind among scientific men, and that is even more important than the renewal of formal associations.

It was the personal relations among scientists that Eddington cared about. Changing organizations like the IRC would take a long time. His fellow Quakers doing relief work in Europe used the same strategy; they knew that nations would not be disarming anytime soon. Instead, Eddington and the Friends sought to build the personal relationships that humanized the enemy and laid the groundwork for peace. Eddington was particularly concerned to reassure Einstein that there was more than hatred in Britain:

> I have been kept very busy lecturing and writing on your theory. My *Report on Relativity* is sold out and is being reprinted. That shows the zeal for knowledge on the subject; because it is not an easy book to tackle. I had a huge audience at the Cambridge Philosophical Society a few days ago, and hundreds were turned away unable to

> get near the room. . . . One feels that things have turned out very fortunately in giving this object-lesson of the solidarity of German and British science even in time of war.

There was some fortune, surely—any project that succeeded despite weather and war needed some. More important, though, was deliberate action. The eclipse expedition became a symbol of German-British solidarity because Eddington chose to craft it that way. Einstein chose to fight against militarism in German science and made that symbolism possible. It is not an accident, not a fluke, that things turned out this way. This was a great moment for science across the gulf of war because certain scientists turned it into one.

CHAPTER 12

The Relativity Circus

"Virtually nothing but Einstein is being talked about here."

ALL THE FUSS was about one number, or maybe two—1.61 or 1.98—and how close they were to Einstein's prediction of 1.75. Years of work and planning, months in the field, weeks of measuring and calculating, all get reduced to those naked numbers. It takes an enormous amount of labor to produce a single scientific fact like "light is bent by gravity." And all that labor tends to be erased afterward, to make it seem as though nature speaks clearly and easily.

We can see how difficult it was to produce that fact by looking at another number: Eddington's annual bicycle rides. His cycling dropped off precipitously in October 1918, just as preparations for the expeditions were picking up steam. In 1918 he rode 2,028 miles; in 1919 just 124 miles. Relativity had taken over his life. That is the kind of effort and commitment it takes to make a fact. Eddington was always a little frustrated by this. His areas of deep scientific interest were the composition and movements of stars, not space and time: "People seem to forget that I am an astronomer, and that relativity is only a side issue." What made the investment in relativity worth it for him was not just the science but the political and social significance of the theory.

It was that significance that consumed Eddington's time even after the presentation of the results. He had to spend some effort defending his new fact's scientific meaning, but he also had to ensure that people *cared* about relativity and Einstein, that they paid attention to the lessons for internationalism and the nature of scientific cooperation. It was those efforts that would change Einstein's life forever. His transformation into an icon of genius was inextricably tied to his emergence as a scientific saint from the chaos of postwar Europe.

At the end of 1919 many scientific societies in Britain set up special meetings to discuss the new revolutionary science. In December the Royal Astronomical Society managed to get both Eddington and Dyson to speak at theirs. Eddington explained relativity and Dyson explained the measurements. Things were significantly different from the November 6 meeting. This time, most who wanted to had been able to inspect the eclipse photographs directly, not just trust the reports. Some people had contacted Dyson with concern about whether the changes in the star images had been too small to measure. He patiently explained that, yes, the changes were very small "but those who are familiar with the measurement of astronomical photographs will know that it is quite possible to measure quantities of this order of magnitude." Translation: astronomers do this every day; trust us. The individual star images were about 4 arc-seconds across; the change was about 1 arc-second. Could they really measure a change smaller than the actual image? Certainly—if you were six feet tall, you would definitely notice if you moved eighteen inches.

Oliver Lodge was there, attending his first-ever RAS meeting. He discussed his usual conceptual objections to relativity and then asked if the results could be rechecked with the gravity from Jupiter (no). Silberstein said the results could only be seen as supporting relativity if one was already biased toward Einstein (he got little agreement). Scientists sent letters so they would be part of the conversation even if they were not present. Alfred Fowler, running the meeting, provided a hint of how British scientists would come to change the way they remembered the war: "We may also find satisfaction in the knowledge that national prejudice was not allowed to interfere with any contribution that we could make to the progress of science." Just months after barring the Germans from the IRC, they were already beginning to celebrate their support of internationalism. Everyone tends to rewrite history—intentionally or not—to make themselves look better; scientists' history about themselves is no different.

There was a frenzy in the scientific community as those interested and able checked the eclipse plates for themselves. Even skeptics began grudgingly coming around. Charles St. John, while accepting the existence of the deflection, hinted that a future theory might explain the observation better than Einstein's. A few people unhappy with relativity, such as Joseph Larmor, made similar suggestions. Most everyone wanted the test redone at the 1922 eclipse, just for thoroughness (and it was). Within a year or so, though, H. H. Turner concluded that the debate over relativity within the scientific community had essentially ceased, with opinion firmly behind Einstein.

Having the evidence for the theory accepted by specialists was crucial but not sufficient for it to triumph. If it was to survive and become part of the scientific canon, it needed to be spread to the next generation of scientists as well—it needed to be taught in formal classes, so science students could make it their own and use it in their own work. Given the huge interest, professors around Britain tried to teach it, though it was often "the blind leading the blind."

The decisive moment was Eddington's first course on relativity at Cambridge, originally planned to be offered in October 1919. The class actually began several weeks later, probably pushed back due to him being too busy with the eclipse data reductions. He lectured twice a week at the famous Cavendish Laboratory, scheduled late in the day so more people could attend. He gave another, more technical, course on relativity in the spring. His lecture notes from that became the basis of one of the first textbooks on relativity, which continued to be used for a generation. The historian Andrew Warwick points out that Cambridge became a center for research in relativity in a way that Leiden did not (even though there were plenty of experts there) because

Eddington made deliberate efforts to pass it on to his students and colleagues.

The students were intensely excited to be learning the theory that seemed to be changing the world. As one student described it: "The thrill of seeing physical science on the march in a new direction, the sense of something stirring, of new adventure, held us tensely expectant even though we might but half comprehend it." We actually have a firsthand account of Eddington's teaching in that class:

> A slight man of average height, in academic gown, reserved almost to the point of shyness, he rarely looks at his class. His keen eyes look at or through the side wall as he half turns from the blackboard and seems to think aloud the significance of the tensors which he has just written on the board. The mathematical theory of relativity is developed *ab initio* before our eyes and the symbols are made to live and take on meaning.

This is in sharp contrast with descriptions of Eddington's public lectures, which were "enriched with literary quality and with his inimitable humour" and where the audience was said to be "breathless." Alice and the denizens of Wonderland appeared often, as did Gulliver, Humpty Dumpty, and the Jabberwock. He also found room for more highbrow references from Shakespeare, Milton, and Chaucer. He rewrote parts of the *Rubáiyát* to describe the Principe expedition:

> *Oh leave the Wise our measures to collate*
> *One thing at least is certain, LIGHT has WEIGHT,*
> *One thing is certain, and the rest debate—*
> *Light-rays, when near the Sun, DO NOT GO STRAIGHT.*

He would address just about any group that asked, even undergraduate clubs (one of which introduced him as a professor of astrology). Many more people attended these lectures than his formal classes, and he earned a reputation as "an expounder to the multitude of the poetry of modern science."

Eddington's public lectures helped spark the broader conversation, which quickly took off in its own directions. A common theme was the sense that relativity had certain political connotations—people took the "revolution in science" headlines quite literally. In 1919, so soon after the Communists came to power in Russia, "revolution" evoked the specter of political overthrow. People explicitly worried about "scientific Bolshevism," and that relativity was paving the way for a great disruption. Literature professor Katy Price points out in her book about this particular cultural moment that even with the war over, this was still a time of serious uncertainty in Britain: rationing was ongoing, there were railway strikes, returning soldiers had a difficult time finding jobs. Einstein's overthrow of Newton seemed to be just one more part of the chaos.

The *New York Times* asked Charles Poor, professor of celestial mechanics at Columbia University, about this. He made the connections quite clearly:

> For some years past, the entire world has been in a state of unrest, mental as well as physical. It may well be that the physical aspects of the unrest, the war, the strikes, the Bolshevist uprisings, are in reality the visible objects of some underlying deep mental disturbance, worldwide in character. . . . This same spirit of unrest has invaded science.

"Einstein" became a shorthand for any kind of broken order or unsettled hierarchy. This was often played for laughs—students not doing their homework, employers refusing to give raises. *Punch* presented "Einsteinized," a piece in which a commuter, his mind "alive with science," becomes unable to navigate a train station due to an overdose of abstract theory. One editorial warned that "the rays of logic emanating from the Mayor's office are bent as badly as Einstein's rays."

Another political aspect was the sense that relativity was antidemocratic. Its complicated mathematics meant it could only be understood by a small group; therefore, it was elitist. A "little knot of experts" had overthrown common sense. This was particularly a problem in the

United States, where newspapers complained about having to trust a theory that laymen could not understand. It seemed anti-American. Perhaps the next bit of wisdom to be overthrown would be the multiplication table?

These worries often combined with continuing concerns about the mysterious, incomprehensible aspects of the theory. Merely the name "relativity" evoked philosophical uncertainty. Einstein was described as "a destroyer of time and space." Talk of a fourth dimension made many people think of the occult. This became an indelible part of the way people thought and talked about both relativity and Einstein himself. Many years later he was at a film premiere with Charlie Chaplin. As the crowds cheered, Chaplin reportedly said, "They cheer me because they all understand me, and they cheer you because no one understands you."

The final political flavor of relativity was the one that Eddington was most concerned with: Einstein as a vindication of internationalism and liberal politics. One British scientist wrote to a German friend, hoping to invite Einstein for a visit:

> Virtually nothing but Einstein is being talked about here, and if he were to come over now, I think he would be celebrated like a victorious general. The fact that a German's theory was confirmed by observations by the English has, as is daily becoming more evident, brought the chance of collaboration between these scientific nations far closer. Thereby Einstein has done an inestimable service to mankind, leaving quite aside the high scientific value of his ingenious theory.

The metaphor of Einstein as victorious general was, while vivid, perhaps somewhat poorly chosen in the wake of the war. Another correspondent with Einstein, trying to describe the "unusual interest" in relativity, explained that in the newspapers, "You are presented as Polish & as Swiss, etc., but especially as one who did *not* sign the ill-fated letter [the Manifesto of 93]. . . . Prof. Eddington was very particularly occupied with your theory and is a kindly man without the modern prejudice that sadly is very strongly developed in some others."

Einstein himself contributed directly to this. Three weeks after the expedition results were announced he wrote a short piece for the *Times*—his first writing for the English-speaking world.

> After the lamentable breach in the former international relations existing among men of science . . . it was in accordance with the high and proud tradition of English science that English scientific men should have given their time and labour . . . to test a theory that had been completed and published in the country of their enemies in the midst of war.

The public seized on this essay like a declaration from an oracle. Everyone wanted to hear more from this genius. This created a cottage industry for anyone willing and able to translate Einstein's work into English. In a bizarre twist, the best-qualified people for this were Anglophone scientists who had been interned in Germany and Austria during the war. Henry Brose, an Australian physicist, had been held in a camp near Berlin. While there, he learned about general relativity and actually started some translation before he was freed. He went back to Oxford after the war, where his translations were put on the road to publication by no less than H. H. Turner.

Robert W. Lawson from the University of Sheffield had been interned in Vienna, where, amazingly, he was allowed to continue doing physics experiments. He wrote directly to Einstein asking for original material he could translate, particularly some aimed at scientists. He wanted to "work toward healing the deep wounds inflicted on the hearts of mankind by this war as quickly as possible and . . . [this would] bring us one big step closer to a welcome collaboration among scientists of different nationalities." His letter closed: "May I in closing congratulate you regarding the recent verification of your prediction related to the 'gravitational deflection' of light? People here have been talking of nothing else for the past few weeks." A British publisher even contacted Willem de Sitter asking for a translation. He declined and recommended Ebenezer Cunningham instead.

The language barrier was a real obstacle, even for correspondence between Eddington and Einstein themselves. Eddington's letters were in English (he apologized for being "unable to write except in my own language"), Einstein's in German (he, too, apologized, saying that "goodwill alone" was insufficient to overcome his linguistic inability). Even after their shared victory, communication was still difficult.

Einstein's fame immediately eclipsed Eddington's, though their names continued to be connected. One physicist rewrote Lewis Carroll's "The Walrus and the Carpenter" into "The Einstein and the Eddington" in honor of their collaboration:

> The time has come, said Eddington,
> To talk of many things;
> Of cubes and clocks and meter-sticks,
> And why a pendulum swings,
> And how far space is out of plumb,
> And whether time has wings. . . .
> You hold that time is badly warped,
> That even light is bent;
> I think I get the idea there,
> If this is what you meant:
> The mail the postman brings today,
> Tomorrow will be sent. . . .
> The shortest line, Einstein replied,
> Is not the one that's straight;
> It curves around upon itself,
> Much like a figure eight,
> And if you go too rapidly
> You will arrive too late.

THE PUBLIC CONVERSATION in Europe went somewhat differently. Much of the reaction in Britain and America came from the suddenness of Einstein's appearance—this essentially unknown figure from an enemy country mysteriously enters their scene by dethroning Isaac

Newton. Everyone wanted to know who he was, where he had come from, and how he had done it.

In Germany he wasn't unknown. He was a well-regarded professor and member of the Prussian Academy (though that was a far cry from being famous). There had been a trickle of newspaper coverage about relativity and the expeditions based on the little information Einstein had. So instead of the sudden avalanche of coverage in British publications, German newspapers took a couple weeks after the announcement of the results to gradually increase their reporting. The driving factor was less the obvious significance of the science and more the realization of the reaction overseas.

It was not until December 14 that the Einstein phenomenon truly broke out across Germany. That was when the *Berliner Illustrirte Zeitung* published a front-page article on Einstein with a dramatic portrait of him looking profound. It was as celebratory as anything that had appeared in Britain, calling him "A New Giant of World History: Albert Einstein, whose research signifies a complete overturning of our view of nature comparable to the insights of Copernicus, Kepler, and Newton." Inside: "A new epoch in human history has now arisen and it is indissolubly bound with the name of Albert Einstein." For hundreds of thousands of people, this was the first time they had heard his name or seen his face.

THE DUTCH NEWSPAPERS were full of translations of the British articles. Ehrenfest narrated it this way: "'Einstein versus Newton! Natural Philosophy Revolutionized,' etc.,—and the startled newspaper ducks flutter up in a hefty bout of quacking. . . . Even Galinka [his daughter] has been swept up by this flurry and quickly laid an artistic egg." He enclosed Galinka's drawing with commentary: in Africa an astronomer is looking at the stars and sun because Einstein has calculated in his house (he is shown throwing the calculations out the window). All the world is excited (even the squirrels), as shown by people running to his house. Einstein responded with delight: "Galinka's little picture is inimitably prettier than all that cackling by the startled flock of newspaper geese."

Nr. 50

28. Jahrgang

Berliner

Illustrirte Zeitung

25 Pfg.

Verlag Ullstein & Co, Berlin SW 68

Eine neue Größe der Weltgeschichte: Albert Einstein,

Berliner Illustrirte Zeitung, *December 14, 1919*

COURTESY OF THE AUTHOR

Good wishes and congratulations flooded in. A fellow member of the Prussian Academy, Carl Stumpf, was effusive: "With all our hearts, we share the elation which must fill you and are proud of the fact that, after the military-political collapse, German science has been able to score such a victory." Einstein was grateful, though not particularly happy at having his work described as a victory for German science. He was pleased, though, to have a huge government grant appear from nowhere to support research in general relativity. The German government was amazed at the sudden international goodwill Einstein was generating and wanted to support it in any way they could. Einstein used his new leverage with the authorities to get a job for Freundlich and to ask permission for an extra room in their building to house his ailing mother. He wrote to Besso about his surprise to find that his name was "in high favor since the English solar eclipse expeditions."

Across Europe, physicists were being hounded for lectures and popular articles about relativity. Both Lorentz and Ehrenfest agreed to produce some. The latter worried himself sick about doing a good job, perhaps with reason—Einstein was quite unhappy with an article Freundlich wrote. Max Born agreed to give a series of lectures on relativity for the stunning sum of 6,000 marks, which was enough to refit all the laboratories at his institute. Arnold Sommerfeld's lectures on it attracted more than a thousand listeners. He commented years later about the "general, somewhat sensational and epidemical interest in the theory of relativity."

AN EPIDEMIC WAS a good metaphor for the spread of relativity fever. It was contagious and grew rapidly. Einstein unexpectedly found himself at the center of nonstop media attention that he would come to call the relativity circus. One of the first signs of what was to come was when a rather confused reporter from the *New York Times* appeared at his apartment wanting to interview the great man. The correspondent knew little about science, only that Einstein was extraordinary in some way. Einstein tried to explain his theory, though the young man apparently thought the "freely falling person" thought experiment involved someone actually toppling off a roof.

Many, many representatives of the press followed in his footsteps. Einstein began complaining about the "riffraff" that had started hounding him. He was baffled by the attention. He didn't think relativity was particularly revolutionary. He was no Copernicus. Relativity, he thought, need not disrupt anyone's beliefs. It "harmonizes with every possible outlook of philosophy and does not interfere with being an idealist or materialist, pragmatist or whatever else one likes." Why would anyone be so concerned?

He knew that it was, at least in part, Eddington's fault: "Due to the newspaper clamor about the solar eclipse, people are pestering me very much. Everyone wants an article, a talk, a photograph, etc.; this business reminds one of 'The Emperor's New Clothes.'" It is unclear how Einstein himself fit into this metaphor—was he the emperor, being celebrated for no good reason? Or was he the unafraid child, the only one willing to point out the absurdity of the situation?

Everywhere he went, the reporters lay in wait. "I'm giving a children's lecture on relativity = appearance of the newspaper lions." It got worse and worse until he complained to Max Born that the publicity was "so bad that I can hardly breathe." Answering the mail was Sisyphean. When he had been traveling for a couple days, Elsa sent him a note warning of the "laundry-basket" of mail that had already arrived.

Early in his fame Einstein sat for an etched portrait by the artist Hermann Struck. Copies flew off the shelves and everyone wanted it autographed by the genius himself. Elsa griped, "Half the world is now buying the Struck picture and is sending it to you so that you can immortalize yourself on it." The volume only increased with time. Years later, when asked about his dog Chico, Einstein replied, "The dog is very smart. He feels sorry for me because I receive so much mail; that's why he tries to bite the mailman."

Einstein's trademark reaction to this sort of frustration was humor. Early on in the circus, he wrote:

> By an application of the theory of relativity to the taste of readers, today in Germany I am called a German man of science, and in England I am represented as a Swiss Jew. If I come to be represented as a bête noire, the descriptions will be reversed and I shall become a Swiss Jew for the Germans and a German man of science for the English.

The press attention seemed to border on worship. "Since the light deflection result became public, such a cult has been made out of me that I feel like a pagan idol. But this, too, God willing, will pass."

He was very, very wrong. His fame has outlived him. Science writer Thomas Levenson has pointed out that Einstein was one of the first modern celebrities. He became famous at just the right time to be a global figure, present on any newspaper around the world. He was also willing to participate, to craft a public persona that the media loved (in contrast, Marie Curie pushed back against her fame). The downside was that the papers recorded his every syllable. He learned this the hard way after the backlash to his off-the-cuff observation in New York that American women led their men around like poodles on a leash.

Eventually accepting his fame, Einstein was affable with reporters and gave answers to whatever inane questions they offered. When asked, he gave his opinion on literature, Prohibition, or just about any subject whether he knew anything about it or not. Along with witty quips, he had a talent for oracular utterances that perfectly balanced obscurity with a promise of insight. Being photogenic helped too—the unkempt hair, the soulful eyes with a glint of humor, the bad posture all made for a great picture. One paper described him as looking "like an artist—a musician. He was. But underneath his shaggy locks was a scientific mind whose deductions have staggered the ablest intellects of Europe." He didn't mind facilitating a good photo—he learned to toss his hat in the air so photographers would have something interesting to shoot.

There was certainly a sense in which he enjoyed the fame (particularly increased attention from women), though he was uncomfortable with the hero worship. A lifetime antiauthoritarian, he found it "distasteful" when a few individuals were credited with "superhuman powers of intellect and character." He was pleased, though, to see that the public made a hero of someone "whose goals lie exclusively in the spiritual and moral domain" (meaning himself) instead of that of money or power. He was also happy to leverage his reputation to support his causes of internationalism and pacifism. His friends knew the fame was exhausting: Ehrenfest tried to lure him to visit by promising, "Here there is nothing but people who are fond of *you* and not just of your cerebral cortex."

SCIENTIFIC FAME AT the end of 1919 was a strange thing. Einstein, the pacifist antinationalist, was suddenly celebrated as the figurehead of

German science. Things could still get stranger. The Royal Swedish Academy of Sciences announced that Fritz Haber had been awarded the Nobel Prize in Chemistry. The prize was for his work on the nitrogen fixation processes that had revolutionized agriculture, but no one was likely to forget that he used those same processes to keep Germany in the war for years or that he had pioneered chemical weapons. He had been isolated from other scientists since the end of the fighting and at one point was on a list of war criminals wanted for extradition. He was devastated and depressed by how the war ended.

Soon, though, Haber dedicated his life to keeping German science going in the postwar economic crisis (he tried to extract gold from seawater to pay war reparations). He never accepted that he had done anything wrong with his invention of gas warfare. After the war a colleague showed him a petition to outlaw chemical weapons—Haber's furious response was that signing that document was unbecoming of a German. Somehow he and Einstein remained friends through all of this. Haber offered Einstein praise of special value coming from a militant toward a pacifist: "In a few centuries the common man will know our time as the period of the World War, but the educated man will connect the first quarter of the century with your name." Einstein did not have many close friends, and he greatly valued the ones he had despite their moral and political differences.

It was still unclear whether postwar Germany would go the way of Einstein or Haber, internationalism or nationalism. Einstein worried that the League of Nations did not seem to be doing much, and certainly that nationalism continued to dominate: "My political optimism has suffered a jolt." He saw the values of internationalism playing out more in private groups than governmental ones. He particularly praised the Quakers for what they had done "to alleviate the misery in central Europe." As long as there were people "willing and able to provide such substantial forces and means to help men without regard to race or political affiliation, we have good reason in spite of it all to believe that the psychological conditions for a useful development of the League of Nations are there." The German government asked him to use his connections with the Society of Friends to help with food aid (in July 1920 a staggering 632,000 children were being fed by the Quakers alone). Einstein was happy to do so, and to emphasize the connections between that work and his. He wrote:

> Whatever great political disappointments we have experienced and still must experience, we must not abandon hope for a just and satisfying order in the world. . . . No other branch of public life is as well suited to revive mutual trust between nations, and even more should be done to make the nation profoundly aware of the blessed work of the Quakers.

The branch of public life Einstein had high hopes for was, of course, his own—science. He said that intellectuals had always been at the forefront of internationalism (the Manifesto of 93 apparently being an exception):

> The most valuable contribution to a reconciliation of the nations and a permanent fraternity of mankind is in my opinion contained in their scientific and artistic creations, because they raise the human mind above personal and national aims of a selfish character. . . . The intellectuals should never weary of emphasizing the internationality of mankind's most beautiful treasures and their corporations should never stoop to foster political passions by public declarations or other demonstrations.

Scientists could repair the wounds of war better than any politician. And he felt he had a perfect example of this: the 1919 eclipse expeditions. He praised the internationalism of the expeditions as showing the way forward for science. "Our English colleagues not only did excellent work but also *behaved superbly in the personal sense*. . . . I see that these men really did rise above the situation." And to de Sitter: "The outcome of the English expeditions pleased me very much and more so the friendly behavior of our English colleagues toward me, despite my still being a half-*Boche*" (referring to an insulting term for Germans).

Haber worried about Einstein's sudden desire to "fraternize" with the British. The world needed to know, Haber said, where this revolution in science came from: "The English and Belgians want to divest the name Albert Einstein of the German character that was hitherto attached to it." Einstein replied that no reasonable person could accuse him of disloyalty—how many job offers had he turned down from other countries so he could stay in Berlin? And, he reminded Haber, it was his German *friends* who he was staying for—not Germany. He also casually mentioned how poorly both his theory and his politics had been received there during the war. Perhaps there was something special about their former enemies after all: "It even has to be said that . . . the English have behaved much more nobly than our colleagues here. They are for the most part Quakers and pacifists. How magnificent their attitude has been toward me and relativity theory in comparison!" Eddington and Russell came to stand in for their whole country, in a mirror image of wartime stereotypes on both sides.

Even with the relativity enthusiasm in Britain, though, wartime echoes there continued to make it difficult for German science to flourish after the war. It was certainly not the case that "most" British scientists were pacifists and Quakers. The lifeblood of scientific communication—journals and papers—was still restricted. Throughout 1919 the RAS, for instance, readily provided back issues of their journals to Belgium and Serbia while refusing to send them to "enemy countries." German and Austrian scientists struggled to get themselves up to date. Planck tried to get one copy of every scientific publication not received during the war—some 13,000 issues. Requests for these were often directed to Eddington, whose sympathies were well known. He did what he could within institutional restrictions, often taking advantage of bureaucratic confusion to bypass the unofficial blockade. Einstein pitched in with his fame, too, hoping to trade German publications for Anglophone ones and getting Lorentz to act as an intermediary.

Sometimes his status was indeed helpful. But there were dark sides to his prominence. One was that he became an excellent target for the surging right wing of German politics. Einstein himself said that as "a

Jew with liberal international views" he was a symbol of everything they hated. Both his science and his character came under attack by the suspiciously well-funded Working Party of German Scientists for the Preservation of Pure Science. This organization, led by the enigmatic Paul Weyland, orchestrated a smear campaign and anti-relativity rallies in Berlin. The group dismissed the enthusiasm for the theory as "mass suggestion."

Its largest anti-Einstein event was held at the Berlin Philharmonic Hall. Einstein actually sneaked in to see what they were saying about him and noted anti-Semitic conspiracy literature being handed out in the foyer. The combination of his absurd fame and the absurd politics was almost too much for him. "This world is a strange madhouse. Currently, every coachman and every waiter is debating whether relativity theory is correct. Belief in this matter depends on political party affiliation."

Powerful people came to Einstein's defense. The German government was particularly concerned about the diplomatic consequences of the most famous scientist in the world coming under attack. Their chargé d'affaires in London warned:

> The attacks on Prof. Einstein and the agitation against the well-known scientist are making a very bad impression over here. At the present moment in particular Prof. Einstein is a cultural factor of the first rank, as Einstein's name is known in the broadest circles. We should not drive out of Germany a man with whom we could make real cultural propaganda.

Einstein appreciated the efforts being made by the higher-ups to keep him in Berlin, even if their sudden concern for him felt faintly ridiculous: "The role I play is similar to that of a saint's relics that a cathedral absolutely has to have." Violence was a real possibility but Einstein decided to stay regardless. Flippantly, he compared the situation to being plagued by bedbugs while sleeping in a good bed.

The anti-Semitism behind the attacks on Einstein was not random. Anti-Jewish sentiment surged in Berlin after the war, mainly aimed at eastern European refugees who were blamed for crime and political chaos. Einstein saw this clearly and spoke up to defend both himself and his fellow Jews. It was this persecution alongside people with whom he shared little other than distant heritage that first pushed him toward Zionism: "These and similar experiences have awakened my Jewish-national feelings." He had little interest in a Jewish homeland per se—he didn't even want to participate in the Jewish community in Berlin—he just wanted a place where people like him might go to be safe. If his fame was good for anything, perhaps it could help here: "I believe that this undertaking deserves energetic collaboration. . . . My name, in high favor since the English solar eclipse expeditions, can be of benefit to the cause by encouraging the lukewarm kinsmen."

All of this took place against a tragic backdrop for Einstein. His mother, Pauline, had been seriously ill throughout 1919. He managed to use his status to bring her to Berlin for her final days. She died in February 1920. Einstein confessed he was completely exhausted by her long decline: "One feels in one's bones the significance of blood ties."

Other family concerns occupied him too. The postwar collapse of the German mark made it more and more difficult for him to support Mileva and his sons. He began to accept invitations to lecture in just about any country that could offer stable currency. Sometimes his outrageous speaking fees were accepted, sometimes not. Practice did not make him a better traveler; Elsa still had to remind him to have his shirts washed and not to wear his shabby traveling suit for formal events. She needled him by casually mentioning the delicious ripe asparagus at home that he would not get to eat. He passed the time on his train and boat rides reading *The Brothers Karamazov*, which he loved. The German government took advantage of his country-hopping to improve their own international relations. Einstein was a godsend for increasing goodwill. A German diplomat in Norway commented, "Admiration for the scientist was extraordinary."

THE ADMIRATION HAD become global, but scientific internationalism remained an ongoing project. Even in Britain, the home of Einstein fever, Eddington was still unusual as an advocate of a general return to normal scientific relations. The framework of the IRC made ordinary scientific interactions quite difficult. Soon after the announcement of the eclipse results a group of German scientists organized a special meeting of the Astronomische Gesellschaft on general relativity, but almost no British scientists were willing to break with the IRC's expected boycott. Eddington decided that even if he couldn't get British participation in any official capacity, he could still participate as an individual. He wrote to the head of the Gesellschaft:

> I hope to show my interest in the Astronomische Gesellschaft by attending the next meeting—an individual step which no one has any right to object to. . . . International Science is bound to win and recent events—the verification of Einstein's theory—has made a tremendous difference in the past month.

He tried to participate in German science as if there had been no disruption from the war, even publishing a paper in the *Zeitschrift für Physik* despite his almost complete inability to write or read German. It began: "This paper is intended to give a full account on the theory of the radiative equilibrium of the stars. It is written primarily because the original papers are not easily accessible in Central Europe in present circumstances."

Eddington must have thought it was a good sign when on November 14, 1919, Einstein was nominated for the Gold Medal of the RAS, the group's highest honor. A month later it was decided that he would indeed receive the award. The decision needed to be formally confirmed at the January meeting but Eddington sent word to Einstein about the medal right away. His friend Ernest Ludlam was heading to Germany to help with the Quaker Emergency Committee's relief work. He could convey the news to Einstein in person, surely better than a letter. A victory for scientific internationalism should be reported as soon as possible.

This turned out to be a bad choice. The January meeting should have been a rubber stamp for Einstein's award. Instead, something went wrong. We do not know exactly what. The minutes of the meeting only record that "the award of the Gold Medal to Professor Einstein was not confirmed." There was a strong implied message, though. For the first time since 1891 the RAS had decided to give no Gold Medal at all that year. Just perhaps, there were some British scientists who were still not comfortable celebrating a German scientist. The original nominations came from H. H. Turner and James Jeans, who, as Eddington said, had been intensely anti-German during the war. Their nomination seemed like a change of heart. Then they did not show up to the confirming vote in January—did they change their mind again? Had it been an elaborate attempt to embarrass Einstein? Eddington penned a heartfelt apology to Einstein:

> I am sorry to say an unexpected thing has happened and at the meeting on Jan. 9 the Council of the RAS rejected the award, which had been carried by quite a large majority at the previous meeting. The facts (which are confidential) are that three names were proposed for the Medal. You were selected by an overwhelming majority in December. Meanwhile the "irreconcilables" took alarm, mustered up their full forces in January, and managed to defeat the confirmation of the award in January. . . . I confess I was very much surprised when the motion was proposed and carried originally (it was proposed by two men who during the war have been violently "patriotic"). . . . I am sure that your disappointment will not be in any way personal; and that you will share with me the regret that this promising opening of a better international spirit has had a rebuff from reaction. Nevertheless I am sure the better spirit is making progress.

The letter closed with Eddington's hopes that Einstein could make a trip to England soon regardless, and perhaps even visit the RAS (he admitted that there might be "some awkwardness after what happened").

Eddington certainly saw the episode as a continuation of wartime animosity, of which there were plenty of examples. Ludlam, too, apologized to Einstein for the incident. Writing on Quaker Emergency

Committee stationery, he offered an explanation for the RAS's embarrassing behavior:

> I find it difficult to believe that English men of science can really be so narrow minded. I think one of the chief difficulties is that scientific men work so hard, and have so much to read, that they have not time to study the real facts in international affairs and accept too easily the opinions of the common press. . . . Perhaps, when you consider the campaign of lies which has lasted for five years—in all countries—you will not judge these poor islanders too harshly.

Perhaps scientists were simply too focused on their technical work to understand the realities of politics. Einstein's response to the "tragicomical" affair tried to soothe his friends' frustration: "The greater will be my pleasure in accepting your invitation [to visit], for now my trip is purely of a private nature. My irritating ignorance of the English language will disturb less." Despite goodwill on both sides, Einstein was still unable to visit Eddington that spring. A face-to-face meeting had to wait for the future.

EINSTEIN WAS NOMINATED for the Gold Medal again the following year, and again rejected. It was not until 1926 that he finally received the award. Eddington told Einstein that he had not much to do with the decision. This was, in a sense, true—he was in Leiden when the vote was taken. But Eddington had created the conditions under which it was possible to welcome a former enemy in Britain at all, from the expedition to popular lectures to battling for the very possibility of international science. His contemporaries commented freely on this. Ernest Rutherford named Eddington as the one responsible for Einstein's fame:

> The war had just ended, and the complacency of the Victorian and Edwardian times had been shattered. The people felt that all their

> values and all their ideals had lost their bearings. Now, suddenly, they learnt that an astronomical prediction by a German scientist had been confirmed by expeditions . . . by British astronomers. An astronomical discovery, transcending worldly strife, struck a responsive chord.

Oliver Lodge said Einstein would have been unknown without Eddington. J. J. Thomson offered a somewhat backhanded compliment that Eddington had "by his eloquence, clearness and literary power persuaded multitudes of people in this country and America that they understand what relativity means."

Einstein's biographers have frequently resorted to religious imagery to help convey his sudden transformation from obscure academic to worldwide authority. He was canonized on November 6; he was another Moses carrying a new scripture; he was a wise man bearing a secularized Christmas message of peace; his lectures were "a place where miracles happen." Eddington then would be some combination of Peter and Paul. The rock on which the church of relativity was built, or the evangelist who brought the good news into hostile lands (even if those are somewhat un-Quakerly images). He was not the prophet, but he made the prophecy possible.

Einstein finally set foot on British soil on June 8, 1921, descending from the White Star liner *Celtic* into Liverpool. He stopped there on his way back to Germany after a long trip to the United States to raise money for a Zionist organization. Freundlich joined him in Great Britain to help with translation (his mother was English). There was great excitement and anticipation about the visit, with lectures, receptions, and dinner parties planned everywhere. Of course, not everyone was happy to see him. Henry Brose sighed at the anti-Germanic screeds still put forward by "a few 'irreconcilables' who, needless to say, have never seen a field of battle."

Einstein's trip would be a whirlwind three days. Young women fainted as he entered the room. He gave lectures in German, with proceeds going to the Imperial War Relief Fund. At Westminster Abbey he placed flowers on Newton's grave. In rooms full of dignitaries he was almost always the youngest person there. At a dinner party the Archbishop of Canterbury asked him what difference relativity should make to the way we thought about morality. He replied, "It makes no difference. . . . It is purely abstract science." Einstein either did not realize or did not care that his host, Lord Haldane, had just written an enormous book claiming the exact opposite.

The newspapers followed every awkward step of the shabby scientist's sockless feet. Depending on which paper you read, Einstein might be described as absentminded, patient, or playful, and wearing an "ill-cut" or possibly "well-cut" morning coat. Everyone commented on his black hair. *Punch* said his mane "takes its own course through space and is not subject to gravitation."

Upon his arrival by train in London he was driven by car to Burlington House, the home of the Royal Astronomical Society, where an overflow crowd waited. As the door was opened for him, he saw a slim figure waiting at the end of the hall. Sharply dressed and preparing to preside over the event as president of the RAS was Arthur Stanley Eddington. They shook hands; we don't know who reached out first. Three years after the cannons went silent, sixteen years after the first paper on relativity, Einstein had won his war. Relativity had triumphed. Einstein, reminiscing later, told Elsa that Eddington was "a splendid chap. He had to endure so much that I would have admired him even without his theories."

WHEN EINSTEIN STEPPED on the stage at the RAS, he had already become something more than himself. He was a walking myth. He had become an icon of science for the entire world, a symbol of the best of what humanity had to offer. William Carlos Williams named him "St. Francis Einstein of the Daffodils." His name was now synonymous with genius. Einstein's mind had become disembodied (literally, after his death, when his brain was removed and eventually taken on a cross-country road trip).

But really he was an intensely *embodied* person, with stomach pains, enjoying cigars, scrunching his toes in the sand, falling in and out of love. Relativity, too, was a tangible thing (for a theory anyway). It came from clocks and rulers and elevators. Its equations were written in well-worn notebooks, crossed out, scrawled over, and remade. It became real on a hot, rainy rock off the coast of Africa.

Today we think of relativity as a fantastically abstract theory, just a few elegant equations on a page. But its emergence was a messy, tangled story. There was no single moment of discovery, no guarantee of fame, only years of struggle and failure and challenge. Einstein had to persist through failure and skepticism; he had to trust his friends. Eddington had to hold to his pacifism against overwhelming pressures; he needed to have faith in both God and physics. Any of a dozen turning points could have waylaid relativity, leaving Einstein no more recognizable a name than Lorentz or Noether, leaving the equations more curious than earthshaking. He might be remembered as one of a dozen or so people who contributed to early quantum theory; relativity might be mentioned as an odd side project of his. We like to forget how hard it is to do science, and how it could have been different. A simple story of science seems more true, more convincing.

We like to simplify the scientists too. The Einstein scholar John Stachel reminds us that the most persistent myth about Einstein was that he was born old. We imagine that he always had gray hair and a lined face, a grandfatherly sage beloved by all humanity. We forget the wartime Einstein, starving, scrappy, a socialist radical battling to make sense of his own ideas, much less persuading anyone else. Einstein was just forty years old in 1919. We project our elderly Einstein back in time because we want him to have always been the great sage.

But he wasn't. Our mythical genius came out of bloody, devastating years of war. It was only in contrast to those horrors that Einstein's triumph was so striking—a victory for pure thought, scientific beauty, and world peace at a time when civilization itself seemed to be in peril. Relativity's sudden explosion, and Eddington's zealous evangelism for it, would never have happened in quieter times. The theory had few applications for decades and, even if it had been confirmed, would likely have languished in dusty journals until cosmologists or GPS engineers realized they needed its delicate adjustments. Without the war, relativity would have been just one more theory, true but obscure. Without the war, Einstein would be just one more name for bored schoolchildren to memorize. Instead, his name is now an idea, an icon, a personification of everything we want science to be.

EPILOGUE

The Legacy of Einstein and Eddington

What kind of politics does science have?

THE 1919 ECLIPSE has lasted a long time. Not the eclipse itself; that was only a handful of minutes. Its legacy has lived on for a century. Every generation has used the story of Einstein and the eclipse to explain what science is, how it works, and what it means. Exactly what those lessons are has changed, and will change. But the 1919 eclipse has become an exemplar for the essence of science—good or bad.

IF YOU ASK a scientist today what makes an idea "scientific," you are likely to get a blank look (that's not a question they need to think about much). If you do get an answer, it will probably be something like this: an idea is scientific if it is *falsifiable*. That is, it is scientific if it can be proved wrong (by an experiment, usually). Science, then, is not about proving good ideas true—it is about proving bad ideas wrong. By eliminating all the inadequate ideas, scientists will gradually get better and better understandings of the world.

This position has the awkward name of *falsificationism*. It is extremely widespread among the people who actually do science. It is the brainchild of the Austrian-born philosopher Karl Popper, one of the most influential philosophers of science of the twentieth century. Like Einstein, Popper is one of those figures who seemed to have been born old—heavy jowls, well-retreated hair, elephantine ears. But to understand his connection to the 1919 eclipse we need to see the young, handsome, dashing Popper. Born in 1902, he was just too young to fight in the war but old enough to see its injustices. That drove him to become a Marxist by age fifteen. When he was seventeen, the 1919 eclipse spread the Einstein phenomenon across the world and he found a new intellectual hero. Young Karl heard the genius speak in Vienna the winter of 1919–1920 and found himself "dazed."

It was not time dilation and curved space-time that startled him. Rather than the science, he was more interested in how Einstein *talked about* his science. What struck him was Einstein's "intellectual modesty"—that the physicist had specified the conditions under which relativity could be refuted. No gravitational redshift, no relativity. No deflection of light, no relativity. This combination of boldness (these are my predictions, go check them) and tentativeness (my theory is only provisional and can be proven wrong at any time) enormously impressed Popper.

He had been dissatisfied with claims that Marxism was scientific for exactly these reasons. Marx's predictions of revolution, it seemed, could only be proved right and never proven wrong. Any world event was claimed by the Marxists as confirmation of Marx's ideas; there was no evidence that could convince them otherwise. Popper noticed something similar about Freudian psychology, one of the other great intellectual frameworks of the age. No matter what sort of dream you had, a Freudian would explain how it supported their theories. It was irrefutable. The Marxists and the Freudians seemed to have immensely powerful theories that could explain anything. Surely that made them scientific?

Popper had been frustrated by this, and Einstein helped him understand why. The mark of a good theory was not that it made predictions; the mark of a good theory was that it made predictions that could be refuted. A scientific theory should declare a rigorous, severe test that would show it is wrong. If light is not deflected by gravity, relativity is not true. Concentrating on evidence *for* a theory, as the Marxists and Freudians did, inevitably would lead one to only look for (and only see) what you want to see, rather than what is really there. That marked them as pseudosciences instead of true science. The ability to falsify a theory was how one found the boundary of the scientific (what philosophers call the demarcation problem).

So the 1919 eclipse was, to Popper, not about demonstrating that Einstein was right. It was a test to see if relativity was wrong. It passed the test and thus one could have provisional confidence in the theory. This became the exemplar of science for Popper's philosophy. Otto Neurath said Popper turned Eddington's experiment into "a scientific model." Popper himself said that all he had done was "to make explicit certain points which are implicit in the work of Einstein." All science, he said, should follow Einstein and Eddington's model.

Popper's falsificationism has become tremendously popular among scientists and science educators, particularly those looking for a clear standard with which to convince courts to keep creationists out of the classroom. It gave a model for how to do their own science (be like Einstein!). It also gave good reasons to keep doing science—since a theory cannot be ultimately proven, there is always more work for scientists to do. That gradual approximation to truth means there is, as Popper wrote, "no point of rest in science."

The 1919 eclipse observations fit nicely into this scheme. Falsificationism demands that the theory be checked again and again, and that is indeed what astronomers did. The Lick Observatory repeated the observations at the 1922 eclipse. W. W. Campbell, perhaps embarrassed by his team's unreliable results from 1918, carried out extensive preparatory work to find exactly the equipment and arrangement that would work best for measuring the deflection. The results strongly confirmed Einstein. Erwin Freundlich finally had the chance to carry out the test himself in 1929 in Sumatra. Each expedition could only show that relativity had been non-disproved once more. It was triumphant . . . until the next test.

Popper's ideas are still regularly invoked by scientists to police radical ideas or shape their fields. Calling an idea unscientific is perhaps the most devastating critique one can level (remember such attacks on relativity), and falsifiability is a convenient way to do that. Nowadays cosmologists talk about multiple universes—but is that a falsifiable idea? Theoretical physicists have been pursuing string theory for decades—but they have yet to propose a Popperian decisive test to see if they are wrong. With Popper's criterion—does this follow the model of 1919?—hypotheses can be discarded without further consideration, the whole trajectories of scientific research can be directed. Creationists have even tried to attack Darwinian evolution by claiming it fails Popper's test of falsifiability (it does not; it is perfectly falsifiable).

Philosophers and historians often point out that despite the appeal and utility of falsificationism, it is deeply flawed. It doesn't actually help define pseudoscience (astrology can be falsified by the experience of any set of twins). And it is not good at describing how scientists actually *do* their work. Despite the phrasing that so impressed Popper, Einstein actually was hoping someone would prove him *right*. Everyone talked about the 1919 results as proving relativity to be true. The later expeditions were indeed important. But the consensus was already that relativity was correct, not that it was a mere conjecture that should be checked.

By the 1950s and '60s other philosophers had begun to point out that deciding whether a theory had actually been falsified was difficult. It is not always clear, as Popper hoped, whether a given experimental result actually refutes a theory. Thomas S. Kuhn's famous *Structure of Scientific Revolutions* argued that one's paradigm (the framework of ideas through which you interpreted the world) could actually change what you thought an experiment's result was. An Einsteinian could look at the photographic plates and see the curvature of space-time; a Newtonian could look and see none.

One version of this problem of interpreting an experimental result, sometimes called the Duhem–Quine thesis, states that any conclusion about what a test is actually testing depends on lots of intermediary knowledge and assumptions. Were Eddington's plates measuring gravitational deflection, refraction in the solar atmosphere, or the uneven heating of a coelostat mirror? We saw those issues explicitly discussed by scientists after the results were first presented. The issues are difficult and absolutely critical to convincing someone that your results mean what you say they do. Data does not speak for itself.

Some skepticism of the idea that the 1919 expeditions had truly confirmed relativity began to appear after World War II. This was when a new generation of scientists had come to prominence who did not remember Einstein's canonization firsthand and when Eddington's reputation had been damaged by his widely panned late-career attempts at unifying physics. They had some distance from which it was easier to ask certain questions.

In 1969 the astrophysicist Dennis W. Sciama wanted to celebrate a new era in relativistic physics in which tools such as radio telescopes could provide results undreamt-of by Einstein. Using radio telescopes is a much more precise way to measure gravitational deflection than eclipses (and you can do it anytime you want). By the fiftieth anniversary of Eddington's observations, the mathematical tools available and the precision expected of an experiment had changed dramatically. The early age of measuring light deflection at a solar eclipse now seemed amateurish. It is an extremely challenging measurement to carry out and even the later expeditions still had a fairly high amount of error. Perhaps Eddington's results had not been as decisive as once thought. According to Sciama their worldwide influence was "partly because the world was amazed that so soon after the Great War the British should finance and conduct an expedition to test a theory proposed by a German."

The physicist C.W.F. Everitt went so far as to completely reject the 1919 results. He wrote that "this was a model of how not to do an experiment." On "cooler reflection" from sixty years' distance it seemed that the data did not support Einstein at all. Perhaps Eddington, Dyson, and Davidson had always intended to prove Einstein right and manipulated the results to that end. Everitt disputed the idea that there was anything reliable on those photographic plates. "Only Eddington's disarming way of spinning a yarn could convince anyone that here was a good check of General Relativity."

Even Stephen Hawking casually dismissed the 1919 results. The errors were, he said, as large as the effect they were trying to measure (like saying you're sure it is Tuesday, but that it might be Monday or Wednesday). His *Brief History of Time* concluded that "their measurement had been sheer luck, or a case of knowing the result they wanted to get, not an uncommon occurrence in science."

Did the photographic plates show what Eddington said they did? A scientist's reflex in a situation like this is to go check again. So in 1979 (for the centenary of Einstein's birth), the original photographs were pulled from the Royal Astronomical Society's archives and checked with modern methods. It is rare for a scientific debate to go back to the original, raw data—but when it does, that is a sign of deep worry. Astronomers are obsessive record keepers for precisely this reason, though. You never know when you will need to go back and look.

The Royal Greenwich Observatory staff reanalyzed the Sobral plates with computerized measuring equipment and checked whether the 1919 analyses had been done properly. For the four-inch telescope, Dyson had reported 1.98 ±0.18. The modern computer reported 1.90 ±0.11. For the astrographic, Dyson had reported the uncorrected value of 0.93, even though if he applied likely corrections for the distorted mirror he would have had 1.52. The computer, though, could apply those corrections much more reliably, getting 1.55 ±0.34. The new analysis gave a combined result of 1.87 ±0.13, solidly close to Einstein's 1.75 prediction. It seemed that Dyson's original analysis had been pretty good, and certainly did not show evidence of tampering in favor of Einstein. Two astronomers wrote a public letter refuting Hawking's complaint, pointing out that the errors were well below the measured value. Rather than not knowing if it was Tuesday, the errors were more like "I know it is Tuesday between lunch and teatime." It is true that these errors are large compared to what comes out of CERN today, where results are uncertain to about 1 in 3 million (called "five sigma"). Partly this is because scientific standards change over time—none of Newton's experiments would pass muster now—and partly because the very first time something is measured, the result is always rough.

Precision becomes much easier once you understand what you are looking for.

The reanalysis of the plates, though, left out a critical part of the events of 1919. Eddington and Dyson had made the case that the Sobral astrographic results should not be taken seriously because of the coelostat problems. They had been worryingly close to the Newtonian half-deflection; if they had been included it would have been difficult to claim a confirmation for Einstein. In 1980, two philosophers, John Earman and Clark Glymour, argued that the Sobral astrographic results should not have been excluded. If those results were dropped, then the Principe results (which were far from perfect) should have been dropped as well. Since Eddington did not do that, he must have been biased—Eddington admitted he thought relativity was true even before the expedition, and he had political reasons for wanting a positive result. Earman and Glymour conclude that Eddington only won the debates because he ended up writing the textbooks afterward. They reassured their readers that, while this might cause "despair on the part of those who see in science a model of objectivity and rationality," it was not a deep problem because we had other reasons to think relativity is true. Indeed, today we have a dizzying number of different experiments confirming relativity—it is one of the best-supported theories of all time. The gravitational redshift can now be seen in any laboratory. The deflection of light is so well established today that it is used as a basic tool for exploring the universe (in the form of "gravitational lensing"). From the movements of galaxies to spinning spheres in orbit to the GPS system in your pocket, physicists keep looking for failures of relativity's predictions and find none.

Earman and Glymour's argument found its way to a much larger audience than typical for an academic paper thanks to a bestselling 1993 book, *The Golem: What Everyone Should Know About Science*. The authors were Harry Collins and Trevor Pinch, two sociologists interested in showing that science is a social construct—that is, its conclusions come from social and cultural processes rather than providing an objective perspective on the physical world. Their second chapter used Earman and Glymour to make an even stronger claim about relativity. It was not just that Eddington was biased; the confirmation of relativity was itself simply a social construct. The 1919 expedition, for Popper the model of how to do science in the most rational and reliable of ways, was now a demonstration of the impossibility of objectivity. Even the greatest of scientific feats—the experiment that made Einstein famous—showed that science was just another series of myths. According to this book, the importance that scientists placed on the expeditions did not come from revelations about the physical world, it was because "science needs decisive moments of proof to maintain its heroic image." It was about the story scientists told one another.

IT IS NOT unusual today to meet physicists who have accepted Collins and Pinch's (really Earman and Glymour's) critique of the 1919 results. They will talk about error bars and bias. This version of the story has become a kind of scientific folklore, passed around at water coolers. *The Golem* was a fantastically successful scholarly publication (more than a dozen printings and many sequels) and its arguments have filtered out into many other books in many different disciplines. Very few physicists repeating this version of the story know where it came from, though. They would probably be quite shocked to learn they were passing along arguments aimed at undermining the very foundations of their field.

Generally scientists have strong negative feelings toward social constructivists like Collins and Pinch. The battles among them even earned a special term—the "Science Wars." In the 1990s there were ferocious (by scholarly standards anyway) debates about whether theories like relativity should be seen as concrete parts of physical reality or as mere sociopolitical negotiations. One part of the question was whether fields like sociology and the humanities could have anything important to say about the natural sciences. Also under debate was the nature of science. What were the forces that made science work? Physical ones like gravity and electricity? Or social ones, like personal bias and political affiliation? And who was allowed to speak about science?

N. David Mermin, a Cornell physicist, wrote a reply to *The Golem* for *Physics Today* that walked an interesting line. He accepted that there were social elements to science (anyone who has set foot in a lab knew that). But he rejected Collins and Pinch's *exclusive* focus on those social elements. Surely, he said, science could be the product of *both* physical and social forces.

Mermin did not follow up on this suggestion, but other scholars have. Daniel Kennefick has stressed that even though Eddington and Dyson made important decisions about what data to accept and what to reject, that does not mean those decisions were *wrong*. Context, he writes, is crucial for understanding any experimental result. The astronomers in 1919 had good reasons for attributing systematic error to the Sobral astrographic but not to the Principe one—and they presented those reasons publicly. The vast majority of scientists qualified to judge those reasons were persuaded that this was the correct choice by the standards of the day. They understood that recognizing the difference between good and bad data was hardly a controversial or suspicious thing. It was an ordinary part of doing science.

It is not always easy to tell the difference between good and bad data. Generally you need lots of experience producing and looking at specific kinds of data. In 1919, it was good to trust people who knew telescopes very well. Conversely, don't trust what a particle physicist has to say about fossils. Direct experience produces a special kind of understanding. The Germans, naturally, have a special, lengthy word for this—*Fingerspitzengefühl*. Literally this is the feeling in your fingertips; more loosely, it refers to that special awareness that comes with long experience. It is how your car mechanic figures out what is wrong with the engine just by listening, or how the chef knows what spice to add a pinch of. It would be very complicated for either one to explain the specific details on which they made their decision, though your car will run better and your meal will be delicious. Science is similar. On May 30, 1919, Davidson and Crommelin glanced at the Sobral astrographic photo and immediately knew something was wrong. Their colleagues in astronomy were able to do the same. Recognizing good from bad was not easy; neither was it mysterious.

Everyone wants a simple explanation for why things turn out as they do. Popper thought the expeditions were extraordinary and made them exemplars of good science. Everitt thought the expeditions were biased and made them exemplars of bad science. Collins and Pinch thought the expeditions were shaped by politics and authority, and made them exemplars of socially constructed science.

Einstein's War has been a story about how none of those are enough. Einstein and relativity's victory involved good science, bad science, politics, and personal authority. Any episode in science does. None of those mean relativity is wrong (it has been confirmed many, many times since then) or that Eddington fudged the numbers (there were good reasons to trust the 1919 results). Science is done by people. That means it will be inherently complicated and often confusing. People will make mistakes, equipment will break, poor decisions will be made because of political or personal bias. But people will also have flashes of insight, they will have friends who make crucial suggestions, they will take up a cause because of political or personal beliefs.

We do not have to be forced into extremes. The presence of human scientists does not make science unreliable. We need to understand, though, what science-done-by-people actually looks like and how it works. That means leaving behind some comforting myths about the dispassionate, purely rational, always-objective nature of science. The deeply human, sometimes chaotic story of relativity is not an exception. It is an exemplar. Science is messy; it is also a powerful way to learn about the real world around us.

Beyond data points and error analyses, most everyone agrees about the broader historical importance of the 1919 expeditions. They were a great victory for the higher values of science. They showed that scientists could rise above petty nationalism, that science could help one escape the shackles of nationalism and war. We saw how Eddington

Einstein and his international allies, September 1923. Back row: Einstein, Paul Ehrenfest, Willem de Sitter. Front row: Eddington, Hendrik Lorentz. EMILIO SEGRÈ VISUAL ARCHIVES

and Einstein intentionally spread this interpretation; they wanted to use the moment to change the way scientists were behaving. They never let up, either. After Dyson died in 1940 in the darkest days of World War II, Eddington used his obituary to again remind everyone that the eclipse expeditions had "opportunely put an end to wild talk of boycotting German science. By standing foremost in testing, and ultimately verifying, the 'enemy' theory, our national Observatory kept alive the finest traditions of science; and the lesson is perhaps still needed in the world today." The expeditions were a model not just in the epistemological sense of Popper, but also in a political and moral sense.

This modeling sometimes becomes self-congratulatory. On the Einstein centenary in 1979 the distinguished British astronomer William McCrea gave a rousing speech celebrating how fortunate it was for Einstein that British science had held to scientific internationalism during and after the Great War. His generation had completely forgotten the brutal battles that had been fought over whether German science would ever be welcome on British shores. Eddington had been successful in portraying the expeditions as triumphs of internationalism, so looking back, it seemed to someone like McCrea that internationalism must have been the obvious and natural way to handle German science. In hindsight everything seems inevitable. Of course relativity would be confirmed; Einstein was a genius. Of course politics would not have gotten in the way; scientists always transcend nationalism.

Even those who attacked the expeditions' scientific value acknowledged its importance for the postwar world. Hawking called them a triumph of international reconciliation in the same sentence that he rejected their data. Physicist Clifford Will in 1986 described the expeditions as something to strive for in his own era:

> In our present time, when cold-war politics sometimes obstructs the free flow of scientific information and interaction, we would do to remember this example: a British government permitting a pacifist scientist to avoid wartime military duty so he could go off and try to verify a theory produced by an enemy scientist.

Again, the internationalism of 1919 was presented as something uncontroversial that all scientists would naturally accept. If Soviet and American scientists couldn't get along, they should look to Einstein and Eddington as an example.

As always, though, "internationalism" is a complicated category that means different things to different people. To celebrate the ninetieth anniversary of the observations a plaque was placed on Principe at the very spot (still accessed by dirt road) where Eddington and Cottingham observed the eclipse. There was some controversy over who should install it—the British or the Portuguese? Was the nationality of the spot itself important or the nationality of the people who worked there? In echoes of the eclipse expedition equipment in 1919, the fifty-kilogram plaque was only allowed through customs because local officials made special arrangements. The plaque is a celebration of internationalism, though it is hardly postcolonial: some of the local children were herded away from the installation, being told it was only for whites.

The right connections between science and politics are not obvious. Some say they should be totally separate—scientists should not be involved with politics and vice versa. During the First World War this idea of separation was almost completely abandoned. Who was allowed to subscribe to a journal was a political question. Who received funding from the government was a political question. What words were used to describe an experimental apparatus was a political question. One might look to this and say that it was war, and nothing was working as it should. It was an anomalous kind of science.

It was not. The war only brought into relief political aspects that have always, and will always, be present in science. Wanting science to be apolitical does not make it so. Instead of ignoring the political aspects of science we would be better off acknowledging them so we can understand them. The question then becomes "What kind of politics does science have?" It is sometimes claimed that science has an inherent connection with certain political frameworks or viewpoints. The sociologist Robert K. Merton, for example, used to argue that science encouraged democracy. You can certainly find connections between scientific values and democratic ones (for example, freedom of speech). But it is not difficult to find connections to, say, anarchism too (for example, there is no formal centralized authority).

During the Great War we saw many different attempts to find these connections. Everyone who signed the Manifesto of 93 thought that the correct politics for science were those of the German Empire. Einstein thought socialism was the most natural politics for science. Eddington argued for internationalism. The practice of science did not carry with it an innate political orientation. Each group, each individual forged their own mixture of scientific practice, national identity, personal beliefs, and past experiences. It was obvious to each scientist that the politics they brought were the best ones for science; what their enemy brought was the worst.

There is not a correct answer here. Being a scientist does not carry with it a natural or default political setting. Politics will be in science regardless, as long as scientists are human. Scientists need to decide for themselves what political values they think are important for the work they do. If you think science needs many different viewpoints and a variety of life experiences, then you should fight for liberal values in science. If you think science suffers from too much government intervention, then you should fight for libertarian values in science. A scientist should not be embarrassed to state political views that they think will make science work better (Einstein certainly wasn't). And there will probably be other scientists who disagree with those views (as Einstein discovered). A scientist who is wondering whether it is acceptable to lobby their representative, or whether they can march in a protest, should remember Eddington at his conscription hearing or Einstein crossing the revolutionary barricades.

Simply having political views does not make it impossible for someone to do science. Scientists are not emotionless machines, nor would we want them to be. We need scientists who care about things and are willing to take actions to support the things they care about. If Eddington had not cared about pacifism, we would not have had the relativity revolution in 1919. It was because people from all across the political spectrum cared about their science that the world had just the right conditions for Einstein's sudden catapulting to fame. The horrors of the war, and pacifists' reactions to them, forged the intricate, fragile network that made relativity what it was. The connections of science to the wider world—politics, religion, culture—are not trivial. How we think about science, how we link it to the other parts of our lives, changes the way science is done. We—scientists and nonscientists—need to choose what values and goals we want to bring to the scientific endeavor. Einstein did.

ACKNOWLEDGMENTS

The publication date on a book creates the illusion that it has a distinct moment of creation (kind of like scientific theories). In reality, of course, it grows slowly and organically. This is more true than usual for *Einstein's War*, which relies on and integrates academic research and writing I have done over two decades. This integration might have, in some places, led to some unavoidable similarities with my earlier publications. My thanks to everyone who has supported my work over the years—your efforts made this book possible, even if you can't see it.

There are a handful of scholars whose work I have relied on particularly heavily in constructing this story, whom I would like to mention even beyond the end notes. Of the nigh-infinite Einstein biographies out there, I have found none that match the balance of detail, clarity, and accessibility found in Albrecht Fölsing's *Albert Einstein: A Biography*, and I have used it as my foundation for this story. Similarly, John Keegan's book *The First World War* strongly framed the way I have thought about the conflict. Hubert Goenner and Giuseppe Castagnetti's research on Einstein's political activism during the war has been extremely helpful, as has the work of Jürgen Renn, Michel Janssen, and everyone else who was part of the *Genesis of General Relativity* series. And this book would have been literally impossible without the army of people who have worked on the Einstein Papers Project over the years and made those letters and documents available both in print and online.

In a book of this type I am unable to provide comprehensive citations of all the relevant scholarly literature as would be expected in an academic book. I apologize to the many, many scholars whose work I have not explicitly mentioned here—so much has been written on both Einstein and the Great War that my notes can only scratch the surface. Where possible I have tried to cite widely accessible secondary sources such as *The Quotable Einstein* or Margaret MacMillan's *Paris 1919* rather than academic publications that may be harder for interested readers to find. Similarly, for the sake of consistency and ease of reference I have generally used published materials for translations from the German.

My deepest thanks to those long-suffering people who read this manuscript front-to-back in various incarnations: Janelle Stanley, Andrew Warwick, Graeme Gooday, Andrew Romig, Matthew Gregory, and Meredith Theeman. Their patience and feedback were essential to the book coming to exist as a coherent whole. David Kaiser and Michael Gordin read huge portions, and I am especially grateful to them for catching many of my stupid mistakes. Thanks to Guy Rader, Bruce Hunt, and everyone in the Department of History at University of Texas at Austin for early feedback on prototype chapters. Professors Damin Spritzer, Jeffrey Johnson, Carsten Reinhardt, Margot Canaday, Luis Campos, and Kitt Price provided invaluable information on specific topics. And I am grateful to Maya and Zoe Stanley for putting up with endless Einstein anecdotes over dinner.

Many people helped out when I was first trying to figure out whether this was a feasible project and how one actually goes about writing a trade book, among them Amanda Petrusich, Kim Phillips-Fein, Brian Keating, and Ken Alder. Many parts of this story were tried out in front of audiences for One Day University across the country—thanks for their attention and to Steven Schragis for making those events happen. I am deeply grateful to Susanne Wofford and NYU's Gallatin School of Individualized Study for providing the time and resources for me to do this work. I would never have written this without the lively conversations with and interdisciplinary atmosphere created by my colleagues, or the tireless work of my research assistants Jacob Ford, Elizabeth Luxenberg, Melody Xu, and Rachel Stern.

Finally, my gratitude to everyone who made the book physically possible. Without my agent Jeff Shreve's confidence in the project none of this would have happened—huge thanks to him and everyone else at the Science Factory. Thanks to my editors Daniel Crewe and Stephen

Morrow for whipping the manuscript into shape, and to Connor Brown, Madeline Newquist, and everyone else at Viking and Dutton for their heroic efforts in getting this on the shelves in time for the 2019 centenary. One of the core themes of this book is the challenges, difficulty, and rewards inherent in the publication of ideas, which makes me especially appreciative of what it took for this to come into being. Thank you, everyone.

NOTES

PROLOGUE

1 **"pale face and long hair":** Alice Calaprice, ed., *The Ultimate Quotable Einstein* (Princeton: Princeton University Press, 2010), 58.
2 **"one of the greatest":** Ibid., 301.
4 **"disturbs fundamentally":** *The Manchester Guardian,* June 10, 1921, quoted in Ronald W. Clark, *Einstein: The Life and Times* (New York: World Publishing, 1971), 271–72.
5 **It did not last:** Clark, *Life and Times,* 272.

CHAPTER 1

7 **"What a paradise this land is":** Samuel Clemens to William Dean Howells, May 4, 1878, in Samuel L. Clemens and William D. Howells, *Mark Twain–Howells Letters: The Correspondence of Samuel L. Clemens and William D. Howells, 1872–1910,* ed. Henry Nash Smith and William M. Gibson (Cambridge, Massachusetts: Belknap Press, 1960).
8 **"his face would turn completely yellow":** *Collected Papers of Albert Einstein* (Princeton: Princeton University Press, 1987–2006), hereafter CPAE, volume 1, "Albert Einstein—A Biographical Sketch by Maja Winteler-Einstein (Excerpt)," xviii.
8 **Hermann read Schiller and Heine:** Abraham Pais, *Subtle Is the Lord: The Science and the Life of Albert Einstein* (Oxford: Oxford University Press, 1982), 36.
8 **"I believe altogether that love":** Albrecht Fölsing, *Albert Einstein* (New York: Penguin, 1997), 26. Alice Calaprice, ed., *The Ultimate Quotable Einstein* (Princeton: Princeton University Press, 2010), 27.
8 **He threw a chair:** CPAE volume 1, "Albert Einstein—A Biographical Sketch by Maja Winteler-Einstein (Excerpt)," xix.
8 **"Your mere presence here undermines":** Calaprice, *The Ultimate Quotable Einstein,* 281.
8 **A favorite entertainment:** CPAE volume 1, "Albert Einstein—A Biographical Sketch by Maja Winteler-Einstein (Excerpt)," xix.
9 **"sacred little geometry book":** Fölsing, *Albert Einstein,* 23.
9 **"could be proved with such certainty":** Quoted in Ibid.
9 **"the fetters of the merely personal":** Quoted in Ibid., 24. Lorraine Daston, "A Short History of Einstein's Paradise Beyond the Personal," in *Einstein for the 21st Century,* ed. Peter Galison Gerald Holton, and Silvan Schweber (Princeton: Princeton University Press, 2008).
9 **He also admitted:** CPAE volume 1, document 22, "My plans for the future," 16.

10 **Einstein levered his diagnosis:** Fölsing, *Albert Einstein*, 29–30.

10 **One of those instructors:** Lewis Pyenson, *The Young Einstein: The Advent of Relativity* (Boca Raton: CRC Press, 1985), 81.

10 **Einstein would read the notes:** Fölsing, *Albert Einstein*, 57.

10 **"I would rather not speculate":** Ibid., 53.

11 **"You are a smart boy":** Pais, *Subtle Is the Lord*, 44.

11 **"roaring, booming, friendly":** Alice Calaprice, ed., *The New Quotable Einstein* (Princeton: Princeton University Press, 2005), 302.

13 **He tried to count the number:** A. Vibert Douglas, *The Life of Arthur Stanley Eddington* (London: Thomas Nelson, 1956), 2.

13 **"some of the greatest astronomers":** 20 June 1898. A. S. Eddington, "A total eclipse of the sun," O.11.22/13, Eddington Papers, Trinity College Library, University of Cambridge. Courtesy of the Master and Fellows of Trinity College, Cambridge.

15 **"indiscriminately bruised the shins":** Douglas, *Arthur Stanley Eddington*, 30.

15 **"With this one friend":** Ibid., 7.

16 **He originally started smoking:** Ibid., 33; Eddington's Notebook, Add. Ms. a. 48, Eddington Papers, Trinity College Library, Cambridge.

16 **"His short skull seems unusually broad":** Calaprice, *Ultimate Quotable Einstein*, 278.

17 **"acted on women as a magnet":** Ibid., 302.

23 **Discouraged but not defeated:** Fölsing, *Albert Einstein*, 68.

23 **Einstein sent a steady stream:** CPAE volume 1, document 136, "Einstein to Mileva Marić, 8? February 1902," 192.

23 **She suffered an attack of scarlet fever:** Abraham Pais, *Einstein Lived Here* (Oxford: Oxford University Press, 1994), 13.

23 **They had to wake the landlord:** Ibid., 11.

24 **"It enforced many-sided thinking":** Fölsing, *Albert Einstein*, 102.

24 **He was marked unfit:** CPAE volume 1, document 91, "Military Service Book, 13 March 1901," 158.

24 **"An ancient, exquisitely cozy city":** CPAE volume 1, document 134, "Einstein to Mileva Marić, 4 February 1902," 191.

29 **"stamp out vermin":** John Norton, "How Hume and Mach Helped Einstein Find Special Relativity," in *Discourse on a New Method: Reinvigorating the Marriage of History and Philosophy of Science*, eds. Mary Domski and Michael Dickson (Chicago: Open Court, 2004), 374.

29 **This sparked several weeks:** Ibid., 367.

29 **He declared that an analysis:** Peter Galison, *Einstein's Clocks, Poincaré's Maps* (New York: W. W. Norton, 2003), 253.

35 **"our work on relative motion":** Pais, *Einstein Lived Here*, 8.

35 **"our theory of molecular forces":** CPAE volume 1, document 101, "Einstein to Mileva Marić, 15 April 1901," 166; CPAE volume 1, document 127, "Einstein to Mileva Marić, 12 December 1901," 185.

37 **"sense of duty and deliberateness":** John Heilbron, *The Dilemmas of an Upright Man: Max Planck as Spokesman for German Science* (Berkeley: University of California Press, 1986), 35.

37 **It is sometimes said:** Ibid., 28.
37 **Those who read the paper:** Andrew Warwick, *Masters of Theory: Cambridge and the Rise of Mathematical Physics* (Chicago: University of Chicago Press, 2003), 404–6.
37 **"absolute, invariant features":** Pais, *Subtle Is the Lord*, 150.
37 **Einstein actually never liked:** Warwick, *Masters of Theory*, 406.

CHAPTER 2

38 **In all, he spent:** A. Vibert Douglas, *The Life of Arthur Stanley Eddington* (London: Thomas Nelson, 1956), 15.
38 **When he was taken:** Eddington's Notebook, Add. Ms. a. 48, Eddington Papers, Trinity College Library, Cambridge.
40 **Eddington became known:** Douglas, *Arthur Stanley Eddington*, 24.
40 **He didn't care much:** Ibid., 18.
41 **"I believe I sewed him":** Ibid., 34.
42 **The Nobel Prize winners:** Albrecht Fölsing, *Albert Einstein: A Biography* (New York: Viking, 1997), 132.
43 **"I admire that man":** Ibid., 216.
43 **Works of genius:** Ibid., 203.
44 **"highly suspicious":** John Stachel, "The First Two Acts," in *The Genesis of General Relativity*, vol. 2, ed. Jürgen Renn (Dordrecht: Springer, 2007), 84.
44 **"the happiest thought of my life":** CPAE volume 7, document 31, "Ideas and Methods," 136.
44 **I was sitting:** Jürgen Renn, "Classical Physics in Disarray," in *Genesis of General Relativity*, vol. 1, ed. Jürgen Renn, n.d., 63.
45 **"of the exact same nature":** Jürgen Renn and Matthias Schemmel, eds., *The Genesis of General Relativity, Volume 1*, Boston Studies in the Philosophy and History of Science (Dordrecht: Springer, 2007), 494.
48 **It had been known:** CPAE volume 5, document 69, "Einstein to Conrad Habicht, December 24, 1907," 47.
49 **"So now I am":** Fölsing, *Albert Einstein*, 246–51.
49 **Despite his father's efforts:** Ibid., 241.
49 **His grooming habits:** Ibid., 262.
50 **He liked to joke:** Ibid., 273.
50 **These were tasks:** Ibid., 294–95.
51 **"I consider myself":** CPAE volume 5, document 389, "Einstein to Elsa Einstein, 30 April 1912," 292.
51 **Participation in international:** Roy M. MacLeod, "The Chemists Go to War: The Mobilization of Civilian Chemists and the British War Effort, 1914–1918," *Annals of Science* 50 (1993): 457.
51 **"meant more than":** Martin J. Klein, *Paul Ehrenfest, Volume I: The Making of a Theoretical Physicist* (New York: Elsevier Science, 1970), 303.
51 **"Lorentz is a marvel":** CPAE volume 5, document 305, "Einstein to Heinrich Zangger, 15 November 1911," 222.
51 **"fatherly kindness":** CPAE volume 5, document 360, "Einstein to Hendrik A. Lorentz, 18 February 1912," 262–63.

52 **"the strangest thing":** Fölsing, *Albert Einstein*, 154.
52 **"One has every right":** Alice Calaprice, ed., *The Ultimate Quotable Einstein* (Princeton: Princeton University Press, 2010), 281.
52 **"the barking of a seal":** Ibid., 294.
53 **"Henceforward space":** Fölsing, *Albert Einstein*, 243.
53 **"Since the mathematicians":** Ibid., 245.
54 **"It is greatly":** Ibid., 308.
54 **"One thing can":** CPAE volume 5, document 281, "Einstein to Erwin Freundlich, 1 September 1911," 201–2.
54 **He saw no:** Fölsing, *Albert Einstein*, 309.
55 **"complete unselfishness":** Douglas, *Arthur Stanley Eddington*, 19.
56 **"scientific maturity":** Alex Soojung-Kim Pang, *Empire and the Sun: Victorian Solar Eclipse Expeditions* (Stanford: Stanford University Press, 2002), 39.
57 **"mummy":** Eddington's Notebook, Add. Ms. a. 48, Eddington Papers, Trinity College Library, Cambridge.
57 **The trains:** Pang, *Empire and the Sun*, 130–31.
58 **There was intense:** Eddington, "Report on an expedition to Passa Quatro, Brazil," MNRAS 73 (1912), 386–90.
59 **"The scene was like fairyland":** Eddington's Notebook, Add. Ms. a. 48, Eddington Papers, Trinity College Library, Cambridge.
60 **He was known:** Douglas, *Arthur Stanley Eddington*, 98.
60 **"Schwarzschild & five mad Englishmen":** 5 August 1913, Eddington to Sarah Ann Eddington, EDDN A 3/1. Eddington Papers, Trinity College Library, Cambridge.
61 **"His interest in women":** Douglas, *Arthur Stanley Eddington*, 30.
61 **"not universally beloved":** Ibid., 90.
61 **He would meet:** Ibid., 109.
61 **Fear of growing:** Catherine Rollet, "The Home and Family Life," in *Capital Cities at War: Paris, London, Berlin 1914–1919: A Cultural History*, vol. 2, eds. Jay Winter and Jean-Louis Robert, Studies in the Social and Cultural History of Modern Warfare (Cambridge: Cambridge University Press, 1997), 316.
61 **The arms race:** John H. Morrow Jr., *The Great War: An Imperial History* (New York: Routledge, 2014), 24.
62 **When Kaiser Wilhelm:** Christopher Clark, *The Sleepwalkers: How Europe Went to War in 1914* (New York: Allen Lane, 2012), 181.
62 **Nonetheless the system:** Morrow, *The Great War*, 27.
62 **"Do you know who":** Clark, *The Sleepwalkers*, 54.

CHAPTER 3

63 **"In a certain sense":** John Norton, "'Nature Is the Realisation of the Simplest Conceivable Mathematical Ideas': Einstein and the Canon of Mathematical Simplicity," *Studies in History and Philosophy of Modern Physics* 31, no. 2 (2000): 137.
63 **"If you want":** Albert Einstein, "On the Method of Theoretical Physics," in *Ideas and Opinions*, ed. Sonja Bargmann (New York: Wings Books, 1954), 270.

64 **And electromagnetism:** Hendrik Lorentz, "Considerations on Gravitation," in *The Genesis of General Relativity* vol. 3, ed. Jürgen Renn (Dordrecht: Springer, 2007), 113.

64 **In particular, Abraham:** Jürgen Renn, "The Summit Almost Scaled," in Ibid., 310.

64 **"That's what happens":** Quoted in Norton, "Nature Is the Realisation of the Simplest Conceivable Mathematical Ideas," 143.

65 **"a stately beast":** Quoted in John D. Norton, "Einstein, Nordström, and the Early Demise of Scalar, Lorentz Covariant Theories of Gravitation," in Renn, *Genesis* vol. 3, 422.

65 **his theory was robust:** Jürgen Renn and Matthias Schemmel, "Introduction," in Ibid., 13–14.

65 **"Every step":** Einstein 1907 (Vol. 2, Doc. 47), "Einstein to Michele Besso, 1912," cited in CPAE volume 4, "Introduction," xv.

66 **"Those who":** John Stachel, "The First Two Acts," in *The Genesis of General Relativity* vol. 1, ed. Jürgen Renn (Dordrecht: Springer, 2007), 99.

68 **"Grossmann, you must":** Abraham Pais, *Subtle Is the Lord: The Science and Life of Albert Einstein* (Oxford: Oxford University Press, 1982), 212.

71 **"Never in my life":** Albrecht Fölsing, *Albert Einstein: A Biography* (New York: Viking, 1997), 315.

71 **He thanked Grossmann:** Thomas Levenson, *Einstein in Berlin* (New York: Bantam Books, 2003), 105.

72 **Einstein's notebook:** See Jürgen Renn and Matthias Schemmel, eds., *The Genesis of General Relativity, Volume 1,* Boston Studies in the Philosophy and History of Science (Dordrecht: Springer, 2007), for a page-by-page commentary on the notebook.

73 **"much too feeble":** Jürgen Renn and Tilman Sauer, "Pathways out of Classical Physics," in *The Genesis of General Relativity Volume 1*, ed. Jürgen Renn (Dordrecht: Springer, 2007), 113–312, 263.

73 **"the theory refutes":** Fölsing, *Albert Einstein*, 317.

73 **"more in the nature":** Ibid.

74 **Not reassuring:** Michel Janssen and Jürgen Renn, "Einstein Was No Lone Genius," *Nature* 527, no. 7578 (November 2015): 298–300.

74 **The cigar-puffing:** Fölsing, *Albert Einstein*, 324.

75 **"You understand":** Ibid., 325.

75 **Patronage of the sciences:** Levenson, *Einstein in Berlin*, 4–5.

75 **The salary would be:** Fölsing, *Albert Einstein*, 328.

75 **Many of the donors:** Jeffrey Johnson, *The Kaiser's Chemists* (Chapel Hill: University of North Carolina Press, 1990), 89–90.

75 **The flower:** Levenson, *Einstein in Berlin*, 1–2.

76 **The famous physiologist:** Hubert Goenner and Giuseppe Castagnetti, "Albert Einstein as Pacifist and Democrat During World War I," *Science in Context* 9, no. 4 (1996): 329–30.

76 **"indulge wholly":** Fölsing, *Albert Einstein*, 330–31.

76 **But the major reason:** CPAE volume 8, document 94, "Einstein to Heinrich Zangger, 7 July 1915," 109–110.

76 **They stayed in Haber's house:** Fritz Stern, *Einstein's German World* (Princeton: Princeton University Press, 2001), 64.

76 **He said he was only willing:** CPAE volume 8, document 23, "Einstein to Mileva Einstein-Marić, 18 July 1914," 33.
77 **She must cease:** CPAE volume 8, document 22, "Memorandum to Mileva Einstein-Marić, with Comments, 18 July 1914," 32–33.
77 **Einstein, crying:** Stern, *Einstein's German World*, 65.
77 **He wrote a note:** CPAE volume 8, document 26, "Einstein to Elsa Einstein, 26 July 1914," 35.
77 **He decided that:** CPAE volume 8, document 27, "Einstein to Elsa Einstein, after 26 July 1914," 36.
77 **"You get so much":** CPAE volume 8, document 6, "Einstein to Adolf Hurwitz and Family, 4 May 1914," 13.
77 **"sheer amount":** CPAE volume 10 (cited as volume 8), document 16a, "Einstein to Zangger, 27 June 1914," 11.
77 **"a certain discipline":** Levenson, *Einstein in Berlin*, 31.
77 **Nonetheless, Planck:** Fölsing, *Albert Einstein*, 335.
77 **"as much respect":** CPAE volume 10 (cited as volume 8), document 16a, "Einstein to Zangger, 27 June 1914," 12.
78 **no less than a complete reworking:** Albert Einstein, "Principles of Theoretical Physics," in *Ideas and Opinions*, ed. Sonja Bargmann (New York: Wings Books, 1954), 222–23.
78 **With Planck's help:** Klaus Hentschel, *The Einstein Tower* (Stanford: Stanford University Press, 1997), 22.
78 **The Academy covered:** Fölsing, *Albert Einstein*, 356.
78 **"celebrate the system":** "The British Association in Australia," *Science* 39, no. 1015 (June 12, 1914): 864–65.
79 **At one astronomical meeting:** A. Vibert Douglas, *The Life of Arthur Stanley Eddington* (London: Thomas Nelson, 1956), 121–22.
79 **Eddington's letters home:** Ibid., 90.
79 **Eddington delivered:** Ibid., 91.
80 **Bystanders recall:** Christopher Clark, *The Sleepwalkers: How Europe Went to War in 1914* (New York: Allen Lane, 2012), 374.
80 **The Austrian authorities:** Ibid., 388.
80 **Ironically, Franz Ferdinand:** Ibid., 393–95.
80 **Germany advised:** Ibid., 417.
81 **"the most insolent document":** Ibid., 456.
81 **The reply arrived:** Ibid., 463.
81 **Germany sent spies:** Ibid., 524.
82 **"honourable expectation":** Ibid., 540.
82 **"sort of a grunt":** Ibid., 541.
83 **"destroy civilization":** John Morrow, *The Great War: An Imperial History* (New York: Routledge, 2014), 30.
83 **He was taken to Odessa:** Hentschel, *The Einstein Tower*, 22.
84 **The chemist William Herdman:** *The Observatory* 479 (October 1914): 397.
84 **"fine prize":** Douglas, *Arthur Stanley Eddington*, 91–92.
84 **This was a real possibility:** *The Observatory* 485 (March 1915): 155.
85 **"quite negligible":** Adrian Gregory, *The Last Great War* (Cambridge: Cambridge University Press, 2008), 16–18.

85 **"a sin against civilization":** Stefan L. Wolff, "Physicists in the 'Krieg der Geister': Wilhelm Wien's 'Proclamation,'" *Historical Studies in the Physical and Biological Sciences* 33, no. 2 (2003): 340.
85 **"a glimpse of Ministers":** Gregory, *The Last Great War*, 13.
85 **The windows of the German embassy:** Alan Wilkinson, *The Church of England and the First World War* (London: Lutterworth, 2014), 12.
85 **"The country calls":** Ibid., 33.
86 **The BEF happily embraced:** Winston Groom, *A Storm in Flanders* (Grove Press, 2003), 13.
86 **"I am musing":** CPAE volume 8, document 34, "Einstein to Paul Ehrenfest, 19 August 1914," 41–42.
86 **"The German people":** John Heilbron, *The Dilemmas of an Upright Man: Max Planck as a Spokesman for German Science* (Berkeley: University of California Press, 1986), 71.
86 **"that all the moral":** Stern, *Einstein's German World*, 44.
87 **"The individual disappeared":** Jon Lawrence, "Public Space, Political Space," in *Capital Cities at War: Paris, London, Berlin 1914–1919; A Cultural History*, vol. 2, eds. Jay Winter and Jean-Louis Robert, Studies in the Social and Cultural History of Modern Warfare (Cambridge: Cambridge University Press, 1997), 283.
87 **"Nowhere is there":** CPAE volume 10 (cited as volume 8), document 34a, "Einstein to Zangger, 24 August 1914," 13.
87 **A week later:** Adrian Gregory, "Railway Stations," in Winter and Robert, *Capital Cities at War*, 28–29.
87 **The Germans had to:** John Keegan, *The First World War* (London: Hutchinson, 1998), 87.
88 **Their 2,000-pound shells:** Ibid., 96.
88 **Even Moltke:** Ibid., 92.
88 **"I went along under cover":** Quoted in Peter Hart, *Fire and Movement* (Oxford University Press, 2014), 94.
89 **During the battle:** Keegan, *The First World War*, 110.
89 **"before the leaves fall":** Groom, *A Storm in Flanders*, 31.
89 **Nearly a quarter-million books:** Keegan, *The First World War*, 93.
89 **Among the troops:** William Van der Kloot, *Great Scientists Wage the Great War* (Oxford: Fonthill Books, 2014), 22–23.
90 **On September 2:** Keegan, *The First World War*, 121.
90 **They were so close:** Kurt Mendelssohn, *The World of Walther Nernst: The Rise and Fall of German Science* (Pittsburgh: University of Pittsburgh Press, 1973), 80.
90 **"be killed":** Holger H. Herwig, *The Marne, 1914: The Opening of World War I and the Battle That Changed the World* (New York: Random House, 2009), 244.
90 **Soon, the only remaining:** Groom, *A Storm in Flanders*, 33.
91 **The death of so many:** Keegan, *The First World War*, 143.
91 **By the time the battle ended:** Ibid., 146.
91 **Their astronomical equipment:** Hentschel, *The Einstein Tower*, 22. Fölsing, *Albert Einstein*, 357. Levenson, *Einstein in Berlin*, 59–60.

91 **"The observations":** CPAE volume 10 (cited as volume 8), document 34a, "Einstein to Zangger, 24 August 1914," 13.

91 **"heroism on command":** Levenson, *Einstein in Berlin*, 60.

92 **The international catastrophe:** CPAE volume 8, document 39, "Einstein to Paul Ehrenfest, December 1914," 46–47.

CHAPTER 4

93 **Irreplaceable cultural treasures:** John Horne and Alan Kramer, *German Atrocities, 1914: A History of Denial* (New Haven: Yale University Press, 2001): 38–40.

93 **HOLOCAUST OF LOUVAIN:** Ibid., 117.

94 **"blind, barbarian" act:** Ibid., 231.

94 **IT IS NOT TRUE THAT OUR TROOPS:** G. F. Nicolai, *The Biology of War* (New York: The Century Co., 1919), ix.

95 **IT IS NOT TRUE THAT FIGHTING:** Ibid.

95 **Einstein's friends:** Stefan L. Wolff, "Physicists in the 'Krieg der Geister': Wilhelm Wien's 'Proclamation,'" *Historical Studies in the Physical and Biological Sciences* 33, no. 2 (2003): 341.

95 **Planck and Klein:** John Heilbron, *The Dilemmas of an Upright Man: Max Planck as a Spokesman for German Science* (Berkeley: University of California Press, 1986), 70.

95 **"as members":** Hubert Goenner and Giuseppe Castagnetti, "Albert Einstein as Pacifist and Democrat During World War I," *Science in Context* 9, no. 4 (1996): 331.

95 **"of real celebrity":** Nicolai, *Biology of War*, xiv.

96 **Instead of saying:** Wolff, "Physicists in the 'Krieg der Geister,'" 343.

96 **"We grieve profoundly":** Daniel Inman, "Theologians, War, and the Universities," *Journal for the History of Modern Theology* 22, no. 2 (2015): 168–89, 176.

96 **The reply had been organized:** Ibid., 175.

97 **"collective insanity":** CPAE volume 9, document 80, "Einstein to Hendrik A. Lorentz, 1 August 1919," 68.

97 **"sexual character of the male":** CPAE volume 10 (cited as volume 8), document 41a, "Einstein to Zangger, 27 December 1914," 13.

97 **His close connections:** Wolf Zuelzer, *The Nicolai Case* (Detroit: Wayne State University Press, 1982), 17–20.

97 **"wonder for the first time":** Ibid., 25.

97 **Never has any:** Nicolai, *Biology of War*, xvii–xix.

98 **"Although I am":** CPAE vol. 8, document 57, "Einstein to Georg Nicolai, 20 February 1915," 69.

98 **Only two others:** Goenner and Castagnetti, "Albert Einstein as Pacifist and Democrat During World War I," 333.

98 **The famed zoologist:** Martin J. Klein, *Paul Ehrenfest, Volume I: The Making of a Theoretical Physicist* (New York: Elsevier Science, 1970), 299.

98 **"be healed":** Zuelzer, *The Nicolai Case*, 345.

98 **Students studied:** Stefan Goebel, "Schools," in *Capital Cities at War: Paris, London, Berlin 1914–1919: A Cultural History*, vol. 2, eds. Jay Winter and

Jean-Louis Robert, Studies in the Social and Cultural History of Modern Warfare (Cambridge: Cambridge University Press, 1997), 211–16.

99 **Einstein was member:** Goenner and Castagnetti, "Albert Einstein as Pacifist and Democrat During World War I," 334–35.

99 **October 29:** Thomas Levenson, *Einstein in Berlin* (New York: Bantam Books, 2003), 85.

100 **This led to:** See John Stachel, "The Hole Argument and Some Physical and Philosophical Implications," *Living Reviews of Relativity* 17, no. 1 (December 2014), 5–66.

101 **"physics too":** Wolff, "Physicists in the 'Krieg der Geister,'" 337–38.

101 **"enemy foreigners":** Ibid., 346.

102 **Foreign terms:** Ibid., 339.

102 **Unjustified English influence:** Ibid., 348.

102 **"a crusade":** Ibid., 353.

103 **Critics argued:** See, for instance, the anonymous pamphlet *Some Arguments for the Maintenance of Voluntary Service* (London: St. Clements Press [1915?]).

103 **At his request:** John Stevenson, *British Society 1914–45* (London: Allen Lane, 1984), 47.

103 **There were not enough doctors:** Michael Robinson, "Broken Soldiers," *History Ireland* 24 (March/April 2016): 30–32, 31.

103 **Cambridge and Oxford:** Stevenson, *British Society*, 52–53.

103 **At the end of August:** Arthur Marwick, *The Deluge: British Society and the First World War* (Basingstoke: Macmillan, 1991), 75.

103 **By October:** Paul Fussell, *The Great War and Modern Memory* (Oxford: Oxford University Press, 1975), 9.

104 **Adm. Charles Penrose:** Nicoletta Gullace, "White Feathers and Wounded Men: Female Patriotism and the Memory of the Great War," *Journal of British Studies* 36, no. 2 (April 1997): 178–206, 193.

104 **Graham Greene:** Royal Society Council Minutes (hereafter RSCM), vol. 10, 258, November 1, 1917. Courtesy of Royal Society.

104 **One of the highest-ranking:** The *Times*, November 1, 1914.

104 **Fewer than half:** James McDermott, *British Military Service Tribunals 1916–1918* (Manchester: Manchester University Press, 2011), 13–14.

104 **So the owner of a munitions factory:** The National Archives of the UK (TNA): Public Record Office (PRO) (hereafter TNA:PRO), MH 47/1: Central Tribunal Minutes, April 6, 1916.

104 **It springs from:** A. Ruth Fry, *A Quaker Adventure* (London: Nisbet and Co., 1926), xvii. See also Hugh Barbour, "The 'Lamb's War' and the Origins of the Quaker Peace Testimony," in *The Pacifist Impulse in Historical Perspective*, ed. Harvey Dyck (Toronto: University of Toronto Press, 1996), 145–58.

105 **We find ourselves:** Rufus Jones, *A Service of Love in War Time* (New York: Macmillan Co., 1920), 3–4.

105 **"Whatever may be":** Ibid., 65–66.

106 **Those philanthropists:** Peter Gatrell and Philippe Nivet, "Refugees and Exiles," in *The Cambridge History of the First World War*, vol. 2, ed. Jay Winter (Cambridge: Cambridge University Press, 2014), 194.

106 **Jonckhèere walked:** *The Observatory* 485 (March 1915): 143–45.

106 **Louvain was again:** Horne and Kramer, *German Atrocities,* 240.

106 **However, the fighting:** Susan Grayzel, "Men and Women at Home," in Winter, *The Cambridge History of the First World War,* 96.

107 **They were uniformed:** Adrian Gregory, "Railway Stations," in Winter and Robert, *Capital Cities at War,* 35.

107 **The green lawns:** Fussell, *The Great War and Modern Memory,* 43.

107 **Further, its domestic agriculture:** Adrian Gregory, "Imperial Capitals at War," *London Journal* 42, no.3 (November 2016), 219-232, 227.

107 **Bread became scarce:** Belinda Davis, *Home Fires Burning: Food, Politics, and Everyday Life in World War I Berlin* (Chapel Hill: University of North Carolina Press, 2000), 24.

107 **Angry Berliners:** Ibid., 27.

108 **The black market:** Adrian Gregory, *The Last Great War* (Cambridge: Cambridge University Press, 2008), 227.

108 **"butter riots":** Davis, *Home Fires Burning,* 1.

109 **Synthetic nitrate:** Fritz Stern, *Einstein's German World* (Princeton: Princeton University Press, 2001), 119.

109 **Einstein was supposed to:** Albrecht Fölsing, *Albert Einstein: A Biography* (New York: Viking, 1997), 354–55.

109 **The government came to view:** Gerald Feldman, "A German Scientist Between Illusion and Reality: Emil Fischer, 1909–1919," in *Deutschland in der Weltpolitik des* 19. und 20. *Jahrhunderts,* eds. Imanuel Geiss and Bernd Jürgen Wendt (Düsseldorf: Bertelsmann Universitätsverlag, 1973), 341–62, 356.

109 **"He died as a soldier":** Stern, *Einstein's German World,* 121.

110 **The faculty of:** Elizabeth Fordham, "Universities," in Winter and Robert, *Capital Cities at War,* 262.

110 **He replied that Chadwick:** Russell McCormmach, *Night Thoughts of a Classical Physicist* (Cambridge, Massachusetts: Harvard University Press, 1982), 144, 210.

110 **King's College:** Fordham, "Universities," 266.

110 **Immigrants were no longer:** Panikos Panayi, "Minorities," in Winter, *The Cambridge History of the First World War,* 222.

110 **Postal censorship:** Niall Ferguson, *The Pity of War* (Basic Books, 1998), 186.

110 **He spent the length:** Wolff, "Physicists in the 'Krieg der Geister,'" 339. McCormmach, *Night Thoughts,* 143, 210.

110 **Albrecht Penck:** Roy MacLeod, "'Kriegsgeologen and Practical Men': Military Geology and Modern Memory, 1914–18," *British Journal for the History of Science* 28, no.4 (1995): 427–50, 431.

CHAPTER 5

112 **It claimed hundreds of thousands:** Ernest Rutherford, "Henry Gwyn Jeffreys Moseley," *Nature* 96, no. 2393 (September 9, 1915): 33–34. John Heilbron, "The Work of H.G.J. Moseley," *Isis* 57, no. 3 (1966): 336–64.

112 **"Had the European War":** Quoted in Daniel Kevles, *The Physicists* (Harvard University Press, 1995), 113.

113 **"a heavy loss":** K. Fajans, *Die Naturwissenschaften* 4 (1916), 381–82.
113 **"pride for their":** Rutherford, "Henry Gwyn Jeffreys Moseley," 33.
113 **"we cannot but recognise":** Ibid., 34.
113 **A single lab:** Roy M. MacLeod, "The Chemists Go to War: The Mobilization of Civilian Chemists and the British War Effort, 1914–1918," *Annals of Science* 50 (1993): 473.
113 **"Killed in Flanders":** "Waste of Brains," *The Times*, December 24, 1916.
114 **No longer:** Alan Wilkinson, *The Church of England and the First World War* (London: Lutterworth, 2014), 212.
114 **A piano:** *The Manchester Guardian*, May 13, 1915.
114 **"vendetta":** Adrian Gregory, *The Last Great War* (Cambridge: Cambridge University Press, 2008), 236.
114 **"No Compromise":** Panikos Panayi, "Minorities," in Winter, *The Cambridge History of the First World War*, 216–41, 227.
115 **I love science:** CPAE volume 8, document 45, "Einstein to Paolo Straneo, 7 January 1915," 57.
115 **"the only thing":** CPAE volume 8, document 84, "Einstein to Heinrich Zangger, 17 May 1915," 97.
115 **"The theoretician is led":** CPAE volume 8, document 52, "Einstein to Hendrik Lorentz, 3 February 1915," 65.
115 **"personal view":** CPAE volume 8, document 43, "From Hendrik A. Lorentz to Einstein, between 1 and 23 January 1915," 53.
116 **"kaleidoscopic mixture":** Quoted in Jürgen Renn and Tilman Sauer, "Pathways Out of Classical Physics," in *The Genesis of General Relativity* vol. 1, ed. Jürgen Renn (Dordrecht: Springer, 2007), 257.
116 **Freundlich's supervisor:** CPAE volume 8, document 54, "Einstein to Erwin Freundlich, 5 February 1915," 66.
117 **"work on gravitation progresses":** Renn and Sauer, "Pathways," 251.
117 **"brought, in a sense":** CPAE volume 8, document 45, "Einstein to Paolo Straneo, 7 January 1915," 57.
117 **"I am working tranquilly":** CPAE volume 8, document 44, "Einstein to Edgar Meyer, 2 January 1915," 56.
117 **"exceedingly modest":** CPAE volume 8, document 75, "Einstein to Tullio Levi-Civita, 14 April 1915," 89.
117 **The most recent issue:** "Notes," *The Observatory* 479 (October 1914): 392.
118 **"Owing to the cutting":** Pickering to Dyson, October 8, 1914, Cambridge University Library, Royal Greenwich Observatory Archives, Papers of Frank Dyson, MS.RGO.8/104.
118 **American astronomers:** Dyson to Pickering, October 20, 1914, Papers of Frank Dyson, op. cit.
118 **Strömgren assured astronomers:** Strömgren to Dyson, November 6, 1914; Dyson to Pickering, November 9, 1914, Papers of Frank Dyson, op. cit.
119 **It was refused again:** Dyson to Postmaster General, November 18, 1914; Post office to Dyson, November 27, 1914, Papers of Frank Dyson, op. cit.
119 **"the appearance of assisting":** Plummer to Dyson, November 26, 1914; Plummer to Dyson, December 3, 1914, Papers of Frank Dyson, op. cit.

119 **"would prefer":** R.T.A. Innes at Johannesburg to Dyson, December 10, 1914, Papers of Frank Dyson, op. cit.

119 **"considerate German":** Lawrence Badash, "British and American Views of the German Menace in World War I," *Notes and Records of the Royal Society of London* 34, no. 1 (July 1979): 95.

119 **A. B. Basset:** Ibid., 94.

120 **At the 1915 BA meeting:** Ibid., 96.

120 **"unimaginative German hands":** Ibid., 96–97.

120 **Nonetheless, he remained:** Ibid., 97.

120 **"abstract, heavy, obscure":** Anne Rasmussen, "Mobilising Minds," in *The Cambridge History of the First World War*, vol. 2, ed. Jay Winter (Cambridge: Cambridge University Press, 2014), 405.

121 **"pull down from":** Badash, "British and American Views," 99–100.

121 **Reports of scientists:** *The Observatory* 489 (July 1915): 306.

121 **"a symbol that the Empire":** *The Observatory* 492 (October 1915): 409.

121 **The 1915 meeting:** Ibid., 413.

121 **It is very sad:** A. S. Eddington to Annie Jump Cannon, July 3, 1915, Annie Jump Cannon Papers, Harvard University Archives HUGFP 125.12 Box 2 HA1UPX. Courtesy of the Harvard University Archives.

122 **"science is thicker than blood":** W. W. Campbell, "International Co-operation in Science," September 15, 1917, attached to Campbell to G. E. Hale, September 19, 1917, Box 9, Folder 6, George Ellery Hale Papers, Archives, California Institute of Technology.

122 **They ran refugee camps:** *Report of the War Victims' Relief Committee of the Society of Friends* (London: Spottiswoode & Co., 1914–1919), vol. 1, 5, vol. 3, 44.

122 **organized to aid enemy citizens:** Ibid., 13.

122 **The Emergency Committee received:** Ibid., 15.

122 **They did not approve:** Rufus Jones, *A Service of Love in War Time* (New York: Macmillan Co., 1920), 252.

123 **His breakthrough of ammonia:** Fritz Stern, *Einstein's German World* (Princeton: Princeton University Press, 2001), 85.

123 **He converted to Christianity:** Ibid., 73–74.

123 **"uncritical acceptance":** L. F. Haber, *The Poisonous Cloud* (Oxford: Clarendon Press, 1986), 2.

124 **Haber's experiments:** Ibid., 27.

124 **Then, they waited:** Guy Hartcup, *The War of Invention: Scientific Developments, 1914–18* (London: Brassey's Defence Publishers, 1988), 96.

124 **creeping forward:** Haber, *Poisonous Cloud*, 34.

124 **The Germans were unprepared:** John Keegan, *The First World War* (London: Hutchinson, 1998), 214.

125 **"troubles the mind":** Haber, *Poisonous Cloud*, 277.

125 **His subordinates grew rapidly:** Hartcup, *War of Invention*, 105.

125 **As a stopgap:** Haber, *Poisonous Cloud*, 45.

125 **Ironically, the first:** Ibid., 31.

125 **The Chemical Society of London:** Roy MacLeod, "Scientists," in Winter, *The Cambridge History of the First World War*, vol. 2, 443.

125 **"The Chemical Society considers":** Lawrence Badash, "British and American Views of the German Menace in World War I," *Notes and Records of the Royal Society of London* 34, no. 1 (July 1979): 110.

126 **Stunned, Haber threw:** Stern, *Einstein's German World*, 122–23.

126 **"Mrs. Haber shot herself":** CPAE volume 8, document 83, "Einstein to Mileva Einstein-Marić, 15 May 1915," 97.

126 **As with Haber:** Suman Seth, *Crafting the Quantum: Arnold Sommerfeld and the Practice of Theory, 1890–1926*, Transformations: Studies in the History of Science and Technology (Cambridge, Massachusetts: MIT Press, 2010), 74–79.

126 **In the fall of 1914:** RCSM, vol. 10 (November 5, 1914): 475.

126 **The War Office did not:** MacLeod, "The Chemists Go to War," 461.

127 **Germany's dominance:** RSCM, April 22, 1915. For an example, see the correspondence between the War Committee and chemical manufacturers trying to replicate German glues. Royal Society Council Documents (CD), CD/67 "Advice to Chemical Manufacturers"—CD/67.

127 **It was found that:** William Henry Perkin, "Presidential Address: The Position of the Organic Chemical Industry," *Journal of the Chemical Society, Transactions* 107 (1915): 557–78.

127 **Britain was almost completely:** MacLeod, "The Chemists Go to War," 459.

127 **Their formal request:** Marwick, *Deluge*, 229.

127 **The military had:** Ibid., 230.

127 **A committee of major scientists:** Peter Alter, *The Reluctant Patron: Science and the State in Great Britain* (New York: Berg, 1987), 96–97.

127 **"On our side":** William Van der Kloot, *Great Scientists Wage the Great War* (Oxford: Fonthill Books, 2014), 93.

128 **The Royal Society Council:** Heilbron, "The Work of H.G.J. Moseley," 336.

128 **The Braggs were so effective:** Arne Schirrmacher, "Sounds and Repercussions of War: Mobilization, Invention, and Conversion of First World War Science in Britain, France and Germany," *History and Technology: An International Journal* 32, no. 3 (October 9, 2016): 269.

128 **In an effort to maintain:** Roy MacLeod, "Sight and Sound on the Western Front," *War and Society* 18 (2000), 23–46, 39.

128 **By the end of the war:** MacLeod, "Scientists," 451.

128 **Much of the gas:** Hartcup, *War of Invention*, 168.

128 **The cylinders:** Ibid., 98–100.

129 **The British system:** MacLeod, "The Chemists Go to War," 466.

129 **Australia sent about half:** Roy M. MacLeod, "The 'Arsenal' in the Strand: Australian Chemists and the British Munitions Effort 1916–1919," *Annals of Science* 46 (1989): 58.

129 **She was denied:** MacLeod, "The Chemists Go to War," 475.

129 **One industrial chemist:** TNA:PRO: MH 47/1: Central Tribunal Minutes.

129 **After the system:** MacLeod, "The Chemists Go to War," 474.

129 **They argued that:** Cambridge Observatory Syndicate Minutes, 1896–1971, December 6, 1915, Cambridge University Archives, UA Obsy A1 iii.

130 **The Society was still:** RSCM, February 22, 1917.

130 **Life for scientists:** Hartcup, *War of Invention*, 35.

130 **They, too, had no clear guidelines:** MacLeod, "The Chemists Go to War," 474.
130 **"standing eye to eye":** Niall Ferguson, *The Pity of War* (Basic Books, 1998), 207.
130 **Erwin Schrödinger:** Walter Moore, *Schrödinger: Life and Thought* (Cambridge: Cambridge University Press, 1989), 83.
130 **A year into the war:** Van der Kloot, *Great Scientists Wage the Great War*, 34.
130 **"no longer knew":** Ibid., 34.
130 **Meitner felt strongly:** Russell McCormmach, *Night Thoughts of a Classical Physicist* (Cambridge, Massachusetts: Harvard University Press, 1982), 171.
131 **"Einstein played the violin":** Hubert Goenner and Giuseppe Castagnetti, "Albert Einstein as Pacifist and Democrat During World War I," *Science in Context* 9, no. 4 (December 1996): 369.
131 **Walther Schücking:** Ibid., 336.
131 **Einstein took it out:** CPAE volume 8, document 86, "Einstein to Heinrich Zangger, 28 May 1915," 100–101.
131 **"until now not politically active":** Goenner and Castagnetti, "Albert Einstein as Pacifist," 372.
131 **"a woman's world":** Emmanuelle Cronier, "The Street," in *Capital Cities at War: Paris, London, Berlin 1914–1919: A Cultural History*, vol. 2, eds. Jay Winter and Jean-Louis Robert, Studies in the Social and Cultural History of Modern Warfare (Cambridge: Cambridge University Press, 1997), 88.
132 **Some 70,000 Eastern European Jews:** Ibid., 96.
132 **Right-wing groups:** Adrian Gregory, "Religious Sites and Practices," in *Capital Cities at War*, vol. 2, 406–17, 406–8.
132 **This should not be mistaken:** McCormmach, *Night Thoughts*, 213.
132 **It was not well received:** John Heilbron, *The Dilemmas of an Upright Man: Max Planck as Spokesman for German Science* (Berkeley: University of California Press, 1986), 74–78.

CHAPTER 6

133 **He speculated about:** A. S. Eddington, "Some Problems in Astronomy XIX: Gravitation," *The Observatory* 484 (1915): 93–98.
134 **Cunningham was:** Andrew Warwick, *Masters of Theory: Cambridge and the Rise of Mathematical Physics* (Chicago: University of Chicago Press, 2003), 409; Andrew C. Thompson, "Logical Nonconformity? Conscientious Objection in the Cambridge Free Churches After 1914," *Journal of United Reformed Church History Society* 5, no. 9 (November 1996): 551.
135 **"would mean that gravitation":** Eddington, "Some Problems," 98.
137 **"There is the problem":** Leo Corry, "The Origin of Hilbert's Axiomatic Method," in *The Genesis of General Relativity*, vol. 4, ed. Jürgen Renn (Dordrecht: Springer, 2007), 771.
138 **Hilbert had refused:** Albrecht Fölsing, *Albert Einstein: A Biography* (New York: Viking, 1997), 364; CPAE volume 8, document 101, "Einstein to Heinrich Zangger between 24 July and 7 August 1915," 115–16.

138 **"Berlin is no match":** CPAE volume 8, document 94, "Einstein to Heinrich Zangger, 7 July 1915," 110.

138 **"clarified very much":** Fölsing, *Albert Einstein*, 364.

138 **With a verbal shrug:** Ibid., 365.

139 **"Hilbert now regrets":** CPAE volume 8, document 94, "Einstein to Heinrich Zangger, 7 July 1915," 110.

139 **A planning meeting:** Hubert Goenner and Giuseppe Castagnetti, "Albert Einstein as Pacifist and Democrat During World War I," *Science in Context* 9, no. 4 (December 1996): 342.

139 **Surely there must be:** CPAE volume 8, document 98, "Einstein to Hendrik A. Lorentz, 21 July 1915," 113.

139 **The powerful would forever:** CPAE volume 8, document 103, "Einstein to Hendrik A. Lorentz, 2 August 1915," 117.

139 **"no education and intellectual":** CPAE volume 8, document 101, "Einstein to Heinrich Zangger, between 24 July and 7 August 1915," 115.

140 **By the fall:** Thomas Levenson, *Einstein in Berlin* (New York: Bantam Books, 2003), 142.

140 **Elsa had been cooking:** CPAE volume 8, document 94, "Einstein to Heinrich Zangger, 7 July 1915," 110.

140 **The German High Command:** Elizabeth Fordham, "Universities," in *Capital Cities at War*, vol. 2, eds. Jay Winter and Jean-Louis Robert (Cambridge: Cambridge University Press, 1997), 252.

140 **A replica of a front-line trench:** Levenson, *Einstein in Berlin*, 118.

140 **Einstein hoped that Germany:** Goenner and Castagnetti, "Albert Einstein as Pacifist and Democrat During World War I," 340.

141 **This was aggravated by:** See CPAE volume 8A, 166 and 277.

141 **"Milk and honey":** CPAE volume 8, document 115, "Einstein to Elsa Einstein, 3 September 1915," 125.

141 **Besso helped mediate:** CPAE volume 8, document 133, "From Michele Besso to Einstein, 30 October 1915," 139–40.

142 **"very vivacious and serene":** Rolland's diary, quoted in Fölsing, *Albert Einstein*, 349.

142 **"It is being harassed":** CPAE volume 8, document 118, "Einstein to Romain Rolland, 15 September 1915," 127.

142 **Humiliatingly, a soldier:** CPAE volume 8, document 120, "Einstein to Heinrich Zangger, 19 September 1915," 129.

143 **"Nature always uses":** *Memoirs of the Royal Academy of Sciences at Paris*, (April 15, 1744), 417–26.

143 **He immediately sat down:** CPAE volume 8, document 122, "Einstein to Hendrik A. Lorentz, 23 September 1915," 131.

143 **But in retracing:** Michel Janssen and Jürgen Renn, "Arch and Scaffold: How Einstein Found His Field Equations," *Physics Today* 68 (2015): 30–36; Michel Janssen and Jürgen Renn, "Einstein Was No Lone Genius," *Nature* 527, no. 7578 (November 2015): 298–300.

143 **"electrifies me enormously":** CPAE volume 8, document 123, "Einstein to Erwin Freundlich, 30 September 1915," 132.

144 **"I must depend":** CPAE volume 8, document 123, "Einstein to Erwin Freundlich, 30 September 1915," 132.
144 **In early October:** Jürgen Renn and Tilman Sauer, "Pathways Out of Classical Physics," in *The Genesis of General Relativity* vol. 1, ed. Jürgen Renn (Dordrecht: Springer, 2007), 113–312.
144 **He agreed to do so:** Goenner and Castagnetti, "Albert Einstein as Pacifist," 346.
144 **"animal hatred":** Ibid., 348.
144 **He called for:** Fölsing, *Albert Einstein*, 367.
146 **"unscrupulous opportunist":** Albert Einstein, "Remarks Concerning the Essays Brought Together in This Co-Operative Volume," in *Albert Einstein: Philosopher-Scientist*, ed. Paul Arthur Schlipp (Evanston, IL: The Library of Living Philosophers, 1949), 683–84.
147 **He described how:** Fölsing, *Albert Einstein*, 372.
147 **"I have immortalized":** CPAE volume 8, document 153, "Einstein to Arnold Sommerfeld, 28 November 1915," 152.
147 **This certainly made for:** Michel Janssen and Jürgen Renn, "Untying the Knot," *The Genesis of General Relativity* vol. 2, ed. Jürgen Renn (Dordrecht: Springer, 2007), 850–51.
147 **"In the last few days":** CPAE volume 8, document 134, "Einstein to Hans Albert Einstein, 4 November 1915," 140.
148 **"I am curious":** CPAE volume 8, document 136, "Einstein to David Hilbert, 7 November 1915," 141.
148 **"the glorification of war":** CPAE volume 8, document 138, "Einstein to Berliner Goethebund, 11 November 1915," 143.
148 **Einstein particularly noted:** CPAE volume 8, document 139, "Einstein to David Hilbert, 12 November 1915," 143.
148 **He would be happy:** CPAE volume 8, document 140, "Hilbert to Einstein, 13 November 1915," 144.
148 **He asked Hilbert:** CPAE volume 8, document 144, "Einstein to David Hilbert, 15 November 1915," 145–46.
149 **His calculation said:** Renn and Sauer, "Pathways," 280.
149 **"Imagine my joy":** CPAE volume 8 (listed in volume 10), document 144a, "Einstein to Heinrich Zangger, 15 November 1915," 19.
149 **In these last months:** CPAE volume 8, document 147, "Einstein to Michele Besso, 17 November 1915," 148.
149 **"Today I am presenting":** CPAE volume 8, document 148, "Einstein to David Hilbert, 18 November 1915," 148.
149 **Einstein had the advantage:** CPAE volume 8, document 149, "From David Hilbert to Einstein, 19 November 1915," 149.
149 **This was probably because:** Renn and Sauer, "Pathways," 280.
150 **"I was beside myself":** CPAE volume 8, document 182, "Einstein to Paul Ehrenfest, 17 January 1916," 179.
150 **"The theory is beautiful":** CPAE volume 8, document 152, "Einstein to Heinrich Zangger, 26 November 1915," 151.
150 **Hilbert credited Einstein:** Jürgen Renn and John Stachel, "Hilbert's Foundation of Physics" in Renn, *Genesis* vol. 4, 911.

150 **Another large part:** Ibid., 857.
151 **"In my personal experience":** CPAE volume 8, document 152, "Einstein to Heinrich Zangger, 26 November 1915," 151.
151 **Hilbert quickly realized:** Renn and Stachel, "Hilbert's Foundation of Physics," 911.
151 **There has been a certain ill-feeling:** CPAE volume 8, document 167, "Einstein to David Hilbert, 20 December 1915," 163.
152 **"rather fishy":** Matthias Schemmel, "The Continuity Between Classical and Relativistic Cosmology in the Work of Karl Schwarzschild," in *The Genesis of General Relativity* vol. 3, ed. Jürgen Renn (Dordrecht: Springer, 2007), 167.
152 **"agreement of two single numbers":** Schemmel, "Continuity," 168.
152 **"As you read":** CPAE volume 8, document 161, "Einstein to Arnold Sommerfeld, 9 December 1915," 159.
152 **"trying to poke holes":** CPAE volume 10 (cited as volume 8), document 159a, "Einstein to Heinrich Zangger, 4 December 1915," 20.
154 **He wanted something new:** CPAE volume 8, document 165, "Einstein to Moritz Schlick, 14 December 1915," 161–62.
154 **He was planning a visit:** CPAE volume 8, document 168, "Einstein to Michele Besso, 21 December 1915," 163.
154 **He did not mention:** Ibid.

CHAPTER 7

156 **"chain of wrong tracks":** CPAE volume 8, document 183, "Einstein to Hendrik A. Lorentz, 17 January 1916," 179.
156 **"walk around without a muzzle":** CPAE volume 8, document 177, "Einstein to Hendrik A. Lorentz, 1 January 1916," 170.
156 **"abominable":** CPAE volume 8, document 183, "Einstein to Hendrik A. Lorentz, 17 January 1916," 179–81.
157 **One of Einstein's friends:** CPAE volume 8, document 247, "Gunnar Nordström to Einstein, 3 August 1916," 241.
157 **The fighting felt close:** Martin J. Klein, *Paul Ehrenfest, Volume I: The Making of a Theoretical Physicist* (New York: Elsevier Science, 1970), 298.
157 **"a supporter of the peace movement":** Hubert Goenner and Giuseppe Castagnetti, "Albert Einstein as Pacifist and Democrat During World War I," *Science in Context* 9, no. 4 (December 1996): 372.
157 **Surely the war:** Albrecht Fölsing, *Albert Einstein: A Biography* (New York: Viking, 1997), 396.
159 **"As you see":** CPAE volume 8, document 169, "From Karl Schwarzschild to Einstein, 22 December 1915," 164.
161 **"the question of light deflection":** CPAE volume 8, document 176, "Einstein to Karl Schwarzschild, 29 December 1915," 170.
162 **"enjoyed and admired":** Andrew Warwick, *Masters of Theory: Cambridge and the Rise of Mathematical Physics* (Chicago: University of Chicago Press, 2003), 476–77.
162 **"is why I would regret":** CPAE volume 8, document 181, "Einstein to Karl Schwarzschild, 9 January 1916," 177.

162 **"foiled assault":** CPAE volume 8, document 186, "Einstein to Arnold Sommerfeld, 2 February 1916," 188; CPAE volume 8, document 207, "Einstein to David Hilbert, 30 March 1916," 205–6.

164 **"We regard":** Report of the 1915 Yearly Meeting, quoted in John W. Graham, *Conscription and Conscience* (London: Allen & Unwin, 1922), 162.

164 **They were formed:** James McDermott, *British Military Service Tribunals 1916–1918* (Manchester: Manchester University Press, 2011), 16.

165 **Some 1.2 million:** Ibid., 24.

165 **The Banbury Local Tribunal:** Adrian Gregory, *The Last Great War* (Cambridge: Cambridge University Press, 2008), 101–2.

167 **"He will carry":** J. D. Symon, *The Universities' Part in the War*, pamphlet (n.p., 1915), Cambridgeshire Public Library C 45.5, 727–28.

167 **"The melancholy of this place":** Quoted in Stuart Wallace, *War and the Image of Germany: British Academics, 1914–1918* (Edinburgh: John Donald Publishing, 1988), 74.

168 **"refused to sheathe":** *Cambridge Magazine*, March 4, 1916, 359.

168 **While an undergraduate:** Warwick, *Masters of Theory*, 451.

169 **"Hitherto I had":** Eddington to de Sitter, June 11, 1916, Leiden UB, AFA FC WdS 14, Leiden University Library, Leiden Observatory Archives, directorate Willem de Sitter. Hereafter WdS.

169 **"I am immensely interested":** Ibid.

169 **"So far as I can":** Eddington to de Sitter, July 4, 1916, WdS.

170 **"I was interested to hear":** Eddington to de Sitter, October 13, 1916, WdS.

170 **"Your theory still seems":** CPAE volume 8, document 243, "From Willem de Sitter to Einstein, 27 July 1916," 239.

170 **"It is a fine thing":** CPAE volume 8, document 290, "From Einstein to Willem de Sitter, 23 January 1917," 279–80.

171 **His general absentmindedness:** Alice Calaprice, ed., *The Ultimate Quotable Einstein* (Princeton: Princeton University Press, 2010), 30.

171 **"very thorough":** CPAE volume 8, document 209a, "Einstein to Elsa Einstein, from Zurich, 6 April 1916," 22.

171 **Einstein's plan:** For example, see CPAE volume 8, document 211, "Einstein to Mileva Einstein-Marić, 8 April 1916," 208–9.

171 **She was also increasingly ill:** Fölsing, *Albert Einstein*, 395.

171 **But our enemies:** William Van der Kloot, *Great Scientists Wage the Great War* (Oxford: Fonthill Books, 2014), 194.

172 **The soap shortage:** Roger Chickering, *The Great War and Urban Life in Germany: Freiburg, 1914–1918*, Studies in the Social and Cultural History of Modern Warfare 24 (Cambridge: Cambridge University Press, 2007), 295–307.

172 **"my aesthetic demands":** CPAE volume 8, document 247a, "Einstein to Heinrich Zangger, 3 August 1916," 27.

172 **"I shut my eyes":** CPAE volume 8, document 232a, "Einstein to Heinrich Zangger, 11 July 1916," 24.

172 **Unable to continue:** Goenner and Castagnetti, "Albert Einstein as Pacifist," 349, 353.

172 **"people will lose faith":** CPAE volume 8, document 264, "Einstein to Werner Weisbach, 14 October 1916," 253.

173 **"state of siege":** Jon Lawrence, "Public Space, Political Space," in *Capital Cities at War: Paris, London, Berlin 1914–1919: A Cultural History*, vol. 2, eds. Jay Winter and Jean-Louis Robert, Studies in the Social and Cultural History of Modern Warfare (Cambridge: Cambridge University Press, 1997), 288.
173 **"among the living":** CPAE volume 8, document 223, "Einstein to David Hilbert, 30 May 1916," 216.
173 **"artistic pleasure from devising":** CPAE volume 6, document 33, "Einstein's Memorial Lecture on Karl Schwarzschild," Victoria Yam, trans.
173 **"He would have been a gem":** CPAE volume 8, document 219, "Einstein to Michele Besso, 14 May 1916," 213.

CHAPTER 8

174 **"We would rather say":** *The Observatory* 503 (August 1916): 337–39.
175 **"Is not the die really cast":** "From an Oxford Note-Book," *The Observatory* 500 (May 1916): 240.
176 **he praised Turner:** RAS MSS Grove Hills 1914, Arthur Eddington to Hills, 2/1 14, 27 January, Royal Astronomical Society, London.
176 **"The dilemma is inexorable":** J. H. Morgan, "German Atrocities: An Official Investigation," quoted in "From an Oxford Note-Book," *The Observatory* 500 (May 1916): 241–42.
176 **"Above all, there is the conviction":** A. S. Eddington, "The Future of International Science," *The Observatory* 501 (June 1916): 271.
177 **Fortunately, most of us know:** Ibid.
178 **"Naturally I do not regard":** Arthur Eddington to Joseph Larmor, June 7, 1916, MS/603, Larmor Papers, Royal Society.
179 **"staring at a sunlit picture of Hell":** Siegfried Sassoon, *Memoirs of an Infantry Officer* (London: Faber and Faber, 1930), 76.
179 **Capt. W. P. Nevill:** Martin Gilbert, *A History of the Twentieth Century, Vol. 1: 1900–1933* (New York: William Morrow, 1997), 408–9; John Keegan, *The First World War* (New York: Vintage, 2000), 308–21.
179 **He casually lit:** J. M. Winter and Blaine Baggett, *1914–1918: The Great War and the Shaping of the 20th Century* (London: BBC Books, 1996), 16.
179 **"by candle light":** Joseph Loconte, "How J.R.R. Tolkien Found Mordor on the Western Front," *New York Times*, June 30, 2016, https://www.nytimes.com/2016/07/03/opinion/sunday/how-jrr-tolkien-found-mordor-on-the-western-front.html, accessed August 16, 2018.
179 **Is it not an actual fact:** "From an Oxford Note-Book," *The Observatory* 502 (July 1916): 23–24.
180 **"highly complicated in form":** A. S. Eddington, "Gravitation and the Principle of Relativity," *Nature* 98 (2461), December 28, 1916, 328–30.
181 **Eddington had hoped:** *The Observatory* 503 (August 1916): 337–39.
181 **"This is objective evidence":** CPAE volume 8, document 278, "Einstein to Hermann Weyl, 23 November 1916," 265.
182 **"After all, we are a university":** Clark Kimberling, "Emmy Noether and Her Influence," in *Emmy Noether, A Tribute to Her Life and Work*, by Emmy Noether,

ed. James W. Brewer and Martha K. Smith, Monographs and Textbooks in Pure and Applied Mathematics 69 (New York: Marcel Dekker Inc., 1982), 14.

182 **"Upon receiving the new work":** CPAE volume 8, document 677, "Einstein to Felix Klein, 27 December 1918," 714.

183 **So his task was:** CPAE volume 8, document 194, "Einstein to Karl Schwarzschild, 19 February 1916," 196; CPAE volume 8, document 226, "E to Hendrik A. Lorentz, 17 June 1916," 221; CPAE volume 8, document 227, "Einstein to Willem de Sitter, 22 June 1916," 223.

183 **Many peace activists:** CPAE volume 8, document 253, "Einstein to Paul Ehrenfest, 25 August 1916," 245.

184 **"A long chain":** CPAE volume 8, document 256, "Einstein to Paul Ehrenfest, 6 September 1916," 248.

184 **"the concurrence in opinion":** CPAE volume 8, document 276, "Einstein to Hendrik A. Lorentz, 13 November 1916," 263.

184 **His wife, Tatyana:** Martin J. Klein, *Paul Ehrenfest, Volume I: The Making of a Theoretical Physicist* (New York: Elsevier Science, 1970), 304.

184 **It is not a coincidence:** Andrew Warwick, *Masters of Theory: Cambridge and the Rise of Mathematical Physics* (Chicago: University of Chicago Press, 2003), 459–60.

184 **the only person in Belgium:** See CPAE volume 8, document 230, "Einstein to Théophile de Donder, 13 June 1916," 226; CPAE volume 8, document 231, "From Théophile de Donder to Einstein, 4 July 1916," 226.

185 **"Einstein began to puff":** Klein, *Paul Ehrenfest*, 303.

185 **"a bit of give and take":** Ibid.

185 **"have to receive me lardless":** CPAE volume 10 (cited as volume 8), document 262b, "Einstein to Elsa Einstein, 7 October 1916," 31.

185 **"The reinvigorating days":** CPAE volume 8, document 268, "Einstein to Paul and Tatyana Ehrenfest, 18 October 1916," 255.

185 **"already come very much alive":** CPAE volume 8, document 270, "Einstein to Michele Besso, 31 October 1916," 257.

185 **Einstein tried to recruit Planck:** CPAE volume 8, document 269, "Einstein to Paul Ehrenfest, 24 October 1916," 256.

185 **"skittish":** CPAE volume 8, document 276, "Einstein to Hendrik A. Lorentz, 13 November 1916," 263.

185 **"very warmly":** CPAE volume 8, document 275, "Einstein to Paul Ehrenfest, 7 November 1916," 263.

186 **This included Elsa's:** CPAE volume 9, document 277, "Einstein to Paul Ehrenfest, 17 November 1916," 265.

186 **Anti-Semitic conspiracy:** Belinda Davis, *Home Fires Burning: Food, Politics, and Everyday Life in World War I Berlin* (Chapel Hill: University of North Carolina Press, 2000), 135.

186 **Mass food protests:** Thomas Levenson, *Einstein in Berlin* (New York: Bantam Books, 2003), 143–44.

186 **In reality, everything was made worse:** Thierry Bonzon and Belinda Davis, "Feeding the Cities," in *Capital Cities at War: Paris, London, Berlin 1914–1919: A Cultural History*, vol. 1, eds. Jay Winter and Jean-Louis Robert, Studies in

the Social and Cultural History of Modern Warfare (Cambridge: Cambridge University Press, 1997), 336.

186 **"contrary to my most":** CPAE volume 10 (cited as volume 8), document 287b, "Einstein to Heinrich Zangger, 16 January 1917," 39–40.

186 **Most sausage available:** Davis, *Home Fires Burning*, 206.

186 **Ersatz sausage:** Stefan Goebel, "Schools," in *Capital Cities at War: Paris, London, Berlin 1914–1919: A Cultural History*, vol. 2, eds. Jay Winter and Jean-Louis Robert, Studies in the Social and Cultural History of Modern Warfare (Cambridge: Cambridge University Press, 1997), 218.

187 **His relatives in southern Germany:** CPAE volume 10 (cited as volume 8), document 287b, "Einstein to Heinrich Zangger, 16 January 1917," 39–40.

187 **Having food sent:** Bonzon and Davis, "Feeding the Cities," 322.

187 **The situation in Berlin:** Catherine Rollet, "The Home and Family Life," in Winter and Robert, *Capital Cities at War*, 333.

187 **"wrath of the evil spirits":** CPAE volume 10 (cited as volume 8), document 287b, "Einstein to Heinrich Zangger, 16 January 1917," 40.

187 **"sickly appearance":** CPAE volume 8, document 298, "Einstein to Paul Ehrenfest, 14 February 1917," 285; CPAE volume 10 (cited as volume 8), document 308a, "Einstein to Heinrich Zangger, 10 March 1917," 45.

187 **At least the pain was better:** CPAE volume 10 (cited as volume 8), document 308a, "Einstein to Heinrich Zangger, 10 March 1917," 45.

187 **"the limiting conditions in the infinite":** Albrecht Fölsing, *Albert Einstein: A Biography* (New York: Viking, 1997), 387.

188 **"If I am to believe":** CPAE volume 8, document 272, "From Willem de Sitter to Einstein, 1 November 1916," 261.

189 **"I am sorry for having placed":** CPAE volume 8, document 273, "Einstein to Willem de Sitter, 4 November 1916," 261.

190 **"The relative velocities":** CPAE volume 6, document 43, "Cosmological Considerations in the General Theory of Relativity," 428.

190 **"justified by our actual knowledge":** Ibid., 432.

190 **As usual when he encountered:** CPAE volume 8, document 300, "Einstein to Erwin Freundlich, 18 February 1917 or later," 287.

191 **"the eternal enigma-giver":** CPAE volume 8, document 306, "Einstein to Michele Besso, 9 March 1917," 293.

191 **"Jehovah did not found":** CPAE volume 8, document 308, "Einstein to Michele Besso, after 9 March 1917," 296.

191 **"rather outlandish":** CPAE volume 8, document 293, "Einstein to Willem de Sitter, 2 February 1917," 281.

191 **"exposes me a bit":** CPAE volume 8, document 294, "Einstein to Paul Ehrenfest, 4 February 1917," 282.

191 **"I argue as I do":** CPAE volume 8, document 308, "Einstein to Michele Besso, after 9 March 1917," 296.

191 **"Now I am no longer plagued":** CPAE volume 8, document 311, "Einstein to Willem de Sitter, 12 March 1917," 301.

191 **"Well, if you do not want":** CPAE volume 8, document 312, "From Willem de Sitter to Einstein, 15 March 1917," 302.

192 **"I do not concern myself":** CPAE volume 8, document 313, "From Willem de Sitter to Einstein, 20 March 1917," 303.
192 **"difference in creed":** CPAE volume 8, document 325, "From Einstein to Willem de Sitter, 14 April 1917," 316; volume 8, document 327, "From Willem de Sitter to Einstein, 18 April 1917."
193 **"probably torpedoed":** CPAE volume 8, document 312, "From Willem de Sitter to Einstein, 15 March 1917," 302.
193 **British grain supplies:** Elizabeth Bruton and Paul Coleman, "Listening in the Dark," *History and Technology: An International Journal* 32, no. 3 (2016): 245.
193 **"When I speak with people":** CPAE volume 8, document 306, "Einstein to Michele Besso, 9 March 1917," 293.
193 **The DFG was no more welcomed:** Hubert Goenner and Giuseppe Castagnetti, "Albert Einstein as Pacifist and Democrat During World War I," *Science in Context* 9, no. 4 (December 1996): 355.
193 **"This agitation is probably connected":** CPAE volume 8, document 204, "Wilhelm Foerster to Einstein, 25 March 1916," 204.
194 **"It is a pity":** CPAE volume 8, document 294, "Einstein to Paul Ehrenfest, 4 February 1917," 282.

CHAPTER 9

195 **her efforts to keep:** Friedrich Herneck, *Einstein at Home*, trans. Josef Eisinger (New York: Prometheus Books, 2016), 118.
195 **"unbelievably loudly":** Alice Calaprice, ed., *The New Quotable Einstein* (Princeton: Princeton University Press, 2005), 346.
195 **"The grub is good":** Albrecht Fölsing, *Albert Einstein: A Biography* (New York: Viking, 1997), 410.
195 **The envelope also included:** CPAE volume 10 (cited as volume 8), document 359a, "Einstein to Elsa Einstein, 30 June 1917," 58.
195 **In Zurich he excitedly wrote:** CPAE volume 10 (cited as volume 8), doc 376c, "Einstein to Elsa Einstein, 31 August 1917," 79.
196 **The military authorities decided:** Hubert Goenner and Giuseppe Castagnetti, "Albert Einstein as Pacifist and Democrat During World War I," *Science in Context* 9, no. 4 (December 1996): 357.
196 **He warned them not to seal:** CPAE volume 10 (cited as volume 8), document 332a, "Einstein to Heinrich Zangger, 4 May 1917," 50.
196 **As an unmarried man:** CPAE volume 8, document 403, "Einstein to Heinrich Zangger, 6 December 1917," 412.
196 **The middle of 1917 saw:** Belinda Davis, *Home Fires Burning: Food, Politics, and Everyday Life in World War I Berlin* (Chapel Hill: University of North Carolina Press, 2000), 206.
196 **Haber, Fischer, and Nernst worked:** Guy Hartcup, *The War of Invention: Scientific Developments, 1914–18* (London: Brassey's Defence Publishers, 1988), 36.
196 **At one point:** Davis, *Home Fires Burning*, 215.
196 **When some forward-thinking:** Ibid., 238.
196 **"All our exalted technological progress":** CPAE volume 8, document 403, "Einstein to Heinrich Zangger, 6 December 1917," 412.

196 **he volunteered to run:** Wolf Zuelzer, *The Nicolai Case* (Detroit: Wayne State University Press, 1982), 22.

196 **As his internationalist views became:** Ibid., 26–32.

197 **He was fined 1,200 marks:** Ibid., 178–79.

197 **He kept pressing Einstein:** CPAE volume 8, document 302, "From Georg Nicolai to Einstein, 26 February 1917," 288–90.

197 **"Thus I raise my voice":** CPAE volume 8, document 303, "Einstein to Georg Nicolai, 28 February 1917," 291.

197 **He immediately regretted:** CPAE volume 8, document 304, "Einstein to Georg Nicolai, 28 February 1917," 291.

197 **Instead, Adler became:** Peter Galison, "The Assassin of Relativity," in *Einstein for the 21st Century*, eds. Peter Galison, Gerald Holton, and Silvan Schweber (Princeton: Princeton University Press, 2008), 185–87.

197 **Einstein quickly wrote a testimony:** Ibid., 189–90.

198 **In May 1917, Adler:** Ibid., 190–95.

198 **Shopkeepers faced steep fines:** Adrian Gregory, *The Last Great War* (Cambridge: Cambridge University Press, 2008), 213.

198 **There was a famous incident:** C. S. Peel, *How We Lived Then, 1914–1918: A Sketch of Social and Domestic Life in England During the War* (London: Bodley Head, 1929), 98.

198 **But tea?:** Gregory, *The Last Great War*, 215.

199 **"Send us chemists":** Roy M. MacLeod, "The Chemists Go to War: The Mobilization of Civilian Chemists and the British War Effort, 1914–1918," *Annals of Science* 50 (1993): 473.

200 **"chemical reserve":** Ibid., 476–77.

200 **That, he said, had been a disaster:** Roy M. MacLeod, "Secrets Among Friends: The Research Information Service and the 'Special Relationship' in Allied Scientific Information and Intelligence, 1916–1918," *Minerva* 37 (1999): 212–14.

201 **H. H. Turner had stopped:** Dyson to Strömgren, 14 November 1917, Papers of Frank Dyson, Cambridge University Library, Royal Greenwich Observatory Archives, MS.RGO.8.150, cited by permission of the Syndics of Cambridge University Library. Hereafter DP.

201 **French and American astronomers:** Bureau de Longitudes to Dyson, 18 May 1917 and Unknown to Dyson 26 November 1917, DP.

201 **One British astronomer supported:** Plummer to Dyson, 26 July 1917, Papers of Frank Dyson, op. cit.

201 **"I am very sensible":** Dyson to Strömgren, 14 November 1917, Papers of Frank Dyson, op. cit.

201 **"I hardly know whether":** F. W. Dyson to Annie Jump Cannon, February 20, 1915, Annie Jump Cannon Papers, Harvard University Archives HUGFP 125.12 Box 2 HA1UPX. Courtesy of the Harvard University Archives.

201 **"great evil with which":** "Report of the Annual Meeting of the Association, Held on Wednesday October 30, 1918," *Journal of the British Astronomical Association* 29, no.1 (1918): 1.

201 **"very skeptical about the theory":** Arthur Eddington to Hermann Weyl, August 18, 1920, ETH-Bibliothek Zürich, Hochularchiv ETHZ, Hs 91: 523.

205 **It proved to be a challenge:** Andrew Warwick, *Masters of Theory: Cambridge and the Rise of Mathematical Physics* (Chicago: University of Chicago Press,

2003), 472; Eddington to Lodge, 7 August 1917; Alfred Lodge to Oliver Lodge, 18 August 1917; Eddington to Lodge, 8 December 1917, MS ADD 89 (Lodge Papers), UCL Library Services, Special Collections.

206 **Eddington gently pointed out:** Eddington to Lodge, 15 January 1918, MS ADD 89 (Lodge Papers), UCL Library Services, Special Collections.

206 **Its yellow-brown clouds:** L. F. Haber, *The Poisonous Cloud* (Oxford: Clarendon Press, 1986), 111.

206 **It rarely killed:** Ibid., 231.

206 **This was three weeks of bombing:** Phil Judkins, "Sound and Fury," in *History and Technology* 32, no. 3 (2016): 239.

206 **The royal family:** *The London Gazette,* July 17, 1917.

207 **In the aftermath:** Panayi, "Minorities," in Winter, *The Cambridge History of the First World War,* 223.

207 **In view of the war:** Lawrence Badash, "British and American Views of the German Menace in World War I," *Notes and Records of the Royal Society of London* 34, no. 1 (July 1979): 112.

207 **"the good faith of the contracting parties":** "From an Oxford Note-Book," *The Observatory* 524 (March 1918): 147.

208 **"We have tried to think":** Ibid., 128.

208 **"plagiarism and piracy":** *Nature* 102, no. 2571 (February 6, 1919): 446–47.

208 **"stubborn, self-righteous":** Paul Fussell, *The Great War and Modern Memory* (Oxford: Oxford University Press, 1975), 12.

208 **Haig was both devoutly Christian:** John Keegan, *The First World War* (New York: Vintage, 2000), 311.

209 **Gas shells were bursting:** Tim Cook, *No Place to Run* (Vancouver: UBC Press, 2001), 138.

209 **"A big shell had just burst":** "100 Years On, Relatives Gather to Remember Passchendaele's Fallen," *The Guardian,* July 31, 2017.

210 **During Passchendaele, conscription:** James McDermott, *British Military Service Tribunals 1916–1918* (Manchester: Manchester University Press, 2011), 22–25.

210 **"punched and pelted":** Nicoletta F. Gullace, *The Blood of Our Sons: Men, Women and the Renegotiation of British Citizenship During the Great War* (New York: Palgrave Macmillan, 2002), 113.

211 **"fire suddenly springs":** A. Vibert Douglas, *The Life of Arthur Stanley Eddington* (London: Thomas Nelson, 1956), 69.

211 **"impish satisfaction":** Ibid., 110.

212 **It could, and would, be checked:** A. S. Eddington, *Report on the Relativity Theory of Gravitation* (London: Fleetway Press, 1918), v.

212 **"Albert Einstein has provoked":** A. S. Eddington, *Space, Time, and Gravitation*, 2nd ed. (Cambridge: Cambridge University Press, 1920), v.

213 **"a strange race of men":** Ibid., 23.

213 **Gulliver regarded the Lilliputians:** Ibid., 23–24.

213 **"if man wishes":** Eddington, *Report*, 53.

214 **"the relativist is sometimes suspected":** Eddington, *Space, Time, and Gravitation*, 27.

214 **There was no Newtonian:** Ibid., 67–68.

214 **It is merely an image:** Eddington, *Report,* 85.

215 **"It suggests that only":** Ibid., 87.

215 **"Let us not be beguiled":** Eddington, *Space, Time, and Gravitation*, 56.

215 **Doubting the fourth dimension:** This metaphor is developed in detail in Eddington, *Space, Time, and Gravitation*, 181.

216 **"a complete history":** Eddington, *Report*, 17.

217 **"Pedantic criticism of":** Ibid., 19.

217 **"we may perhaps suspend":** Ibid., 57.

217 **If you could catch:** Eddington, *Space, Time, and Gravitation*, 111.

218 **"Whether the theory ultimately":** Ibid., v.

218 **"the purview of physics":** Eddington, *Report*, xi.

218 **Most people were still cautious:** *Nature* 103, no. 2575 (March 6, 1919): 2.

218 **Regardless of skepticism:** For example, see "The Theory of Relativity," *American Mathematical Monthly* 28, no. 4 (April 1921): 175.

219 **"We consoled ourselves":** Peel, *How We Lived Then*, 94–96.

219 **Planck hovered constantly:** Fölsing, *Albert Einstein*, 410–12.

220 **the institute's meager funds:** Ibid., 411.

220 **"I do not believe in":** CPAE volume 8, document 381, "Einstein to Michele Besso, 22 September 1917," 374.

220 **He speculated that the economic:** CPAE volume 10 (cited as volume 8), document 372a, "Einstein to Heinrich Zangger, 21 August 1917," 76.

220 **"steered by hard facts":** CPAE volume 8, document 374, "Einstein to Romain Rolland, 22 August 1917," 368.

220 **Perhaps suggesting one:** CPAE volume 8, document 376, "From Romain Rolland to Einstein, 23 August 1917," 371.

221 **Einstein, as always valuing:** CPAE volume 8, document 445, "From Fritz Haber to Einstein, 29 January 1918," 453, and CPAE volume 8, document 446, "Einstein to Fritz Haber, 29 January 1918," 454.

221 **He had been living with Elsa:** Fölsing, *Albert Einstein*, 418.

221 **We don't know the details:** CPAE volume 8, document 545, "From Ilse Einstein to Georg Nicolai, 22 May 1918," 564.

221 **He did not consider the whole episode:** Thomas Levenson, *Einstein in Berlin* (New York: Bantam Books, 2003), 154–56.

221 **If Einstein were to be awarded:** CPAE volume 8, document 449, "Einstein to Mileva Einstein-Marić, 31 January 1918," 456, and CPAE volume 10 (cited as volume 8), doc 475a, "From Mileva Einstein-Marić to Einstein, 5 March 1918," 89.

222 **Planck then chided Einstein:** CPAE volume 8, document 479, "From Max Planck to Einstein, 12 March 1918," 494.

222 **This immediately resulted in:** Fölsing, *Albert Einstein*, 417.

222 **"The state of mind":** CPAE volume 7, document 7, "Motives for Research," 1918, 42–45.

222 **The supreme task:** Ibid., 44.

223 **He emphasized its boldness:** Fölsing, *Albert Einstein*, 210–12.

CHAPTER 10

225 **"a 'slice of life'":** Quoted in R. C. Sherriff, *No Leading Lady* (London: Victor Gollancz, 1968), 45.

225 **What remains in my memory:** J. E. Edmonds, *Military Operations France and Belgium, 1918: Volume 1, The German March Offensive and Its Preliminaries* (London: Imperial War Museum Department of Printed Books, 1995 [1935]), 470.

226 **"We've no chance":** "1918: Year of Victory," National Army Museum Online, https://www.nam.ac.uk/explore/1918-victory, accessed October 16, 2018.

226 **"write encouragingly to friends":** Paul Fussell, *The Great War and Modern Memory* (Oxford: Oxford University Press, 1975), 17.

226 **One serious problem was:** John Keegan, *The First World War* (London: Hutchinson, 1998), 433.

226 **The Allies, however:** Ibid., 437.

226 **The protesters were no longer:** Thomas Levenson, *Einstein in Berlin* (New York: Bantam Books, 2003), 180.

226 **Perhaps it could even:** CPAE volume 8, document 521, "Einstein to David Hilbert, before 27 April 1918," 540, and CPAE volume 8, document 522, "Einstein to David Hilbert, before 27 April 1918," 541.

226 **He wanted no one:** Albrecht Fölsing, *Albert Einstein: A Biography* (New York: Viking, 1997), 415.

227 **It would only hurt:** CPAE volume 8, document 530, "From David Hilbert to Einstein, 1 May 1918," 547.

227 **"grave problems":** CPAE volume 8, document 531, "From Ernst Troeltsch to Einstein, 1 May 1918," 548.

227 **He had wanted to:** CPAE volume 8, document 548, "Einstein to David Hilbert, 24 May 1918," 568.

227 **"German work":** Hubert Goenner and Giuseppe Castagnetti, "Albert Einstein as Pacifist and Democrat During World War I," *Science in Context* 9, no. 4 (December 1996): 367.

227 **"by way of thinking":** Ibid., 370.; CPAE volume 8, document 560, "Einstein to Adolf Kneser, 7 June 1918," 581.

228 **Ilse Einstein, seized by:** Wolf Zuelzer, *The Nicolai Case* (Detroit: Wayne State University Press, 1982), 230–32.

228 **By summer 1918, Einstein:** Belinda Davis, *Home Fires Burning: Food, Politics, and Everyday Life in World War I Berlin* (Chapel Hill: University of North Carolina Press, 2000), 21.

228 **Hoping to spare Albert:** CPAE volume 8, document 514, "Einstein to Heinrich Zangger, 22 April 1918," 535.

228 **"the envy of all Berlin":** CPAE volume 8, document 517, "Einstein to Auguste Hochberger, before 24 April 1918," 537.

228 **Some part of him:** CPAE volume 8, document 514, "Einstein to Heinrich Zangger, 22 April 1918," 535.

228 **He joked to Max Born:** Fölsing, *Albert Einstein*, 418.

229 **This was at best:** Ibid., 419.

229 **Pirates, it was thought:** Lord Walsingham, "German Naturalists and Nomenclature" *Nature* 102 (September 5, 1918): 4.

230 **"honestly well disposed":** Ibid.

230 **Prussianism would dissappear:** W. J. Holland, "Shall Writers upon the Biological Sciences Agree to Ignore Systematic Papers Published in the German Language Since 1914?" *Science* 48, no. 1245 (November 8, 1918).

230 **"doubt the necessity":** Robert M. Yerkes, ed., *The New World of Science: Its Development During the War* (New York: Scribner's, 1920), 409.
230 **"bitter arguments":** Ibid., 410.
231 **There is nothing metaphysical:** A. S. Eddington, *Report on the Relativity Theory of Gravitation* (London: Fleetway Press, 1918), 29.
232 **"matter-of-fact":** A. S. Eddington, "Einstein's Theory of Gravitation," *The Observatory* 510 (1917): 93.
232 **If he could persuade:** "Report of the Meeting of the Association Held on Wednesday, November 27, 1918" *Journal of the British Astronomical Association* 29, no. 2 (1918–1919): 36–37.
232 **There was talk of eliminating:** Peter Brock, *Twentieth-Century Pacifism* (London: Van Nostrand Reinhold Co., 1970), 16. See RSCM Vol. 11, 1914–1920, April 25, 1918, 304.
233 **In making application:** Observatory Syndicate Minutes, 1896–1971, March 12, 1918, UA Obsy A1 iii.
234 **Their presence was powerful:** James McDermott, *British Military Service Tribunals 1916–1918* (Manchester: Manchester University Press, 2011), 56.
234 **in addition to questioning:** *Cambridge Daily News,* March 6, 1916.
234 **"You are exploiting God":** John W. Graham, *Conscription and Conscience* (London: Allen & Unwin, 1922), 71.
234 **"unpatriotic, a slacker, a weakling":** W.H.W., "A Guide to the Conscientious Objector and the Tribunals," 6. Note that contemporary writers sometimes also used "CO" to stand for "commanding officer."
234 **You have presumably heard:** *Cambridge Daily News,* March 4, 1916.
234 **The Earl of Malmesbury:** Denis Hayes, *Conscription Conflict* (London: Sheppard Press, 1949), 208.
235 **"Liberals, Socialists and Pacifists":** Co-ordinating Committee for Research into the Use of the University for War, *Cambridge University and War* (Cambridge: unknown publisher [pamphlet]), 19. Hereafter CUW.
235 **Things are coming near:** CUW, 37–38.
235 **Only then were they told:** Hayes, *Conscription Conflict*, 260, and Graham, *Conscription*, 81; The No-Conscription Fellowship, "Two Years' Hard Labour for Refusing to Disobey the Dictates of Conscience" (London: NCF, [1918]), 70.
235 **"the lot of that class":** Graham, *Conscription*, 82.
236 **Then, refusing to fight:** Wormwood Scrubs was the destination of most imprisoned COs, to the point where the largest Quaker Meeting in London during the war was held inside the prison. Thomas Kennedy, *British Quakerism* 1860–1920 (Oxford: Oxford University Press, 2001), 349.
236 **Once he finished his sentence:** After the war, the government expressed some misgivings (though not publicly) about rearresting men for what was essentially the same offense. See TNA:PRO MH 47, "Report of the Central Tribunal Appointed Under the Military Service Act 1916" (London: H.M.S.O., 1919), 20.
236 **Eddington personally transcribed:** Minutes: Cambridge, Huntington and Lynn Monthly Meeting, January 9, 1918, R59/26/5/4. Courtesy of Cambridgeshire Archives.
236 **I should explain first:** Eddington to Lodge, 22 July 1918, MS ADD 89 (Lodge Papers), UCL Library Services, Special Collections.
238 **The chairman briefly mentioned:** *Cambridge Daily News,* June 14, 1918.

239 **Unusually, they gave him:** *Cambridge Daily News*, June 28, 1918.
239 **This caused trouble for many:** *Cambridge Daily News,* March 18, 1916.
239 **Nonetheless, he wanted to be recognized:** Niall Ferguson, *The Pity of War* (Basic Books, 1998), 186.
239 **Sometimes COs simply stacked rocks:** "Report of the Central Tribunal," 11–12. TNA: PRO MH 47. Participants with special or technical skills were supposed to be put to use in their fields. Apparently, this did not ever happen. See TNA:PRO MH 47/1, June 28, 1916.
239 **Some COs asked:** Graham, *Conscription*, 232.
240 **He would not abandon:** S. Chandrasekhar, "The Richtmeyer Memorial Lecture—Some Historical Notes," *American Journal of Physics* 37, no. 6: 579–80. A similar version of the story is told in "Verifying the Theory of Relativity," in *Notes and Records of the Royal Society of London*, 1976, vol. 30, 249–60.
242 **Eddington then left the hall:** *Cambridge Daily News*, July 12, 1918.
242 **The rate of typhoid:** Leo van Bergen, "Military Medicine," in *The Cambridge History of the First World War*, vol. 3, ed. Jay Winter (Cambridge: Cambridge University Press, 2014), 303.
242 **Half of the medical:** Ibid., 301–6.
242 **In February 1915:** Jay Winter, "Shell Shock," in Winter, *The Cambridge History of the First World War*, 315.
243 **Historians have pointed out:** Anne Rasmussen, "The Spanish Flu," in Ibid., 337.
243 **Military doctors tried:** Ibid., 345–49.
243 **"fashionable illness":** Ibid., 335.
243 **"There was no doubt":** Ibid., 334.
243 **In one month Einstein's Berlin:** Levenson, *Einstein in Berlin*, 187.
244 **Elsa had been taking lozenges:** CPAE volume 9, document 7, "Einstein to Heinrich Zangger, 28 February 1919," 8.
244 **The government now recommended:** Levenson, *Einstein in Berlin*, 183–87.
244 **A British soldier wrote:** Dairy of Robert Cude, 8–9 August 1918. IWM CUDE R MM, 177. The National Archives, http://www.nationalarchives.gov.uk /pathways/firstworldwar/battles/counter.htm.
244 **I saw Wylie:** "2010 Private Edward William Wylie," Australian Red Cross Society Wounded and Missing Enquiry Bureau Files 1DRL/0428, Australian War Memorial Research Centre.
245 **The admiral in charge:** Keegan, *The First World War*, 446.
246 **Lenin sent agents:** Levenson, *Einstein in Berlin*, 189.
246 **On October 19:** CPAE volume 8B, document 640, "From Max Planck to Einstein, 26 October 1918," footnote 1, 930–31.
246 **"Think of the oath":** CPAE volume 8, document 640, "From Max Planck to Einstein, 26 October 1918," 684.
246 **Planck, with whom Einstein:** Suman Seth, *Crafting the Quantum: Arnold Sommerfeld and the Practice of Theory, 1890–1926*, Transformations: Studies in the History of Science and Technology (Cambridge, Massachusetts: MIT Press, 2010), 177; John Heilbron, *The Dilemmas of an Upright Man: Max Planck as Spokesman for German Science* (Berkeley: University of California Press, 1986), 82.
247 **"canceled because of revolution":** Fölsing, *Albert Einstein*, 421.

247 **"Militarism and the privy-councilor":** CPAE volume 8, document 652, "Einstein to Paul and Maja Winteler-Einstein, 11 November 1918," 693.
247 **His friends at the BNV:** Goenner and Castagnetti, "Albert Einstein as Pacifist," 361.
247 **"I am enjoying":** CPAE volume 8, document 663, "Einstein to Michele Besso, 4 December 1918," 703.
248 **"the most precious asset":** Goenner and Castagnetti, "Albert Einstein as Pacifist," 362.
248 **"Force breeds only bitterness":** CPAE volume 7, document 14, "On the Need for a National Assembly, 13 November 1918," 76.
248 **Democracy, he thought:** Goenner and Castagnetti, "Albert Einstein as Pacifist," 359.
248 **We do not know exactly how much:** Ibid., 362.
248 **"We do not keep to":** Ibid., 364.
249 **This matched nicely:** Ibid., 365.
249 **Skirmishes in the streets:** Levenson, *Einstein in Berlin*, 199.
249 **In a letter to his sons:** CPAE volume 8, document 667, "Einstein to Hans Albert and Eduard Einstein, 10 December 1918," 707.
249 **On December 7:** Levenson, *Einstein in Berlin*, 200.
249 **"The military religion":** CPAE volume 8, document 663, "Einstein to Michele Besso, 4 December 1918," 703.
249 **"I find everything unspeakably dire":** Seth, *Crafting the Quantum*, 178.
249 **"of the firm conviction":** CPAE volume 8, document 665, "Einstein to Arnold Sommerfeld, 6 December 1918," 705–6.
250 **The BNV held a meeting:** Goenner and Castagnetti, "Albert Einstein as Pacifist," 366.
250 **Huge bonfires were lit:** Elizabeth Fordham, "Universities," in *Capital Cities at War*, vol. 2, eds. Jay Winter and Jean-Louis Robert (Cambridge: Cambridge University Press, 1997), 307–8.
250 **Suffering and unrest:** Cambridge Friends Meeting Minutes November 1918, Cambridgeshire County Records R59/26/5/5.
250 **"squeezing Germany until the pips squeak":** *The Times*, November 24, 1918.
251 **"misconduct":** Graham, *Conscription*, 311.
251 **About seventy conscientious objectors:** Ibid., 322; Charles L. Mowat, *Britain Between the Wars* (Chicago: University of Chicago Press, 1955), 6.
251 **They journeyed into far:** Edward Thomas, *Quaker Adventures* (London: Fleming H. Revell, 1935), 1–22.
251 **While chemists are testing:** A. Ruth Fry, *A Quaker Adventure* (London: Nisbet and Co., 1926), 355.
251 **"closely connected":** Ibid., 330–31.

CHAPTER 11

252 **"In journeying to observe":** *The Observatory* 457 (January 1913): 62.
253 **He was frustrated both:** CPAE volume 9, document 10, "Einstein to Ehrenfest, 22 March 1919," 10.

253 **Finally, he was ordered:** Albrecht Fölsing, *Albert Einstein: A Biography* (New York: Viking, 1997), 424–25.
253 **"well-fed citizens":** CPAE volume 9, document 3, "Einstein to Hedwig and Max Born, 19 January 1919," 3.
254 **Unimpressed with the state of Germany:** CPAE volume 9, document 17, "Einstein to Pauline Einstein and Maja Winteler-Einstein, 4 April 1919," 15.
254 **There wasn't much interest:** Fölsing, *Albert Einstein*, 426.
254 **The conference officially opened:** Margaret MacMillan, *Paris 1919* (New York: Random House, 2002), 26, 46, 63.
255 **Answering these questions:** Alex Soojung-Kim Pang, *Empire and the Sun: Victorian Solar Eclipse Expeditions* (Stanford: Stanford University Press, 2002), 41.
256 **Much of the information:** Rory Mawhinney, "Astronomical Fieldwork and the Spaces of Relativity: The Historical Geographies of the 1919 British Eclipse Expeditions to Principe and Brazil," forthcoming in *Historical Geography*, 5.
256 **Eddington would go to Principe:** "Obituary Notices: Fellows: Cottingham, Edwin Turner," *Monthly Notices of the Royal Astronomical Society* 101: 131.
256 **Cortie was known:** "Obituary Notices: Fellows: Cortie, Aloysius L," *Monthly Notices of the Royal Astronomical Society* 86: 175.
256 **Dyson trusted him implicitly:** Richard Woolley, "Charles Rundle Davidson. 1875–1970," *Biographical Memoirs of Fellows of the Royal Society*, vol. 17 (November 1971), 193–94.
257 **It had captured good fields:** F. W. Dyson, A. S. Eddington, and C. Davidson, "A Determination of the Deflection of Light by the Sun's Gravitational Field, from Observations Made at the Total Eclipse of May 29, 1919," *Philosophical Transactions of the Royal Society of London Series A*, 220 (1920): 295.
258 **Normally, overhauling them:** RAS Papers 54, Council Minutes, Minutes of the JPEC Subcommittee, vol. 2, 14 June 1918.
259 **"this in itself calls":** *The Observatory* 537 (March 1919): 119–22.
259 **Hoping to get:** RAS Papers 54, Council Minutes, Minutes of the JPEC Subcommittee, vol. 1, 10 January 1919 and vol. 2, 14 February 1919.
260 **"false trichotomy":** John Earman and Clark Glymour, "Relativity and Eclipses: The British Eclipse Expeditions of 1919 and Their Predecessors," *Historical Studies in the Physical Sciences* 11, no. 1 (1980): 84; Alistair Sponsel, "Constructing a 'Revolution in Science': The Campaign to Promote a Favourable Reception for the 1919 Solar Eclipse Experiments," *British Journal for the History of Science* 35, no. 127 (December 2002): 442.
260 **There were other possible:** "RAS Meeting," *The Observatory* 526 (May 1918): 215.
260 **"Eddington will go mad":** A. S. Eddington, "Forty Years of Astronomy," in Joseph Needham, *Background to Modern Science* (Cambridge: The University Press, 1938), 117–44, 141–42, and A. Vibert Douglas, *The Life of Arthur Stanley Eddington* (London: Thomas Nelson, 1956), 40.
261 **By summer they would:** Sponsel, "Constructing a 'Revolution in Science,'" 444–47.
261 **Berliner admitted that:** CPAE volume 9, document 19, "From Arnold Berliner to Einstein, 9 April 1919," 17.

262 **"free your laboratory":** Lawrence Badash, "British and American Views of the German Menace in World War I," *Notes and Records of the Royal Society of London* 34, no. 1 (July 1979): 113.
262 **His son Erwin:** Fritz Stern, *Einstein's German World* (Princeton: Princeton University Press, 2001), 46–47.
262 **"He is rooted to":** CPAE volume 9, document 52, "Einstein to Heinrich Zangger, 1 June 1919," 44.
262 **"The country is like":** CPAE volume 9, document 16, "Einstein to Aurel Stodola, 31 March 1919," 15.
263 **Of Booth's:** Mawhinney, "Astronomical Fieldwork," 7.
263 **This was some of the first:** 11 March 1919, Eddington to Sarah Ann Eddington, EDDN A4/1, Eddington Papers, Trinity College Library, Cambridge.
263 **"unlimited sugar":** Ibid.
263 **The ship stopped:** 15–16 March 1919, Funchal, Eddington to Sarah Ann Eddington, EDDN A4/2, Eddington Papers, Trinity College Library, Cambridge.
263 **He was instrumental:** Mawhinney, "Astronomical Fieldwork," 8; E. Mota, P. Crawford, and A. Simoes, "Einstein in Portugal: Eddington's Expedition to Principe and the Reactions of Portuguese Astronomers," *The British Journal for the History of Science* 42, no. 2 (2009): 256.
264 **Finally free of rationing:** 15–16 March 1919, Funchal, Eddington to Sarah Ann Eddington, EDDN A4/2.; March 27, 1919, Funchal, Eddington to Sarah Ann Eddington, EDDN A4/3 Eddington Papers, Trinity College Library, Cambridge.
264 **"as he was neither beautiful":** 5 May 1919, Principe, Eddington to Winifred Eddington, EDDN A4/8, Eddington Papers, Trinity College Library, Cambridge.
264 **only place to get proper tea:** 27 March 1919, Funchal, Eddington to Sarah Ann Eddington, EDDN A4/3, Eddington Papers, Trinity College Library, Cambridge.
264 **"I expect Mother":** 5 May 1919, Principe, Eddington to Winifred Eddington, EDDN A4/8, Eddington Papers, Trinity College Library, Cambridge.
264 **The other passengers:** Katy Price, *Loving Faster than Light: Romance and Readers in Einstein's Universe* (Chicago: University of Chicago Press, 2012), 119.
265 **Many of those companies:** Catherine Higgs, *Chocolate Islands: Cocoa, Slavery, and Colonial Africa* (Athens, Ohio: Ohio University Press, 2012), 16; Carol Off, *Bitter Chocolate* (Toronto: Random House Canada, 2006), 49.
265 **The Principe authorities:** Higgs, *Chocolate Islands*, 143.
265 **And as the light brightened:** Ibid., 59.
266 **They were met by both:** Morize, "From Port to Planation," 14. RAS Papers 54, Council Minutes, Minutes of the JPEC Subcommittee, vol. 2, 14 June 1918 and vol. 2, 14 February 1919.
266 **Police patrolled to keep:** Morize, "From Port to Plantation," 15.
266 **Most evenings the group:** 29 April–2 May 1919, Eddington to Sarah Ann Eddington, EDDN A4/7, Eddington Papers, Trinity College Library, Cambridge.

266 **Eddington and Cottingham:** Mota, Crawford, and Simoes, "Einstein in Portugal," 258.
267 **The plantation owner:** 29 April–2 May 1919, Eddington to Sarah Ann Eddington, EDDN A4/7, Eddington Papers, Trinity College Library, Cambridge.
267 **"It was a very fine sight":** Higgs, *Chocolate Islands*, 61.; 29 April–2 May 1919, Eddington to Sarah Ann Eddington, EDDN A4/7, Eddington Papers, Trinity College Library, Cambridge.
267 **He asked after Punch:** 5 May 1919, Principe, Eddington to Winifred Eddington, EDDN A4/8, Eddington Papers, Trinity College Library, Cambridge.
268 **Arthur Schuster:** Pang, *Empire and the Sun*, 69.
268 **In a superstitious age:** A. S. Eddington, *Space, Time, and Gravitation*, 2nd ed. (Cambridge: Cambridge University Press, 1920), 113.
269 **"the most perfect discipline":** Pang, *Empire and the Sun*, 75.
270 **A small observatory:** Cambridge University Library, Royal Greenwich Observatory Archives MS.RGO.8.150.
270 **But they rapidly diminished:** Dyson, Eddington, and Davidson, "A Determination of the Deflection of Light by the Sun's Gravitational Field," 299.
270 **"Eclipse splendid":** *The Observatory* 540 (June 1919): 256.
270 **"were a few gleams":** 21 June and 2 July 1919, Eddington to Sarah Ann Eddington, EDDN A4/9, Eddington Papers, Trinity College Library, Cambridge.
270 **"almost took away":** Eddington, "Forty Years," 141.
270 **"almost painful":** Pang, *Empire and the Sun*, 72.
270 **At that moment the astronomers:** 14 July x 30 October 1919, "Account of an Expedition to Principe," A. S. Eddington, EDDN C1/3, Eddington Papers, Trinity College Library, Cambridge.
270 **"We had to carry out":** 21 June and 2 July 1919, Eddington to Sarah Ann Eddington, EDDN A4/9, Eddington Papers, Trinity College Library, Cambridge.
270 **He had to pause:** Dyson, Eddington, and Davidson, "A Determination of the Deflection of Light by the Sun's Gravitational Field," 314.
271 **There is a marvelous spectacle:** Eddington, *Space, Time, and Gravitation*, 115.
271 **"Through cloud. Hopeful.":** *The Observatory* 540 (June 1919): 256.
271 **"not entirely from impatience":** Eddington, "Forty Years," 142.
272 **May 30, 3 am:** Dyson, Eddington, and Davidson, "A Determination of the Deflection of Light by the Sun's Gravitational Field," 309.
272 **We took 16 photographs:** 21 June 1919, Eddington to Sarah Ann Eddington, EDDN A4/9, Eddington Papers, Trinity College Library, Cambridge.
273 **"large as astronomical measures go":** Eddington, "Forty Years," 142.
274 **consequently I have:** 21 June 1919, Eddington to Sarah Ann Eddington, EDDN A4/9, Eddington Papers, Trinity College Library, Cambridge.
274 **"I knew that Einstein's theory":** Eddington, "Forty Years," 142.
274 **He called this:** Douglas, *Arthur Stanley Eddington*, 40–41.
274 **"Cottingham, you won't have":** Ibid., 40.

275 **He would likely arrive:** 21 June 1919, Eddington to Sarah Ann Eddington, EDDN A4/9, Eddington Papers, Trinity College Library, Cambridge.
275 **She had successfully:** Fölsing, *Albert Einstein*, 427.
275 **No one was allowed:** Ibid., 427–28.
275 **He would often work:** Abraham Pais, *Subtle Is the Lord: The Science and Life of Albert Einstein* (Oxford: Oxford University Press, 1982), 301.
275 **He had heard via:** Fölsing, *Albert Einstein*, 438.
275 **Their army would be shrunk:** MacMillan, *Paris 1919*, 158.
276 **"We will search their pockets":** Ibid., 189.
276 **Disgusted, he resigned:** Ibid., 182; Jay Winter and Antoine Prost, *The Great War in History: Debates and Controversies, 1914 to the Present*, Studies in the Social and Cultural History of Modern Warfare 21 (Cambridge: Cambridge University Press, 2005), 41.
276 **"Just as well that":** Thomas Levenson, *Einstein in Berlin* (New York: Bantam Books, 2003), 242.
276 **"Here the political wave":** CPAE volume 9, document 52, "Einstein to Heinrich Zangger, 1 June 1919," 43–44.
276 **He continued to try:** CPAE volume 9, document 76, "From Hendrik A. Lorentz to Einstein, 26 July 1919," 64–65, and CPAE volume 9, document 108, "Einstein to Lorentz, 21 September 1919," 92–93.
277 **"We do not forget":** *The Observatory* 542 (August 1919): 297.
277 **"We look back at that meeting":** Ibid., 301.
277 **Curtis measured the plates:** Ibid., 298.
277 **"some evidence":** Ibid., 299.
278 **"It chances that certain astronomers":** Ibid., 306.
278 **They saw themselves as inaugurating:** Brigitte Schroeder-Gudehus, "Challenge to Transnational Loyalties: International Scientific Organizations after the First World War," *Science Studies* 3, no. 2 (April 1973): 93–118.
278 **"Our grand governess":** Daniel J. Kevles, "Into Hostile Political Camps," *Isis* 62, no. 1 (1971): 57–59.
279 **Davidson worked with:** Daniel Kennefick, "Not Only Because of Theory: Dyson, Eddington, and the Competing Myths of the 1919 Eclipse Expedition," in *Einstein and the Changing Worldviews of Physics*, ed. Christopher Lehner (Boston: Birkhäuser, 2012), 201–32.
280 **Einstein's predicted deflection:** Dyson, Eddington, and Davidson, "A Determination of the Deflection of Light by the Sun's Gravitational Field," 320–28.
280 **Measuring those provided:** Ibid., 299–309.
280 **"I am glad":** Eddington to Dyson, 3 October, 1919, Papers of Frank Dyson, MS.RGO.8/150/138.
281 **I do not like:** Eddington to Dyson, 21 October 1919, Papers of Frank Dyson, MS.RGO.8/150/143.
281 **"They disclose something":** Sponsel, "Constructing a 'Revolution in Science,'" 445.
282 **Other scientists proposed:** "Astronomy at the British Association," *The Observatory* (October 1919): 363–65.

282 **Perhaps most important:** Sponsel, "Constructing a 'Revolution in Science,'" 456–57.

282 **He reported back:** CPAE volume 9, document 127, "From Hendrik A. Lorentz to Einstein, 7 October 1919," 109.

283 **"Eddington found stellar shift":** CPAE volume 9, document 110, "From Hendrik A. Lorentz to Einstein, 22 September 1919," 95.

283 **Full of enthusiasm:** Ilse Rosenthal-Schneider, *Reality and Scientific Truth: Discussions with Einstein, Von Laue, and Planck* (Detroit: Wayne State University Press, 1980). Another version can be found in Ilse Rosenthal-Schneider, "Albert Einstein 14 March 1879–18 April 1955," *Australian Journal of Science* (August 1955): 19.

283 **"I knew I was right":** Levenson, *Einstein in Berlin*, 215.

283 **In another version:** Alice Calaprice, ed., *The Ultimate Quotable Einstein* (Princeton: Princeton University Press, 2010), 94.

284 **I cannot postpone:** CPAE volume 9, document 113, "Einstein to Pauline Einstein, 27 September 1919," 98; volume 9, document 121, "Einstein to Max Planck, 4 October 1919," 105.

284 **"congratulate you heartily":** CPAE volume 9, document 185, "From Willem de Sitter to Einstein, 1 December 1919," 158.

284 **"This is certainly one":** CPAE volume 9, document 127, "From Hendrik A. Lorentz to Einstein, 7 October 1919," 109.

284 **Ehrenfest hoped to get:** CPAE volume 9, document 123, "From Paul Ehrenfest to Einstein, 5 October 1919," 106–7.

284 **Einstein immediately sent:** CPAE volume 7, document 23, "A Test of the General Theory of Relativity," 97, dated October 9, 1919, published October 10, 1919, in *Die Naturwissenschaften* 7 (1919): 776.

284 **All doubts have now been spent:** CPAE volume 9, document 131, "From the Zurich Physics Colloquium, 11 October 1919," 113.

284 **Einstein's responding verse:** CPAE volume 9, document 139, "Einstein to the Zurich Physics Colloquium, 16 October 1919," 118.

284 **"So all is going better":** CPAE volume 9, document 148, "From Heinrich Zangger to Einstein, 22 October 1919," 126–28.

285 **This evening at the colloquium:** CPAE volume 9, document 149, "Einstein to Max Planck, 23 October 1919," 128.

285 **"How hard this magnificent":** CPAE volume 9, document 160, "Einstein to Paul Ehrenfest, 8 November 1919," 136.

285 **"The result is now definite":** CPAE volume 9, document 151, "Einstein to Pauline Einstein, 26 October 1919," 130.

285 **"My theory has been verified":** CPAE volume 10 (cited as volume 9), document 148b, "Einstein to Elsa Einstein, 23 October 1919," 138.

286 **Alistair Sponsel estimates:** Sponsel, "Constructing a 'Revolution in Science,'" 460.

286 **The whole atmosphere of tense interest:** A. N. Whitehead, *Science and the Modern World* (New York: Macmillan, 1947), 15.

286 **"After a careful study":** "Joint eclipse meeting of the Royal Society and the Royal Astronomical Society," *The Observatory* 42, no. 545 (November 1919): 391.

287 **This meant one could say:** Ibid., 393.
287 **This is the most important result:** Ibid., 394.
288 **"The conclusion is so important":** Ibid.
288 **He asked if Eddington:** Ibid., 395.
288 **H. F. Newall acknowledged:** Ibid., 396.
288 **The discovery made at the eclipse expedition:** Ibid., 397.
289 **"Professor Eddington, you must be":** As recalled in Subrahmanyan Chandrasekhar, *Eddington: The Most Distinguished Astrophysicist of His Time* (Cambridge: Cambridge University Press, 1983), 83.
289 **It was written by:** Sponsel, "Constructing a 'Revolution in Science,'" 463–64.
290 **"During the war":** *The Times*, November 8, 1919.
290 **"besieged by inquiries":** Ibid.
290 **There was a genuine sense:** CPAE volume 9, document 174, "From Adolf Friedrich Lindemann to Einstein, 23 November 1919," 145.
291 **A patriot fiddler-composer:** *Punch*, November 19, 1919, 422.
291 **"hundreds were turned away":** CPAE volume 9, document 186, "From Arthur S. Eddington to Einstein, 1 December 1919," 159.
291 **"The theoretical researches":** A. S. Eddington, "Einstein's Theory of Space and Time," *Contemporary Review* 116 (1919): 639.
291 **We cheered the Eclipse Observers':** "From an Oxford Note-Book," *The Observatory* 546 (December 1919): 456.
293 **"astronomer-adventurer":** Joshua Nall, "Constructing Canals on Mars: Event Astronomy and the Transmission of International Telegraphic News," *Isis* 108 (June 2017): 280–306, and Pang, *Empire and the Sun*, 52–53.
293 **"The Great Result":** "From an Oxford Note-Book," 452.
293 **"It is said that Professor Eddington":** Ibid., 453.
294 **All England has been talking:** CPAE volume 9, document 186, "From Arthur S. Eddington to Einstein, 1 December 1919," 158–59.
294 **I have been kept very busy:** Ibid.

CHAPTER 12

296 **"People seem to forget":** A. Vibert Douglas, *The Life of Arthur Stanley Eddington* (London: Thomas Nelson, 1956), 115.
297 **"but those who are familiar":** "From an Oxford Note-Book," *The Observatory* 546 (1920): 37–38.
298 **"We may also find satisfaction":** Ibid., 44.
298 **A few people unhappy:** *Nature* 104 (December 25, 1919): 412.
298 **Most everyone wanted:** F. Schlesinger to Dyson, 16 February 1920, Papers of Frank Dyson, MS.RGO.8/150/123.
298 **Within a year or so:** "From an Oxford Note-Book," *The Observatory* 560 (January 1921): 234–35. See Jeffrey Crelinsten, "William Wallace Campbell and the 'Einstein Problem': An Observational Astronomer Confronts the Theory of Relativity," *Historical Studies in the Physical Sciences* 14, no. 1 (1983): 1–91, and Jeffrey Crelinsten, *Einstein's Jury* (Princeton: Princeton University Press, 2006).

298 **"the blind leading the blind":** Douglas, *Arthur Stanley Eddington*, 42.
298 **The historian Andrew Warwick:** Andrew Warwick, *Masters of Theory: Cambridge and the Rise of Mathematical Physics* (Chicago: University of Chicago Press, 2003), 480–99.
299 **"The thrill of seeing":** Douglas, *Arthur Stanley Eddington*, 52.
299 **A slight man:** Ibid., 51.
299 **"enriched with literary quality":** Ibid., 107; Warwick, *Masters of Theory*, 482.
299 **He also found room:** Douglas, *Arthur Stanley Eddington*, 117–18.
299 **Oh leave the wise:** Ibid., 44.
299 **"an expounder to the multitude":** Ibid., 105.
300 **A common theme:** Marshall Missner, "Why Einstein Became Famous in America," *Social Studies of Science* 15, no. 2 (1985): 270.
300 **Katy Price points out:** Katy Price, *Loving Faster than Light: Romance and Readers in Einstein's Universe* (Chicago: University of Chicago Press, 2012), 18, and *The Observatory* 557 (October 1920): 375.
300 **For some years past:** Abraham Pais, *Subtle Is the Lord: The Science and Life of Albert Einstein* (Oxford: Oxford University Press, 1982), 310.
300 ***Punch* presented:** Price, *Loving Faster than Light*, 30, 40.
300 **"little knot of experts":** Ibid., 31; Missner, "Why Einstein Became Famous in America," 277.
301 **Perhaps the next bit:** Albrecht Fölsing, *Albert Einstein: A Biography* (New York: Viking, 1997), 447.
301 **"a destroyer of time and space":** Missner, "Why Einstein Became Famous in America," 271–72.
301 **"They cheer me":** Fölsing, *Albert Einstein*, 457.
301 **Virtually nothing but Einstein:** CPAE volume 9, document 182, "From Arnold Berliner to Einstein, 29 November 1919," 155–56.
301 **"You are presented as":** CPAE volume 9, document 174, "From Adolf Friedrich Lindemann to Einstein, 23 November 1919," 146.
302 **After the lamentable breach:** *The Times*, November 28, 1919.
302 **He went back to Oxford:** Heidi König, "General Relativity in the English-Speaking World: The Contributions of Henry L. Brose," *Historical Records of Australian Science* 17, no. 2 (December 2016): 2006.
302 **"May I in closing":** CPAE volume 9, document 177, "From Robert W. Lawson to Einstein, 26 November 1919," 152.
302 **He declined and recommended:** CPAE volume 9, document 185, "From Willem de Sitter to Einstein, 1 December 1919," 158.
303 **Eddington's letters were in English:** CPAE volume 9, document 186, "From Arthur S. Eddington to Einstein, 1 December 1919," 159; volume 9, document 216, "Einstein to Arthur S. Eddington, 15 December 1919," 184–85.
303 **The time has come, said Eddington:** Douglas, *Arthur Stanley Eddington*, 116–17.
304 **In Germany he wasn't:** Fölsing, *Albert Einstein*, 449.
304 **a front-page article:** Ibid., 452; Thomas Levenson, *Einstein in Berlin* (New York: Bantam Books, 2003), 219–20; David E. Rowe, "Einstein's Allies and Enemies: Debating Relativity in Germany, 1916–1920," in *Interactions:*

Mathematics, Physics and Philosophy, 1860–1930, ed. Vincent F. Hendricks et al. (Dordrecht: Springer, 2006), 221.

304 **For hundreds of thousands:** Rowe, "Einstein's Allies and Enemies," 223.

304 **"Einstein versus Newton!":** CPAE volume 9, document 175, "Paul Ehrenfest to Einstein, 24 November 1919," 147.

304 **"Galinka's little picture":** CPAE volume 9, document 189, "Einstein to Paul Ehrenfest, 4 December 1919," 161–62.

305 **"With all our hearts":** Pais, *Subtle Is the Lord*, 306.

305 **Einstein used his new leverage:** CPAE volume 9, document 194, "Einstein to Konrad Haenisch, 6 December 1919," 165–66.

306 **He wrote to Besso:** CPAE volume 9, document 207, "Einstein to Michele Besso, 12 December 1919," 178.

306 **The latter worried himself sick:** CPAE volume 10 (cited as volume 8), document 152a, "Einstein to Elsa Einstein, 28 October 1919," 140.

306 **"general, somewhat sensational":** Suman Seth, *Crafting the Quantum: Arnold Sommerfeld and the Practice of Theory, 1890–1926* (Cambridge, Massachusetts: MIT Press, 2010), 190.

306 **One of the first signs:** Fölsing, *Albert Einstein*, 451.

306 **Einstein began complaining about:** CPAE volume 9, document 198, "Einstein to Max Born, 8 December 1919," 169.

306 **"harmonizes with every":** Rowe, "Einstein's Allies and Enemies," 198.

306 **"Due to the newspaper clamor":** CPAE volume 9, document 233, "To Heinrich Zangger, 24 December 1919," 197.

307 **"I'm giving a children's lecture":** CPAE volume 10 (cited as volume 9), document 17, "Einstein to Elsa Einstein, 17 May 1920," 163.

307 **"so bad that I can hardly breathe":** Alan Friedman and Carol Donley, *Einstein as Myth and Muse* (Cambridge: Cambridge University Press, 1985), 11.

307 **"Half the world":** CPAE volume 10 (cited as volume 9), document 20, "From Elsa Einstein to Einstein, 20 May 1920," 165.

307 **"The dog is very smart":** Alice Calaprice, ed., *The Ultimate Quotable Einstein* (Princeton: Princeton University Press, 2010), 234.

307 **By an application:** Albert Einstein, "What Is the Theory of Relativity?" *The Times*, November 28, 1919.

307 **"Since the light deflection":** CPAE volume 9, document 242, "Einstein to Heinrich Zangger, 3 January 1920," 204–5.

307 **He became famous:** Levenson, *Einstein in Berlin*, 224–28.

308 **When asked, he gave:** Ibid., 221.

308 **Being photogenic helped:** Rowe, "Einstein's Allies and Enemies," 199.

308 **"like an artist":** Friedman and Donley, *Einstein as Myth and Muse*, 18.

308 **There was certainly a sense:** Rowe, "Einstein's Allies and Enemies," 222.

308 **He was also happy:** Fölsing, *Albert Einstein*, 457–58.

308 **"Here there is nothing":** CPAE volume 9, document 84, "Paul Ehrenfest to Einstein, September 8 1919," 84.

309 **He had been isolated:** Fritz Stern, *Einstein's German World* (Princeton: Princeton University Press, 2001), 125.

309 **Soon, though, Haber dedicated:** Ibid., 134.

309 **After the war:** L. F. Haber, *The Poisonous Cloud* (Oxford: Clarendon Press, 1986), 291–92.
309 **"In a few centuries":** Stern, *Einstein's German World*, 60.
309 **"My political optimism":** CPAE volume 9, document 187, "Einstein to Adriaan D. Fokker, 1 December 1919," 159.
309 **"willing and able to provide":** CPAE volume 7, document 41, "'On the Quaker relief effort', after 11 July 1920," 192.
309 **The German government asked:** CPAE volume 10, document 74, "From German Central Committee for Food Relief to Einstein, 9 July 1920, Re: Quaker relief," 207.
310 **Whatever great political:** CPAE volume 7, document 40, "Einstein to the German Central Committee for Foreign Relief, 11 July 1920," 191.
310 **The most valuable contribution:** CPAE volume 7, document 47, "On the contributions of intellectuals to international reconciliation," published after 29 September 1920, 201.
310 **Scientists could repair:** CPAE volume 7, document 69, "Impact of science on the development of pacifism" (before December 9, 1921), 488–91.
310 **"Our English colleagues":** CPAE volume 9, document 187, "Einstein to Adriaan D. Fokker, 1 December 1919," 159.
310 **"The outcome of the English":** CPAE volume 9, document 208, "Einstein to Willem de Sitter, 12 December 1919," 179.
311 **"The English and Belgians":** CPAE volume 12, document 87, "From Fritz Haber to Einstein, 9 March 1921," 70.
311 **"It even has to be said":** CPAE volume 12, document 88, "Einstein to Fritz Haber, 9 March 1921," 71.
311 **"enemy countries":** RAS Papers 2, Council Minutes, volume 11, 14 February 1919; 9 April 1920; 11 February 1921.
311 **Planck tried to get:** John Heilbron, *The Dilemmas of an Upright Man: Max Planck as Spokesman for German Science* (Berkeley: University of California Press, 1986), 89.
311 **He did what he could:** RAS Letters 1919, Arthur Eddington to Andrew Crommelin, 28 December; RAS Letters 1921, Arthur Eddington to W. H. Wesley, 30 January.
311 **Einstein pitched in:** CPAE volume 14 (cited as volume 7), document 36a, "An Exchange of Scientific Literature, between 24 March and 4 April 1920," 3; *Neue Zürcher Zeitung*, 4 April 1920; CPAE volume 10, document 26, "Einstein to Lorentz, 22 May 1920," 169.
311 **an excellent target:** Levenson, *Einstein in Berlin*, 246–50.
312 **"a Jew with international views":** CPAE volume 7, document 45, "My Response. On the Anti-Relativity Company,"197.
312 **"mass suggestion":** Rowe, "Einstein's Allies and Enemies," 217.
312 **Einstein actually sneaked in:** Ibid., 227; Fölsing, *Albert Einstein*, 460–62.
312 **"This world is a strange madhouse":** CPAE volume 10, document 148, "Einstein to Grossmann, 12 September 1920," 271.
312 **The attacks on Prof. Einstein:** Fölsing, *Albert Einstein*, 464.
312 **"The role I play":** CPAE volume 10, document 148, "Einstein to Marcel Grossmann, 12 September 1920," 271.

312 **Flippantly, he compared:** Levenson, *Einstein in Berlin*, 250.
313 **He had little interest:** CPAE volume 7, document 57, "How I Became a Zionist," 234; CPAE volume 10, document 238, "Einstein to Jewish Community of Berlin, 22 December 1920," 338.
313 **"I believe that this undertaking":** CPAE volume 9, document 207, "Einstein to Michele Besso, 12 December 1919," 178.
313 **"One feels in one's bones":** Pais, *Subtle Is the Lord*, 303.
313 **The postwar collapse of the German mark:** Fölsing, *Albert Einstein*, 475.
313 **She needled him:** CPAE volume 10, document 10, "Elsa Einstein to Einstein, 9 May 1920," 157.
313 **"Admiration for the scientist":** Fölsing, *Albert Einstein*, 479–85.
314 **I hope to show my interest:** Eddington to Strömgren. November 1919, quoted in Hertzsprung-Kapteyn, "J. C. Kapteyn," *Space Science Reviews* 64 (1993): 81.
314 **He tried to participate:** A. S. Eddington, "Das Strahlungsgleichgewicht der Sterne," *Zeitschrift für Physik* 7 (1921): 531. It is not clear who translated the article.
314 **"This paper is intended":** August 1921, Manuscript of "Radiative Equilibrium of the Stars," EDDN C1/2, Eddington Papers, Trinity College Library, Cambridge.
314 **A month later it was:** RAS Papers 2, Council Minutes, volume 11, 14 November 1919; 12 December 1919.
315 **"the award of the Gold Medal":** RAS Papers 2, Council Minutes, volume 11, 9 January 1920.
315 **There was a strong implied:** R. J. Tayler, ed., *History of the Royal Astronomical Society*, vol. 2 (Oxford: Blackwell Scientific Publications, 1987), 20.
315 **I am sorry to say:** Eddington to Einstein, January 21, 1920, AEA ALS 9-264.
316 **I find it difficult:** Ludlam to Einstein, January 23, 1920, AEA ALS 9-266.
316 **"tragicomical":** CPAE volume 9, document 293, "Einstein to Arthur S. Eddington, 2 February 1920," 245.
316 **The war had just ended:** S. Chandrasekhar, *Truth and Beauty* (Chicago: University of Chicago Press, 1987), 115.
317 **"by his eloquence":** Douglas, *Arthur Stanley Eddington*, 104.
317 **"a place where miracles happen":** Pais, *Subtle Is the Lord*, 305, 310; Fölsing, *Albert Einstein*, 456.
317 **Einstein finally set foot:** Ronald Clark, *Einstein: The Life and Times* (New York: World Publishing, 1971), 270.
317 **"a few 'irreconcilables'":** König, "General Relativity in the English-Speaking World," 188.
317 **He gave lectures in German:** Fölsing, *Albert Einstein*, 508.
318 **Einstein either did not realize:** Price, *Loving Faster than Light*, 35.
318 **"takes its own course":** Ibid., 37.
318 **"a splendid chap":** CPAE volume 14, document 127, "Einstein to Elsa Einstein, 1 October 1923," 123.
319 **The Einstein scholar:** John Stachel, "The Young Einstein," in *Einstein from "B" to "Z"* (Boston: Birkhäuser, 2002), 21.

EPILOGUE

321 **Every generation has used:** For examples see Stephen G. Brush, "Prediction and Theory Evaluation: The Case of Light Bending," *Science* 246 (1989): 1124–129; Deborah G. Mayo, *Error and the Growth of Experimental Knowledge* (Chicago: University of Chicago Press, 1996), 133–37, 278–93; and Robert Hudson, "Novelty and the 1919 Eclipse Experiments," *Studies in the History and Philosophy of Modern Physics* 34 (2003): 107–29.

322 **That drove him:** Karl Popper, "On Reason and the Open Society," *Encounter* 38, no. 5 (1972): 13.

322 **"dazed":** Roberta Corvi, *An Introduction to the Thought of Karl Popper* (New York: Routledge, 1996), 4.

322 **"intellectual modesty":** Richard Bailey, *Education in the Open Society—Karl Popper and Schooling* (New York: Routledge, 2000), 12.

323 **"a scientific model":** Malachi Haim Hacohen, *Karl Popper: The Formative Years 1902–1945* (Cambridge: Cambridge University Press, 2000), 95.

323 **"to make explicit certain points":** Dario Antiseri, *Popper's Vienna* (Aurora, Colorado: Davies Group Publishers, 2006), 25.

323 **Popper's falsificationism has become:** Michael Gordin, "Myth 27: That a Clear Line of Demarcation Has Separated Science from Pseudoscience," in *Newton's Apple and Other Myths About Science*, ed. Ronald Numbers and Kostas Kampourakis (Cambridge, Massachusetts: Harvard University Press, 2015), 219–25.

323 **"no point of rest in science":** Popper, "On Reason and the Open Society," 17.

323 **The results strongly confirmed:** Jeffrey Crelinsten, *Einstein's Jury* (Princeton: Princeton University Press, 2006).

325 **"partly because the world":** Dennis William Sciama, *The Physical Foundations of General Relativity* (New York: Doubleday, 1969), 69.

325 **"cooler reflection":** C. W. Francis Everitt, "Experimental Tests of General Relativity: Past, Present and Future," in *Physics and Contemporary Needs*, vol. 4, ed. Riazuddin (New York: Springer, 1980), 533.

325 **"Only Eddington's disarming way":** Ibid., 534.

326 **"their measurement had been":** Stephen Hawking, *A Brief History of Time* (New York: Bantam, 1988), 32.

326 **The new analysis gave:** G. M. Harvey, "Gravitational Deflection of Light," *The Observatory* 99 (December 1979): 195–98.

326 **Two astronomers wrote:** P. A. Wayman and C. A. Murray, "Relativistic Light Deflections," *The Observatory* 109 (October 1989): 189–91.

327 **"despair on the part":** John Earman and Clark Glymour, "Relativity and Eclipses: The British Expeditions of 1919 and Their Predecessors," *Historical Studies in the Physical Sciences* 11 (1980): 49–85.

327 **Earman and Glymour's argument:** H. M. Collins and Trevor Pinch, *The Golem* (Cambridge: Cambridge University Press, 1993).

328 **It was not just that Eddington was biased:** Ibid., 27–56.

328 **"science needs decisive moments":** Ibid., 52.

329 **Surely, he said, science:** N. David Mermin, "What's Wrong with This Sustaining Myth?" *Physics Today* 49, no. 3 (March 1996): 11–13; N. David Mermin, "The Golemization of Relativity," *Physics Today* 49, no. 4 (April 1996): 11–13.

329 **Context, he writes:** Daniel Kennefick, "Testing Relativity from the 1919 Eclipse—A Question of Bias," *Physics Today* 62 (2009): 37–42, and Daniel Kennefick, "Not Only Because of Theory: Dyson, Eddington, and the Competing Myths of the 1919 Eclipse Expedition," in *Einstein and the Changing Worldviews of Physics*, ed. Christopher Lehner (Boston: Birkhäuser, 2012), 201–32.

330 **colleagues in astronomy:** Matthew Stanley, *Practical Mystic* (Chicago: University of Chicago Press, 2007), 122–23 and 268 *n*181.

331 **"opportunely put an end":** Eddington, "Sir Frank Dyson, 1868–1939," *Royal Society Obituary Notices of Fellows* 3 (1940), 167.

332 **On the Einstein centenary:** W. H. McCrea, "Einstein: Relations with the RAS," *Quarterly Journal of the Royal Astronomical Society* 20, no. 3 (1979): 251–60.

332 **Hawking called them:** Hawking, *A Brief History*, 32.

332 **In our present time:** Clifford Will, *Was Einstein Right?* (New York: Basic Books, 1993), 76.

333 **The plaque is a celebration:** Gisa Weszkalnys, "Principe Eclipsed," *Anthropology Today* 25, no. 5 (October 2009), 8–12. Also see Richard Ellis, Pedro Ferreira, Richard Massey, and Gisa Weszkalnys, "90 Years On—The 1919 Eclipse Expedition at Principe," *Astronomy and Geophysics* 50, no. 4 (August 2009): 412–15.

INDEX